Rapid Thermal Processing
Science and Technology

Rapid Thermal Processing
Science and Technology

Edited by

Richard B. Fair
MCNC
Center for Microelectronic Systems Technologies
Research Triangle Park, North Carolina

ACADEMIC PRESS, INC.
Harcourt Brace Jovanovich, Publishers
Boston   San Diego   New York
London   Sydney   Tokyo   Toronto

This book is printed on acid-free paper. ∞

Copyright © 1993 by Academic Press, Inc.

All rights reserved.
No part of this publication may be reproduced or
transmitted in any form or by any means, electronic
or mechanical, including photocopy, recording, or
any information storage and retrieval system, without
permission in writing from the publisher.

ACADEMIC PRESS, INC.
1250 Sixth Avenue, San Diego, CA 92101-4311

United Kingdom Edition published by
ACADEMIC PRESS LIMITED
24–28 Oval Road, London NW1 7DX

Library of Congress Cataloging-in-Publication Data

Fair, Richard B.
    Rapid thermal processing : science and technology / Richard B. Fair.
       p.   cm.
    Includes bibliographical references and index.
    ISBN 0-12-247690-5
    1. Semiconductors--Heat treatment. 2. Semiconductor doping.
I. Title
TK7871.85.F299   1993
621.3815'2--dc20
                                              93-6882
                                                CIP

Printed in the United States of America
93 94 95 96 97   BB   9 8 7 6 5 4 3 2 1

Contents

Contributors viii

Rapid Thermal Processing—A Justification 1
Richard B. Fair

I.	Manufacturing Issues in the Gigachip Age	1
II.	The Parameter Budget Crisis	3
III.	Conclusions	10
	References	10

Rapid Thermal Processing-Based Epitaxy 13
J. L. Hoyt

I.	Introduction to Silicon Epitaxy	15
II.	Characteristics of Rapid Thermal Processing-Based Silicon Epitaxy	25
III.	Growth of Strained Silicon–Germanium Alloys	31
IV.	Summary	40
	References	41

Rapid Thermal Growth and Processing of Dielectrics 45
Hisham Z. Massoud

I.	Equipment Issues in Rapid Thermal Oxidation	47
II.	Rapid Thermal Oxidation Growth Kinetics	51
III.	Rapid Thermal Processing of Oxides	58
IV.	Electrical Properties of Rapid Thermal Oxidation/ Rapid Thermal Processing Oxides	62
V.	Conclusions	72
	Acknowledgments	72
	References	73

4 Thin-Film Deposition ... 79
Mehmet C. Öztürk

- I. Equipment ... 81
- II. Thin-Film Deposition Processes ... 83
- III. *In Situ* Processing—Applications ... 106
- IV. Equipment Issues ... 111
- V. Summary ... 118
  - References ... 118

5 Extended Defects from Ion Implantation and Annealing ... 123
Kevin S. Jones and George A. Rozgonyi

- I. Introduction ... 123
- II. Defect Formation Kinetics ... 133
- III. Defect Annealing Kinetics ... 155
- IV. Summary ... 162
  - References ... 163

6 Junction Formation in Silicon by Rapid Thermal Annealing ... 169
Richard B. Fair

- I. Rapid Thermal Annealing of Ion-Implanted Junctions ... 174
- II. Dopant Activation ... 213
- III. Summary and Conclusions ... 220
  - References ... 221

7 Silicides ... 227
C. M. Osburn

- I. Introduction ... 228
- II. Formation of Silicides ... 240
- III. Properties of Silicides and Silicided Junctions ... 264
- IV. Applications of Silicides and Process/Device Considerations ... 282
- V. Summary ... 292
  - References ... 292

8 Issues in Manufacturing Unique Silicon Devices Using Rapid Thermal Annealing ... 311
B. Lojek

- I. Impact of Patterned Layers on Temperature Nonuniformity during Rapid Thermal Annealing ... 314

Contents

II.	Bipolar Transistor Processing	325
III.	MOS Transistor Processing	337
IV.	Conclusion	344
	References	346

9 Manufacturing Equipment Issues in Rapid Thermal Processing 349
Fred Roozeboom

I.	Historical Survey of Rapid Thermal Processing	351
II.	Fundamental Thermophysics in Rapid Thermal Processing	352
III.	General Rapid Thermal Processing System Components	359
IV.	Survey of Commercial Rapid Thermal Processing Equipment	381
V.	Temperature Nonuniformity, System Modeling, and Effective Emissivity	390
VI.	Noncontact *In Situ* Real-Time Process Control Options	401
VII.	Recent Developments and Future Trends in Rapid Thermal Processing	407
VIII.	Technology Roadmap and Concluding Remarks	414
	References	417

Index 425

Contributors

Number in parentheses indicate the pages on which the authors' contributions begin.

RICHARD B. FAIR (1, 169), *MCNC, Center for Microelectronic Systems Technologies, P.O. Box 12889, Research Triangle Park, North Carolina 27709 and Department of Electrical Engineering, Duke University, Durham, North Carolina 27706*

J. L. HOYT (13), *Solid State Electronics Laboratory, McCullough Building, Room 226, Stanford, California 94305*

KEVIN S. JONES (123), *Department of Materials Science and Engineering, University of Florida, Gainesville, Florida 32611*

B. LOJEK (311), *Motorola Inc., Advanced Technology Center, 2200 West Broadway Road, Mesa, Arizona 85202*

HISHAM Z. MASSOUD (45), *Semiconductor Research Laboratory, Department of Electrical Engineering, Duke University, Durham, North Carolina 27706*

C. M. OSBURN (227), *MCNC, Center for Microelectronic Systems Technologies, P.O. Box 12889, Research Triangle Park, North Carolina 27511 and Department of Electrical and Computer Engineering, North Carolina State University, Raleigh, North Carolina 27695*

MEHMET C. ÖZTÜRK (79), *North Carolina State University, Department of Electrical and Computer Engineering, Raleigh, North Carolina 27695-7911*

FRED ROOZEBOOM (349), *Philips Research Laboratories, P.O. Box 80,000, NL-5600 JA Eindhoven, The Netherlands*

GEORGE A. ROZGONYI (123), *Department of Materials Science and Engineering, North Carolina State University, Raleigh, North Carolina 27695*

1 Rapid Thermal Processing— A Justification

Richard B. Fair
MCNC
Center for Microelectronic Systems Technologies
Research Triangle Park
North Carolina
and
Department of Electrical Engineering
Duke University, Durham
North Carolina

I. Manufacturing Issues in the Gigachip Age	1
II. The Parameter Budget Crisis	3
A. Thermal Budget	4
B. Ambient Control Budget	5
C. Mechanical Budget	6
D. Contamination Budget	6
E. Electrical Budget	9
III. Conclusions	10
References	10

I. Manufacturing Issues in the Gigachip Age

The cost of manufacturing submicron, ultralarge scale integration (ULSI) chips is scaling upwards in at least inverse proportion to the downward scaling of device feature sizes. These expenses are driven by a manufacturing budget crisis associated with the technological and complexity limits of integrated circuit (IC) design, costs of research and development to address manufacturing issues, and facility capitalization [1]. Thus, examination of the rate of progress in microelectronics over the past 30 years suggests that the primary challenge in reaching the gigachip age in the year 2001 will be the semiconductor industry's ability to change the cost trend lines; that is, to change the economics of how ICs are developed and manufactured [2].

The scale of integration of dynamic random-access memory (DRAM) chips has continued to increase by four times every three years! And there is evidence that nationalistic efforts in Japan, Europe, and the United States are attempting to accelerate the integration-versus-time trend curves in the face of shrinking profit margins in order to achieve world dominance. However, there is great risk in these investments. Indeed, the system applications that would utilize higher density DRAMs are not keeping pace with chip availability. It is clear that the semiconductor manufacturers are trying to drive the end-user market! Thus, accelerated leapfrog programs to produce gigachips may have an inadequate market for timely DRAM sales [3]. It is expected that 85% of gross annual sales of 1G DRAMs will be required to pay for research and development and manufacturing costs, assuming normal market growth [4]. This estimate is based on 1G DRAM research and development investment growing to 10-15 times that of the 1M DRAM, 1.2 to 1.3 times more processing steps per generation, 0.9 times fewer chips per wafer per generation in spite of larger diameter wafers, and 40-50 times larger investment in production equipment!

Several approaches have been suggested for reducing the costs associated with developing and manufacturing gigachips, including internationalizing the technology through global partnerships. Requirements on contamination, process control (manufacturing parameter budgets), and cost of manufacturing floor space are driving a paradigm shift to a microprocessing methodology. Thus, single-wafer processing environments with highly controlled, ultraclean ambients clustered together in specialty process modules are being considered. In single chamber machines it is necessary to extract a silicon wafer out of a carrier and present it to the process chamber. Wafer transport among modules can be done best in a modest vacuum (10^{-4} to 10^{-5} torr). *In situ* vacuum processing equipment accounts for 40% of the total equipment today. Rapid thermal processing (RTP) using lamp heating will move thermal processes into cluster tools. Dry cleaning will also move vacuum processes to cluster tools. It is even projected that *in situ* lithography is possible. Thus it is possible that 80% of gigachip equipment could be *in situ* vacuum-based clusters controlled by a factory information system [2].

Rapid thermal processing is a key technology in the cluster tool, single-wafer manufacturing approach. With RTP a single wafer is heated quickly at atmospheric or low pressure under isothermal conditions. The processing chamber is made of either quartz, silicon carbide, stainless steel, or aluminum with quartz windows. The wafer holder is often made of quartz and contacts the wafer at a minimum number of places. A temperature measurement system is placed in a control loop to set wafer temperature. The RTP system is interfaced with a gas handling system and a computer that controls system operation. The small thermal mass inherent in this

Rapid Thermal Processing—A Justification

Table I Technology comparisons.

Furnace	RTP
Batch	Single wafer
Hot wall	Cold wall
Long time	Short time
Small dT/dt	Large dT/dt
High cycle time	Low cycle time
Temperature measurement —environment	Temperature measurement —wafer
Issues	Issues
—small thermal budget	—uniformity
—particles	—repeatability
—atmosphere	—throughput
—vertical furnaces	—stress
	—measurements
	—automation

processing system along with stringent ambient and particle control allow for reduced processing times and improved control in the formation of *pn* junctions, thin oxides, nitrides and silicides, thin deposited layers, and flowed glass structures. In essence RTP provides a controlled environment for thermally activated processes that is increasingly difficult for existing batch furnace systems to achieve. In addition, RTP is fully consistent with the advanced microprocessing-for-manufacturing paradigm.

A comparison between batch furnace and RTP technologies is shown in Table I. In order to achieve short processing times, one trades off a new set of challenges including temperature and process uniformity, temperature measurement and control, wafer stress, and throughput.

A road map showing the introduction of RTP into manufacturing DRAMs is depicted in Table II. Each of the key processing areas that will be impacted by RTP is listed along with estimated timing for the transition from batch to single-wafer technology. Details regarding these processing areas are provided in the subsequent chapters of this book. A discussion of the manufacturing requirements that are driving the expanded use of RTP follows.

II. The Parameter Budget Crisis

The challenge to the semiconductor industry to maintain future viability is to develop equipment and processes that will mass produce ULSI chips with tight tolerance, high reliability, and low costs. This challenge translates to technological problems associated with patterning, doping, interconnections

Table II Rapid thermal processing technology road map.

	1985	1990		1995		2000	
Design rule (μm)	2	1.3	0.8	0.5	0.3	0.2	0.15
DRAM equivalent	256k	1M	4M	16M	64M	256M	1G
Wafer size (in.)	5	5	6–8	8	8 –10		12
t_{ox} (nm)	35	25	20	15	12	10	8
x_j (μm)	0.3	0.25	0.2	0.15	0.10	0.08	0.06
Repeatability (°C)	±15	±5		±2		±1	
Uniformity (°C)	±7–10	±3–5		±2–3		<±2	
Accuracy (°C)	±20$^+$	±5–10		±3–5		<±3	
Thermal budget x_j		Furnace			RTA		
t_{ox}		Furnace			RTO		
Thin dielectrics		LPCVD			RTCVD-ONO		
Silicides	Furnace		RTA				
Epitaxy		LPCVD, MBE, APCVD, etc.			RTCVD		
Polysilicon		LPCVD			RTCVD-SiGe		
Metals		LPCVD, sputtered, evaporated				RTCVD	
Cleaning		Wet chemistry			RTC		
Processing		Batch		Single wafer	Cluster tools		
Atmospheric budget	0.1/cm^2	0.01/cm^2		0.005/cm^2		0.001/cm^2	
	>0.5 μm	>0.3 μm		>0.2 μm		>0.1 μm	

defect densities, mechanical and structural aspects of handling large-diameter silicon wafers, contamination, and thermal requirements. All of these problems can be discussed in terms of process parameter targets and control tolerances or budgets. The allowed processing parameter budgets are set by device performance and manufacturing requirements, and it is through quantifying these budgets that the requirements for advanced manufacturing are set [1].

A. THERMAL BUDGET

The process thermal budget refers to the allowed time at temperature that can be tolerated to control dopant impurity diffusion and oxide growth [5]. In addition there is a manufacturing thermal budget that deals with temperature control and uniformity across a wafer. Critical concerns for ULSI manufacturing include wafer temperature control during ion implantation, implantation damage annealing, sheet resistance variation of doped layers, oxide thickness control, and absolute and repeatable temperature measurements.

Rapid Thermal Processing—A Justification

The requirements on equipment to meet the thermal budget needs of junction formation and oxidation include the following:

- a mechanical budget that satisfies the structural and mechanical aspects of processing large-diameter silicon wafers;
- a temperature uniformity budget across each wafer from run to run;
- an atmospheric budget for gases used in the furnace;
- a particle contamination budget;
- a time budget that deals with processing times as small as a few seconds and throughputs measured in hundreds of wafers per hour; and
- absolute, repeatable temperature measurements of wafers.

Improper processing conditions and wafer handling can lead to the nucleation of structural defects in silicon such as slip dislocations or wafer warpage, nonuniform oxide thickness, and irregular silicide contact interfaces. These effects produce concomitant problems with device junction leakage, lithography excursion, dielectric and contact nonuniformities, nonuniform junction sheet resistance, etc. Many such problems have been solved in large, multiwafer furnace annealing systems with large thermal masses. However, these systems have difficulty meeting all the time, particulate, and atmospheric budgets of ULSI technology. An alternative is to go to single-wafer systems using RTP.

Rapid thermal processing uses transient radiation sources such as arc lamps and graphite heaters to produce short-time, high-temperature, isothermal wafer processing. RTP can also be accomplished with continuous heat sources where the wafer is moved rapidly in and out of the vicinity of the heat source.

B. AMBIENT CONTROL BUDGET

The smallest fabricated dimension in a MOSFET is the gate oxide thickness, which is grown by thermal oxidation. Ultrathin oxide growth requires careful process control and oxidation furnace optimization. In large-diameter batch furnace tubes, control of the furnace ambient is difficult because of backstreaming of air from the large open ends of the tubes. If control of the partial pressures of the oxidant gases (dry or wet oxygen) were the only variable in achieving 70 ± 3.5 Å oxides, then these partial pressures would have to be maintained at $\pm 6\%$ of nominal [1]. Trace amounts of water must also be minimized to the ppb range [6].

Silicon surface control prior to the growth of thin oxide layers is also important because of the fact that a native oxide grows on bare silicon at room temperature. Thus, 70-Å film growth can be controlled if the Si surface is HF cleaned, leaving the surface Si bonds terminated with H [7].

Desorption of the H at 300°C in a highly pure Ar gas ambient followed by the formation of one monolayer of oxide passivates the surface for subsequent gate oxide growth [8].

For ambient budget control, single-wafer RTP processing chambers offer a microenvironment approach that can satisfy the stringent requirements for ultrathin oxide growth using rapid thermal oxidation (RTO).

C. Mechanical Budget

The mechanical budget for ULSI manufacturing impacts on the patterning budget and the budgets for wafer defects and dopant profile control. Included in this budget are mechanical systems for alignment, wafer film stresses, wafer handling, and wafer flatness.

The patterning budget, overlay registration accuracy, Δ_t, is an important concern for ULSI. There are numerous contributions, Δ_i, to Δ_t that, if mutually independent, can be summed together in quadrature [9]:

$$\delta_t = \left(\sum_i \Delta_i^2\right)^{1/2} \tag{1.1}$$

Mechanical contributions to Δ_t include the following:

- Alignment system errors—due to limitations of the systems in lithographic printers for registering alignment marks or the masks to the wafer alignment marks.
- Wafer processing errors—due to changes in wafer feature dimensions from mechanical stresses of deposited films, high temperature processing, and etching tolerances.
- Mask and wafer mounting errors—due to deformations of the mask as mounted in the exposure tool or local changes in wafer flatness during chucking of properly selected, low-warpage wafers [10].

D. Contamination Budget

Scaling transistors to smaller dimensions has a profound effect on the manufacturing yield and reliability of integrated circuits. Processing complexity (i.e., the numbers of lithography levels) increases as devices become smaller. This added complexity is a result of the need for additional levels of metal to interconnect the increased number of subcircuits on a chip. Each added metal layer requires two or more film layers and two masks. Processing complexity is also increased by the need to overcome those material or circuit parameters that do not scale with decreasing device dimension. Such parameters include the metal–semiconductor work function, silicon conductivity, and circuit operating voltages. A doubling of mask levels and films is expected as the technology is scaled down from 2 to 0.25 μm. In addition a similar doubling is expected in the number of process steps for manufacturing a chip [11].

These trends make devices more susceptible to contamination introduced by particulate and chemical impurities. Beside the increased amount or processing and, thus, exposure to impurities, smaller devices are susceptible to smaller defects and smaller amounts of chemical impurities that may cause chip loss [11]. For example, smaller devices have larger perimeter-to-area ratios, and defects along pattern edges are more likely to cause problems. Thinner oxides are vulnerable to smaller particles. Smaller devices biased with voltages that are not scaled produce higher internal electric fields that aggravate hot electron effects and oxide breakdown.

Particles can cause yield loss through the presence of random defects in the patterning of film levels. Chip yield is expressed in terms of the defect density through various statistical models such a Poisson distribution [12]:

$$\text{Yield} = e^{-A\rho} \qquad (1.2)$$

where A is the chip area and ρ is the density of defects per unit area. On the basis of the device design parameter trends and forecasts of the allowable killer surface particle sizes and densities to achieve a *total* allowed defect density of $0.25/\text{cm}^2$, it has been shown that 0.001 particles/cm^2 per step are required. The objective of 0.25 defects/cm^2 provides a yield of 78% for a 1-cm^2 chip for a Poisson distribution. Both the defect density and the killer defect size decrease as device dimensions decrease. These results are based on the rule of thumb that killer defects are at least 1/3 the size of a lithographic feature or 1/2 of a film thickness. With a gate oxide thickness of 70 Å, a 35-Å particle could be fatal! As a result, scaling device dimensions means that improved means must be found for controlling particle sizes and numbers in the processing environment.

The particle budget crisis is illustrated in Fig. 1. Measured particle densities are plotted versus particle size for airborne particles in a state-of-the-art semiconductor clean room [13], in bulk gases [14–16], in different semiconductor chemicals [17], and the minimum reported size distribution in deionized water. All the distributions in Fig. 1 show increasing particle densities with decreasing particle size. And this is the environment in which smaller devices will be made. Device scaling by a factor of two takes place in a processing environment in which the number of potentially fatal particulates in the air increases by four to eight times! The impact on yield may be devastating. Under these conditions a process that yields 25% and is limited by particle contamination would yield nothing after scaling down dimensions by a factor of two [11].

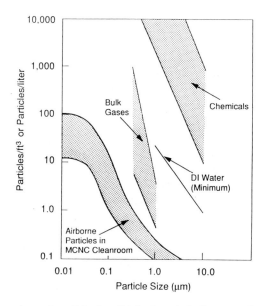

FIGURE 1. Comparison of particle size distributions in bulk gases, chemicals, deionized water and in processed air in the MCNC semiconductor clean room. The particle densities increase rapidly with decreasing size, which is the environment in which traditionally processed silicon devices are being scaled (after Fair, Ref. 1).

In contrast to what is depicted in Fig. 1 the National Advisory Committee on Semiconductors the United States has come out with consensus targets for particle densities in chemicals, in gases, and clean rooms. The target for Microtech 2000 in chemicals is 2000 particles/L of size greater than 0.02 μm, a reduction by a factor of 1000. Bulk gas targets for 0.02 μm particles are 0.02 particles/ft^3, and for airborne particles of size greater than 0.02 μm, the target is 1/ft^3! By extrapolation from current clean room practices and chemical purification methods, these targets will come only at great expense, if at all. In addition, the target for chemicals is much too high, paving the way for dry processing to eliminate wet chemistry completely. For example, it has been demonstrated that the number of particles added in an aqueous native oxide clean-up step is five times greater than in a vapor HF process [2]. Such gains in particle control will drive new strategies in silicon processing such as *in situ* dry cleaning.

Particulate and contamination control possible with *in situ* processing is also expected to be important in chemical vapor deposition (CVD) of thin layers. Rapid thermal CVD (RTCVD) processing is also consistent with the need for multiple, sequential processes such as RTO, rapid thermal nitration, and CVD polysilicon and etching. Other requirements for

RTCVD films that have not been quantified for gigachip manufacturing include film conformity and planarity, film stress, integrity, and the degree to which such films absorb or evolve water.

E. ELECTRICAL BUDGET

The electrical budget refers to the control of electrical device parameters that are determined by device scaling rules. Included in this budget are specifications for contacts, interconnections, electric-field levels, and process-induced electric charge. The electrical budget is shown in Table III.

For ULSI it is essential that low-resistance contacts to semiconductor junctions be made with high yield. These contacts must also be reliable, serving as barriers to unwanted metal reactions with silicon. Barrier layers of titanium/tungsten and titanium nitride have proven to be good choices.

Low-resistance contacts are imperative. If contact dimensions are halved, contact resistance increases by a factor of four. Thus, specific contact resistances must be decreased by factors of 10 or better.

Refractory silicides of transition metals have been used to improve contact resistances. Recent work with Al-TiW-TiSi$_2$ contacts to shallow n^+ junctions has been reported [19]. By performing sputter etching of the TiSi$_2$ surface to remove any oxides prior to TiW deposition in the same vacuum environment, specific contact resistances below 2×10^{-8} ohm·cm^2 can be achieved. Thus, the contact resistance budget is driving the use of *in situ* vacuum processing in an RTP, single-wafer module.

Table III Electrical budgets.

- Contacts
  - High yield
  - Low resistance (10^{-8} ohm·cm^2)
  - Reliable
- Interconnects
  - High conductivity (higher is better)
  - Compatible
  - Multilevel (2–4)
  - Low electromigration ($J = 5 \times 10^5$ A/cm^2)
  - Good step coverage
  - Low stress
  - Low interlevel capacitance
- V_t control
- Hot carrier injection (limiting E-fields)
- Radiation damage

The intrinsic electrical parameter budget of submicron semiconductor devices becomes increasingly bounded as device dimensions decrease. For example, submicron MOSFETs must be designed and manufactured without exceeding physical limits imposed by drain-junction avalanche break down, bulk punchthrough, short-channel effects, and hot-electron effects [20–22]. In addition, device design for one-transistor DRAM cells is constrained by noise-margin requirements that are dominated by the threshold-voltage mismatch at the input of the on-chip sense amplifier [23]. While there are several sources of threshold-voltage mismatch, one that will become much more important as device dimensions approach 0.1-μm minimum feature size is the variation due to channel doping distribution statistics. Indeed, channel doping by ion implantation causes scattering in the location of the impurity atoms in the silicon, causing a one-sigma variation in V_t of 15 mV for a 0.1 μm channel length [24]. However, this is a control problem that can be solved by rapid thermal epitaxy and atomic layer doping in a single-wafer processing module. Such new doping technologies will also impact on controlling subthreshold conduction current between source and drain of a MOSFET. This current can vary in a $\frac{1}{2}$-μm device by a factor of ten with a 500-Å change in channel profile depth. In addition, such process parameter variations would also change the subthreshold slope (change in gate voltage to produce a change in drain current) by 25 mV/decade [25].

III. Conclusions

In the chapters to follow, the state of the art in technology and understanding of RTP is presented. The collective wisdom of the authors who have contributed to this book is intended to form a basis on which further developments in RTP for manufacturing can build. While it is recognized by all who work in this field that new developments in equipment and processing knowledge will evolve rapidly, many of the concepts in this book are fundamental. Thus, while some of the details may change, the scientific underpinnings are established and are portrayed here in order to educate both the users and the developers of RTP technologies.

References

1. R. B. Fair, *Proc. IEEE* **78**, 1687 (1990).
2. P. Chatterjee, in *Extended Abstracts of International Symposium on Semiconductor Manufacturing Technology*, Tokyo (1992), p. 93.
3. R. B. Fair, in *Extended Abstracts of International Symposium on Semiconductor Manufacturing Technology*, Tokyo (1992), p. 73.
4. H. Komiya, in *Extended Abstracts of International Symposium on Semiconductor Manufacturing Technology*, Tokyo (1992), p. 85.

5. C. M. Osburn and A. Reisman, *J. Supercomputing* **1**, 149 (1987).
6. E. A. Irene and R. Ghez, *J. Electrochem. Soc.* **124**, 1757 (1977).
7. N. Yabumoto, K. Saito, M. Morita, and T. Ohmi, *Japan J. Appl. Phys.* **30**, 1419 (1991).
8. T. Ohmi, M. Morita, A. Teramoto, K. Makihara, and T. S. Teng, *Appl. Phys. Lett.* **60**, 2126 (1992).
9. R. F. Watts, in *VLSI Technology*, 2nd ed. (S. M. Sze, ed.), McGraw-Hill, New York (1988), p. 141.
10. J. R. Maldonado, in *Proceedings of the 2nd Workshop on Radiation-Induced and/or Process-Related Electrically Active Defects in Semiconductor-Insulator Systems* (A. Reisman, ed.), MCNC, Research Triangle Park (1989), p. 204.
11. C. M. Osburn, H. Berger, R. Donovan, and G. Jones, *Proc. Instit. Environ. Sci.* **31**, 45 (1988).
12. B. T. Murphy, *Proc. IEEE* **52**, 1537 (1963).
13. B. R. Locke, R. P. Donovan, D. S. Ensor, and C. M. Osburn, in *Aerosols* (Y. H. Benjamin, D. Liu, Y. H. Pui, and H. J. Fissan, eds.), Elsevier Science Publishing Co., New York (1984), p. 669.
14. S. D. Cheung and R. P. Roberge, *Microcontamination* **5**, 45 (1987).
15. J. M. Davidson and T. P. Ruane, *Microcontamination* **5**, 35 (1987).
16. R. M. Thorogood, A. Schwartz, W. T. McDermott, and C. D. Holcomb, *Microcontamination* **4**, 28 (1986).
17. G. Sielaff and N. Harder, *Microcontamination* **2**, 57 (1984).
18. D. Hall, *Semicond. Inter.* **182**, (1984).
19. K. Shenai, P. A. Piacente, and B. J. Baliga in *Advanced Materials for VLST* (M. Scott, Y. Arkasaka, and R. Rief, eds.). Vol. 88-19, Electrochem Soc., Pennington, NJ (1988), p. 155.
20. Y. El-Mansy, *IEEE Trans. Electron Dev.* **ED-29**, 567 (1982).
21. E. Takeda, G. A. C. Jones, and H. Ahmed, *IEEE Trans. Electron Dev.* **ED-32**, 322 (1985).
22. E. Sun, T. Moll, J. Berger, and B. Alders, *IEDM Tech. Dig.* 478 (1983).
23. W. H. Lee, T. Osakama, K. Asdada, and T. Sugano, *IEEE Trans. Electron Dev.* **35**, 1876 (1988).
24. S. Asai, *Symposium on Advanced Science and Technology of Si Materials*, Kona, Hawaii (1991), unpublished.
25. K. M. Cham and S. Y. Chiang, *IEEE Trans. Electron Dev.* **ED-31**, 964 (1984).

2 Rapid Thermal Processing–Based Epitaxy

J. L. Hoyt
Solid State Electronics Laboratory
Stanford, California

I. Introduction to Silicon Epitaxy 15
   A. Role of Silicon Epitaxy in Integrated Circuit Technology 15
   B. Conventional Epitaxial Growth Processes 17
   C. Important Properties of Epitaxial Silicon 18
   D. Advanced Epitaxial Growth Techniques 20
   E. Rapid Thermal Processing–Based Epitaxial Reactors 23
II. Characteristics of Rapid Thermal Processing–Based Silicon Epitaxy 25
   A. Growth Kinetics 25
   B. Doping Profile Abruptness 27
   C. Material Quality 29
III. Growth of Strained Silicon–Germanium Alloys 31
   A. Device Applications 33
   B. Growth Basics 33
   C. Heterojunction Bipolar Transistors 35
   D. Misfit Dislocations 37
IV. Summary 40
References 41

Epitaxial growth of silicon and silicon related materials using rapid thermal processing (RTP) techniques is a rapidly growing field. Compared to other applications of rapid thermal processing such as annealing or oxidation, epitaxial growth is one of the most progressive, and hence less well-developed areas. Work in this area began in the mid-1980s with the development of the limited reaction processing (LRP) technique by Gibbons and Gronet at

Stanford University [1]. This technique combines rapid thermal processing and chemical vapor deposition (CVD). In early work on LRP, the wafer temperature, rather than the flow of reactive gases, was used to initiate and terminate layer growth. Limited reaction processing has been used to grow multilayer structures consisting of thin layers of n- and p-type Si [2], silicon/oxide/polysilicon structures [3], and thin $Si_{1-x}Ge_x$ layers, with thicknesses in the range of tens to hundreds of angstroms [4]. The technique has also been applied to the epitaxial growth of III–V compounds [5, 6].

In addition to the pioneering work on LRP at Stanford, research on epitaxial growth using rapid thermal processing techniques appeared in the literature in the late 1980s and early 1990s under the name of rapid thermal chemical vapor deposition (RTCVD) [7, 8], rapid transient epitaxy [9], rapid thermal processing chemical vapor deposition (RTPCVD) [10], as well as various photon and plasma assisted single-wafer processes [11]. One feature of such work is the growth of multiple layers without removing the wafer from the process chamber, thereby reducing the potential for interfacial contamination. Another key attribute is the ability to optimize the growth temperature for each layer in a complicated structure, since wafer temperature can be changed as readily as gas flows. Thermal exposure of the substrate is inherently minimized. The equipment employs lamp heating of individual wafers. The absence of a thick graphite susceptor allows for rapid changes in wafer temperature, and reduces memory effects from layer to layer and wafer to wafer. Such reactors are designed to minimize wall deposition, which reduces memory effects as well as particulate problems associated with flaking. In this chapter we lump together all CVD techniques that involve single-wafer epitaxy, in which rapid changes of wafer temperature can be achieved by lamp heating, under the heading of "rapid thermal processing applied to epitaxy." This is a natural grouping for a book on rapid thermal processing, since the differences between the various techniques mentioned above are more subtle than the distinction between RTP-based epitaxy and other growth techniques such as molecular beam epitaxy or ultrahigh vacuum chemical vapor deposition, which are discussed in Section I D.

Among the various physical processes involved in silicon integrated circuit fabrication, the growth of thin crystalline silicon layers on silicon wafers ("epitaxy") is the most demanding in terms of the requirements placed on the processing environment and equipment. In epitaxial growth, a perfect replication of the crystal structure of each atomic layer is required. However, there is a natural tendency for defects to form, particularly at low growth temperatures. The growth of epitaxial silicon layers with high electrical quality and precise thickness and doping control is a challenge that imposes constraints on all epitaxial growth equipment. Constraints related to characteristics such as ambient purity, vacuum compatibility, wall deposition, and processing time imply that a rapid thermal processor suitable for epitaxial growth will look different from one that is designed for annealing or silicide formation. Hence, this chapter begins with a brief review of general considerations for silicon-based epitaxy. Requirements for conventional and advanced epitaxy are discussed, and the various forms of advanced epitaxy are compared. Section II reviews the characteristics of RTP-based silicon epitaxy. The third section discusses a particular application for which RTP is well suited, namely epitaxial growth of Si and $Si_{1-x}Ge_x$ layers for heterojunction bipolar transistors.

I. Introduction to Silicon Epitaxy

This section briefly reviews conventional epitaxial growth applications, and the important properties of silicon epitaxial layers are listed. Various advanced epitaxial growth techniques are introduced in Section I D. For a general discussion of silicon epitaxial growth, the reader is referred to Chapter 2 of *VLSI Technology* [12].

A. ROLE OF SILICON EPITAXY IN INTEGRATED CIRCUIT TECHNOLOGY

Silicon epitaxy provides a means of controlling doping profiles beyond what can be achieved using diffusion and ion implantation. In its original application, epitaxy solved competing device requirements for low collector resistance and capacitance, as well as high breakdown voltages in bipolar transistors [13]. A lightly doped epitaxial silicon layer, which provides a lower base–collector breakdown voltage, is grown upon a heavily doped "buried layer" or substrate, which lowers the collector resistance and improves frequency performance. The evolution of epitaxial silicon technology as applied to bipolar transistors is shown in Fig. 1. The doping profile in Fig. 1a illustrates an *npn* bipolar transistor of the 1970s, including a 6 μm-thick n^- epitaxial Si layer grown on a heavily doped buried layer. The *p*-type base and n^+ Si emitter were typically formed by diffusing impurities from the surface into the lightly doped epitaxial Si. A typical bipolar transistor of the 1980s is shown in Fig. 1b, with a thin ion-implanted base layer. The thickness of the Si epitaxial layer, and in particular the thickness of the lightly doped collector region, is scaled considerably compared to transistors of the 1970s. A hypothetical doping profile for a fully scaled bipolar transistor is shown in Fig. 1c. All three regions (collector, base, and emitter) are formed by an advanced epitaxial growth technique.

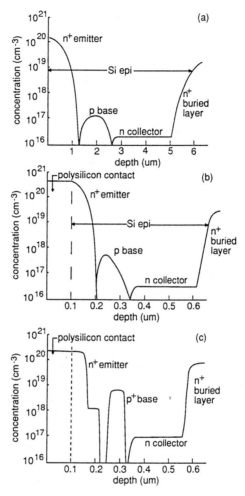

FIGURE 1. The evolution of vertical doping profiles in Si bipolar transistors: (a) 1970s, (b) 1980s, and (c) advanced transistor.

Epitaxial layers are also used in metal oxide semiconductor (MOS) technology to reduce alpha particle and latch-up problems [14], and in bipolar complementary MOS (BICMOS) applications. Selective epitaxial growth offers the potential to improve the performance of bipolar and MOS circuits [15]. However, the majority of applications of epitaxial silicon still consist of a single lightly doped layer, with thickness in the range of 1 to 10 μm. Highly advanced applications seek to provide arbitrary doping profiles, with thicknesses ranging from tens of angstroms up to several micrometers.

B. CONVENTIONAL EPITAXIAL GROWTH PROCESSES

Chemical vapor deposition of epitaxial silicon is usually performed in a quartz reaction chamber with a number of wafers placed flat against a silicon carbide coated, graphite susceptor. Deposition takes place with the wafers held at elevated temperature while a gas mixture consisting of purified hydrogen and a silicon source gas is flowed into the reactor. Silane and dichlorosilane are typical silicon source gases. Hydrides such as phosphine and diborane can be added to dope the layers during growth. Hydrogen provides the required gas velocity, dilutes unwanted impurities, and participates in various surface reactions. The details of the chemical reactions involved in silicon epitaxial growth are still not completely understood. However, the overall CVD process is generally modeled as a series combination of a mass transport process that is weakly temperature dependent, and a surface reaction process that is exponentially dependent on wafer temperature [16]. In this case, the growth rate G is given by an expression of the form $G \propto (1/g_m + 1/g_s)^{-1}$, where g_m and g_s represent the gas phase mass transport and surface reaction rates, respectively. An Arrhenius plot of silicon growth rate for various silicon sources is shown in Fig. 2 [17]. The exponential region corresponds to surface reaction rate limited growth ($g_m \gg g_s$), while the relatively flat region above 1000 °C is indicative of mass transport limitations ($g_m \ll g_s$).

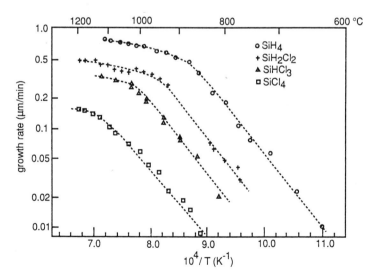

FIGURE 2. Arrhenius plot of epitaxial silicon growth rate using various Si sources. From Eversteyn [17].

Epitaxial reactors are designed to operate at either atmosphere (760 torr) or reduced (typically 20 to 100 torr) pressures. As indicated in Section I C, there is a tendency for the properties of epitaxial silicon layers to improve as the growth pressure is reduced. In most cases, the graphite susceptor is heated inductively or by infrared lamps, and the quartz and the adjoining apparatus are water and/or air cooled. Growth temperatures are typically in the 1150 °C range. However, there is a trend towards low-temperature epitaxy, which generally refers to growth between 750 and 1000 °C [18]. This trend is driven by device requirements for thinner layers, better dopant profile control, and selective epitaxial growth [15].

C. Important Properties of Epitaxial Silicon

Epitaxial layers generally have lower carbon and oxygen concentrations than silicon substrates grown by the Czochralski process. However, bulk silicon wafers are free of dislocations, stacking faults, and other structural defects that can be introduced during silicon epitaxial growth. The following is a list of important characteristics for epitaxial silicon as applied to integrated circuit fabrication [19]. Each characteristic is briefly described.

- The *defect density* observed after etching with various solutions that reveal crystallographic defects is an important characteristic. Layers should be at least 1 μm thick for adequate delineation. Typical etching solutions are the Wright, Sirtl, and Schimmel etches [20]. Defects such as dislocations, stacking faults, and hillocks can be revealed. A Nomarski micrograph of an epitaxial defect after Wright etching is shown in Fig. 3. Defect densities less than 1/cm^2 are required for high-density circuits. Defect density generally increases as growth temperature is reduced and as growth pressure is increased [18].
- *Contaminant levels*, including metals, carbon, and oxygen, must be kept to a minimum. The presence of high levels of contamination can be determined by secondary ion mass spectrometry (SIMS) [21]. For epitaxial silicon, the contaminant levels should be below the detection limit of SIMS for these elements in silicon, which is roughly 10^{16} cm^{-3} for metals such as K, Na, and Cu, and about 5×10^{17} and 5×10^{16} cm^{-3} for oxygen and carbon respectively [22]. Metals and other deep traps can be measured by highly sensitive techniques such as deep level transient spectroscopy [23].
- High *minority carrier lifetimes* indicate the absence of electron and hole recombination and generation centers, which can degrade device performance. For lightly doped material, lifetimes should be in the 100 to 500 μsec range. Lifetimes tend to be reduced as the growth temperature is lowered.

FIGURE 3. An epitaxial defect as seen by Nomarski microscopy after Wright etching.

- The *surface appearance* is an indication of the general epitaxial silicon quality and the properties of the starting substrate. Layers should be free of haze, speckle, and roughness under bright UV and white light illumination.
- The *thickness and resistivity uniformity* are also important properties. Gas flow, reactant depletion, and temperature variations across the wafer all play a role in determining uniformity. The resistivity is a function of both the doping level and the layer thickness.
- *Autodoping* refers to the unintentional carryover of dopants from underlying layers ("vertical autodoping"), and from hot parts of the reactor or other regions of the wafer ("lateral autodoping") [24]. Autodoping for As and Sb is reduced at low pressures and temperatures. Autodoping limits dopant transition widths and device applications.
- *Dopant transition widths and thin layer control* are functions of the temperature–time exposure, surface chemistry, and gas pressures. Abrupt transitions are required in order to produce profiles of arbitrary shape in advanced epitaxy.
- Requirements for both n *and* p-*type doping ranges* are being expanded as the number and type of applications for epitaxial layers increases. For conventional epitaxy, the majority of applications require doping in the range of 10^{14} to 10^{17} cm^{-3}. Advanced applications require high n and p-type doping levels.
- *Slip* refers to the displacement of planes of atoms as a result of stress associated with temperature gradients in the wafer during epitaxial growth [25]. Slip is a strong function of the wafer support, the heating mechanism, and the temperature–time exposure. Slip at the edge of a silicon wafer after epitaxial silicon growth is shown in the Nomarski micrograph in Fig. 4.
- *Pattern shift* refers to the tendency for patterns on the wafer prior to epitaxial growth to be shifted laterally and/or distorted at the surface of the epitaxial layer. Pattern shift is reduced by low-pressure growth [12].
- *Surface particulate contamination* associated with the gas source or the reactor parts can degrade subsequent device processing, or nucleate defects if it occurs prior to epitaxial growth.

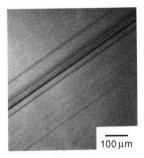

FIGURE 4. Nomarski micrograph of slip at the edge of a silicon wafer after epitaxial growth.

- *Selective epitaxial growth* refers to growth in exposed silicon regions patterned in a silicon dioxide masking layer. Silicon can be grown selectively with a dichlorosilane source, or by using silane with the addition of hydrogen chloride gas.
- The *throughput* is the number of wafers that can be processed per unit time. The throughput of a reactor influences the economic feasibility in commercial applications.
- The capability for *multilayer deposition* (see Fig. 1c) and for heteroepitaxial growth (see Section III) are important properties of advanced epitaxy.

D. ADVANCED EPITAXIAL GROWTH TECHNIQUES

One of the driving forces for advanced epitaxy is lower cost per wafer. Factors such as capital equipment, energy consumption, raw materials, and labor figure into the cost of epitaxy. If these factors are roughly constant, then the cost per wafer scales inversely with the throughput of the reactor [26]. For a conventional reactor, throughput depends primarily on the number of wafers that can fit on the susceptor. As the diameter of the wafers increases, the usable area on the susceptor decreases, so that the susceptor must be scaled to very large dimensions. Hence power and gas consumption and the cost per wafer increase considerably. Single-wafer epitaxy offers one solution to this problem, and commercial single-wafer equipment is available [27]. For single-wafer epitaxy, throughput depends very little on wafer diameter, but rather on processing time. For a machine based on RTP, the process time can be minimized, especially if the equipment includes separate chambers for automatic wafer loading and removal. In addition to cost and efficiency considerations, maintaining layer thickness uniformity as wafer diameters are scaled beyond 6 in. is more feasible in a single-wafer machine, compared to a batch reactor.

Rapid Thermal Processing–Based Epitaxy

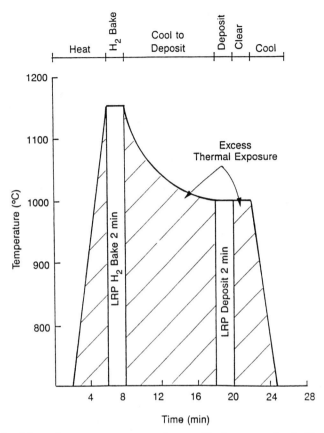

FIGURE 5. Schematic temperature versus time profile for conventional silicon epitaxy (hatched regions) compared to the LRP growth process. From Gronet [19].

There are also fundamental device considerations driving the development of advanced silicon epitaxy. The vertical scaling of devices is a strong driving force for the development of advanced, blanket-wafer epitaxy, as illustrated in Fig. 1 for bipolar devices. The lateral scaling of devices is an important driving force for the development of selective epitaxial growth [15]. Both types of scaling offer improvements in device and circuit performance. If devices are to be scaled it is important that the total temperature–time exposure of the wafer be minimized to prevent dopant diffusion during processing. The reduction in temperature–time exposure afforded by an RTP-based technique is illustrated schematically in Fig. 5. The hatched regions show the excess exposure characteristic of conventional epitaxy, compared to the LRP growth process [20].

1. Molecular Beam Epitaxy

Silicon molecular beam epitaxy (MBE) is a physical deposition process. References [28] and [29] contain detailed discussions of silicon-based MBE. Growth is performed at pressures in the range of 10^{-8} to 10^{-10} torr. In conventional MBE, fluxes of silicon and dopant species obtained by evaporation are directed toward a silicon wafer. Substrate temperatures range from 400 to 800 °C. Because of the low growth temperature, dopant diffusion is minimized and deposition of layers with thicknesses on a monolayer scale is possible. A number of new advances in solid state physics and materials science have been achieved using structures fabricated by MBE. However, several factors have inhibited acceptance in integrated circuit fabrication, included high defect densities (typically in the 10^4 cm^{-2} range), difficulties with dopant incorporation, and low throughput. In addition, metal contamination has historically limited minority carrier lifetimes, although improvements have been made in this area [30]. MBE using gas sources, rather than evaporation from a solid source, is a promising new area of research [31].

2. Ultrahigh Vacuum Chemical Vapor Deposition

There are several groups studying chemical vapor deposition of Si in ultrahigh vacuum environments [32–34]. In the low-temperature, batch process developed by Meyerson, wafers are held vertically in a quartz boat, along the axis of a hot-wall, quartz furnace [32]. The geometry is similar to that used in a conventional oxidation tube. The background pressure of the load-locked deposition system is in the 10^{-11} torr range, and the growth pressure is maintained in the 1 to 10 mtorr range by a turbomolecular pump. There is no hydrogen carrier gas. Uniform deposition is achieved in spite of the close stacking of the wafers because of the combination of long mean free path and low sticking coefficient of the precursor molecules in the source stream. The low deposition temperature of this process (about 550 °C) makes it very attractive for dopant profile control, and impressive minority carrier device results have been obtained using p-type $Si_{1-x}Ge_x$ layers [35]. However, there are serious difficulties associated with n-type doping.

3. Other Advanced Growth Techniques

Other advanced techniques currently under investigation include atomic layer epitaxy (which seeks to achieve self-limited growth [36]), plasma enhanced CVD [37], and photon-assisted CVD [38]. Each of these techniques applies special chemistry or energy enhancement to achieve controlled growth at reduced temperatures. However, information about the material and electrical quality of epitaxial silicon grown by such techniques is still incomplete. Reactors designed for the growth of multiple layers of III-V compounds have been built and researched for many years under the name metallorganic chemical vapor deposition (MOCVD) [39]. In these reactors, fast gas switching designs are often employed to achieve abrupt interfaces. A III-V MOCVD reactor has been modified to grow thin layers of epitaxial silicon with abrupt doping transitions [40]. Although these authors do not use RTP, combinations of gas and temperature switching provide enhanced processing flexibility [5].

E. RAPID THERMAL PROCESSING-BASED EPITAXIAL REACTORS

Epitaxial reactors incorporating RTP technology are available commercially, although much of the work is still being pursued in research laboratories. A rapid thermal processing system suitable for advanced epitaxy has a number of design constraints not usually associated with systems designed for other types of processing. These considerations include control of the ambient purity (oxygen and water levels in the part per billion range), vacuum compatibility, optimization of gas flow patterns, and minimization of wall deposition. Temperature uniformity is critical for obtaining uniform deposition rates at temperatures below 1000 °C, and for eliminating slip in wafers that are heated above this temperature. Systems with individual and dynamic lamp control are under development. A highly schematic view of an advanced, RTP-based epitaxial reactor is shown in Fig. 6.

One of the most critical engineering issues associated with RTP systems in general is the measurement and control of wafer temperature. This topic is discussed in detail in Chapter 9 of this book, and reviews can be found in Refs. 41 and 42. Basic physical mechanisms such as infrared (IR) radiation pyrometry, thermal expansion, IR transmission, and acoustic wave propagation can be used with varying degrees of success to infer wafer temperature. Certain requirements for epitaxial growth place additional constraints on the wafer temperature measurement technique. The low water and oxygen content of the reactor ambient must not be compromised by the addition of apparatus associated with temperature measurement. Metallic or ionic contamination of the silicon layers may result if metal parts or special optical materials come into contact with hot, corrosive

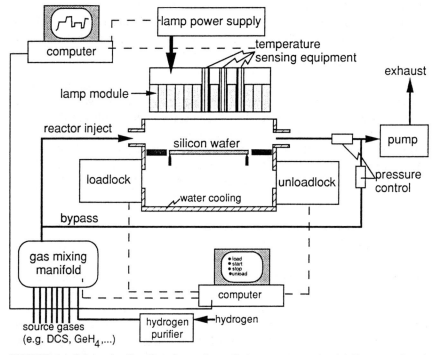

FIGURE 6. Schematic diagram of an advanced, low-pressure epitaxial Si reactor based upon RTP technology.

gases during growth. The use of pyrometry must include special techniques to avoid or correct for deposition on the optical window, variations in the optical properties of the various layers being deposited on the wafer (e.g., Ge and Si), and interference from the lamps and other hot bodies within the field of view of the pyrometer. Wall deposition can be reduced by the use of a chlorinated silicon source species, and careful temperature control and purging of the window region. Infrared transmission is a relatively simple, noncontact technique that can be used through quartz, and hence does not require modification of an existing quartz tube design [43]. A laser diode and detector are used to measure the IR transmission of the silicon substrate, which varies with temperature. A schematic diagram of a set-up in use in an RTCVD reactor [43] is shown in Fig. 7. Using laser diode sources at 1.3 and 1.55 μm, this technique can cover the temperature range from 400 to 850°C. The growth of several micrometers of silicon has little effect on this temperature measurement technique. However,

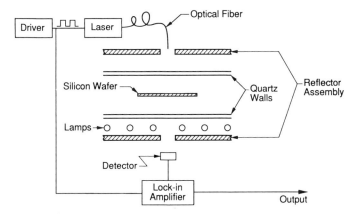

FIGURE 7. Schematic diagram of an IR transmission setup used for temperature measurement in an epitaxial reactor. Modulation is used to increase the signal-to-noise ratio. From Sturm et al. [43].

for growth of $Si_{1-x}Ge_x$, errors associated with absorption in the growing layer limit the thickness and germanium fraction for which the technique is applicable. To date, this temperature measurement technique has been demonstrated for thicknesses and germanium fractions up to roughly 1000 Å and 45% Ge [44].

II. Characteristics of Rapid Thermal Processing–Based Silicon Epitaxy

The characteristics of epitaxial silicon layers grown using RTP techniques have been reported by several groups [20, 45–47]. The following sections discuss growth kinetics, doping profile abruptness, and material quality.

A. Growth Kinetics

Regolini et al. have reported detailed studies of the kinetics of silicon epitaxial growth in an LRP reactor at growth pressures in the 1 to 10 torr range using silane (SiH_4), dichlorosilane (SiH_2Cl_2), and disilane (Si_2H_6) source gases, mixed with various combinations of hydrogen (H_2), nitrogen (N_2), and hydrogen chloride (HCl) [48–50]. An Arrhenius plot for epitaxial silicon growth rate using (a) SiH_4, (b) SiH_2Cl_2, and (c) SiH_4/HCl source gases is shown in Fig. 8 [50]. In each case, the carrier gas is hydrogen, at a flow rate of 2 standard liters per minute (slm). The activation energies for these three curves are 1.7, 2.6, and 3.2 eV respectively. Note that the growth

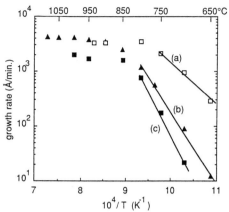

FIGURE 8. Silicon growth rate in an LRP reactor operating at 2 torr, with 2 slm of hydrogen, and (a) 40 sccm of SiH_4, (b) 80 sccm of SiH_2Cl_2, and (c) 20 sccm of SiH_4 and 2 sccm of HCl. After Regolini et al. [50].

rate using silane is higher than that for dichlorosilane, and that the difference is magnified at low temperatures. Curve (c) shows that in the presence of HCl the growth rate from a silane source is dramatically reduced. In addition, the growth rate from a dichlorosilane source decreases linearly with HCl flow [49]. On the basis of a detailed study, these authors propose a model for epitaxial growth from a dichlorosilane source that is slightly different from the chemistry previously reported in the literature. Most of the studies in the literature have focused on growth at atmospheric pressure, or in the 20 to 100 torr range. The model proposed by Regolini et al. assumes that the kinetics are controlled by the main decomposition reaction

$$SiH_2Cl_2 \rightarrow SiHCl + HCl \qquad (1)$$

(dissociation energy 2.7 eV), which is different from the widely accepted reaction

$$SiH_2Cl_2 \rightarrow SiCl_2 + H_2 \qquad (2)$$

(dissociation energy 1.4 eV). The observed energetics and dependence on the HCl flow are better explained by Eq. 1. In addition, the HCl decomposition product in Eq. 1 is believed to be the main reason that full selectivity is obtained for growth using dichlorosilane over the entire temperature range from 1100 to 650 °C, without the addition of HCl gas.

Rapid thermal processing also offers researchers the capability of studying fundamental aspects of chemical vapor deposition in a pressure and time regime not previously accessible in the laboratory. For example, using an RTCVD reactor connected to an ultrahigh vacuum (UHV) analysis chamber, researchers at IBM have been able to "freeze in" surface species present during growth by rapidly cooling the sample [51]. For wafer temperatures up to 575 °C, the hydrogen coverage can be completely frozen on the surface by rapid cooling and pump-down of the reactor. At higher temperatures, the coverage is only partially maintained. After the quenching process, the hydrogen coverage was studied *in situ* by operating the reactor as a thermal desorption spectrometer. This technique was used to study the growth kinetics from 450 to 700 °C, using a silane source and a growth pressure in the millitorr range. The authors conclude that below about 560 °C the growth rate is limited by hydrogen desorption, while above this temperature the growth kinetics is controlled by the decomposition rate of silane on the silicon surface. The rapid quenching afforded by an RTP reactor equipped with rapid pumping capability enables researchers to study the silicon surface using high vacuum (less than 10^{-5} torr) analysis equipment, such as a mass spectrometer. Hence, this type of analysis is no longer limited to UHV-based growth techniques such as MBE or UHVCVD.

B. Doping Profile Abruptness

In order to obtain a thin epitaxial layer on a heavily doped substrate or buried layer, the doping transition width must be small. The transition width will be limited by outdiffusion, which depends upon the thermal exposure of the epitaxial growth process, by autodoping, which is a function of gas residence time and dopant evaporation rates, and by surface segregation, which refers to site exchange between surface silicon atoms and subsurface dopant atoms during growth. Early results on LRP-grown layers revealed that intrinsic epitaxial silicon layers can be grown on n^+ silicon substrates with dopant transition widths comparable to those obtained by MBE [1]. In general, the transition widths are below the depth resolution of routine SIMS analysis, which is approximately 150 Å/decade for the leading edge of an Sb signal using 15-keV Cs^+ bombardment during SIMS sputtering [52]. The SIMS data in Fig. 9 illustrates the doping profile for an $i/p^+/i$ epitaxial silicon structure grown by LRP at 900 °C. The boron transition slopes for this SIMS measurement are roughly 100 Å/decade.

More accurate determination of the actual doping profile abruptness requires special SIMS techniques [40, 53]. For LRP-grown $i/p^+/i$ structures, a detailed SIMS analysis has been performed by measuring SIMS profiles using various O_2^+ bombardment energies [53]. The true decay length associated with the doping transition regions can be estimated by extrapolating the measured decay lengths to zero bombardment energy. For

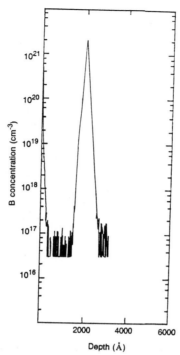

FIGURE 9. SIMS profile for an $i/p^+/i$ epitaxial silicon structure grown by LRP at 900 °C. Analysis was performed using 5-keV O_2^+ ions. The SIMS depth resolution limits the measurement of the pulse leading and trailing edges to approximately 100 Å/decade in this case. From Gronet et al. [2].

LRP growth at 900 and 1000 °C, the transition regions for boron pulses are consistent with diffusion broadening during growth. For boron pulses grown at 900 °C, doping transition widths of approximately 40 Å/decade were found. Lowering the growth temperature decreases the doping transition width, and boron transitions of 20 to 30 Å/decade have been obtained at 800 °C. As the growth temperature is lowered still further, for example for $Si_{1-x}Ge_x$ growth at 625 °C, even thinner pulses can be grown (on the order of tens of angstroms).

In general, *n*-type doping during silicon epitaxial growth is a more difficult task than *p*-type doping. During growth of *n*-type silicon, processes such as autodoping, diffusion, surface segregation, and defect generation have limited the ability to obtain thin, abrupt, heavily doped epitaxial layers. Impressive results for the growth of $i/n/i$ structures have recently been reported using dichlorosilane and phosphine in a single-wafer reactor operating at atmospheric pressures [54].

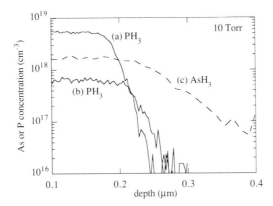

FIGURE 10. SIMS data comparing (a) and (b) phosphorus, and (c) arsenic turn-on transients during LRP growth in the temperature range 800 to 850 °C. The strong tendency for surface segregation of arsenic is believed to produce the delay in achieving a steady-state concentration of arsenic in the silicon. The dopant gas to dichlorosilane ratios are (a) 1.9×10^{-4}, (b) 7.5×10^{-5}, and (c) 6.3×10^{-5}.

Under similar CVD growth conditions, phosphorus doping appears to result in more abrupt doping transitions than arsenic. This may be associated with the strong tendency for arsenic atoms to segregate to the growing surface [55]. A delay in achieving a steady state concentration of dopant can result in a slow turn-on of the doping as the layer is grown. The turn-on characteristics for arsenic- and phosphorus-doped layers grown under similar condition are compared in Fig. 10. The dopant-to-silicon ratios in the gas phase are roughly 10^{-5} in this experiment. It appears that a steady-state dopant concentration in the silicon is obtained more rapidly for phosphorus than for arsenic. Using very high arsine-to-dichlorosilane ratios, reasonably abrupt turn-on transients have been reported by RTCVD [56].

C. MATERIAL QUALITY

The suitability of LRP-grown epitaxial silicon for device applications has been assessed by a variety of characterization techniques [20, 45]. Very low etch-pit densities have been obtained. Interfacial and bulk metal, carbon, and oxygen contamination levels are below the SIMS detection limits. Photocurrent decay techniques have been used to measure minority carrier recombination lifetime in the 30 to 80 μsec range in this material [45]. DLTS analysis of multiple interrupted-growth interfaces indicates the absence of both electron and hole traps, within the sensitivity of the measurement technique, which is approximately 10^{11} cm^{-3} [20].

1. Interrupted-Growth Interfaces

Using RTP techniques, temperature and/or gas-flow switching of the growth may be employed. In temperature-switched growth there are short interrupt periods between layers (on the order of tens of seconds to several minutes), during which the wafer is cool and the gas flows are being established in the reactor. Growth is then initiated by rapidly heating the wafer to the desired growth temperature [1]. The potential for impurity incorporation at interrupted-growth interfaces is an important issue in the case of temperature-switched growth [6]. In most reactors, the primary contaminant species is water vapor. Hence oxides are expected to form on the silicon surface during the interrupt period, when the wafer is cool. Apparently, these oxides are removed by various chemical reactions that take place during the growth of the first few atomic layers of epitaxial silicon [20]. For temperature-switched growth using dichlorosilane, interfacial carbon and oxygen concentrations below the SIMS detection limit have been obtained. SIMS carbon and oxygen depth profiles are shown in Fig. 11 for both (a) a Si/Si/Si multilayer structure grown at

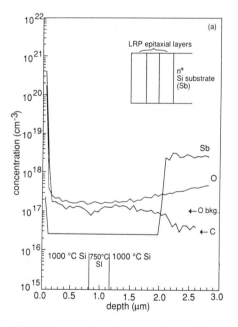 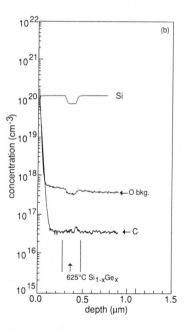

FIGURE 11. SIMS carbon and oxygen depth profiles in an LRP-grown (a) Si/Si/Si structure and (b) Si/Si$_{1-x}$Ge$_x$/Si structure. In (b) the germanium fraction in the alloy layer is roughly 15%. The arrows indicate the SIMS background levels. The dip in the background oxygen level in the Si$_{1-x}$Ge$_x$ is associated with the lower secondary ion yield for oxygen in that material.

1000/750/1000 °C, and (b) a Si/Si$_{1-x}$Ge$_x$/Si structure with epitaxial layers grown at 850/625/850 °C [57]. The incorporation of oxygen in Si$_{1-x}$Ge$_x$ is discussed in more detail in Section III B.

Regolini *et al.* have performed capacitance–voltage (CV) analysis on epitaxial layers grown using silane, dichlorosilane, and disilane source gases in an LRP reactor [46]. The background doping was *n*-type in each case, with levels as low as 10^{13} cm^{-3} obtained using disilane. Measurements on samples containing multiple epitaxial interfaces indicate that there are very few shallow donor states at interrupted-growth interfaces. Interfacial charge densities as low as 10^9 cm^{-2} and minority carrier lifetimes in the 300 to 400 μsec range were deduced from the CV analysis [46].

2. In Situ *Grown* pn *Junction Diodes*

Majority [3] and minority carrier devices [58] exhibiting good performance have been fabricated using LRP-grown epitaxial layers. Since the interrupted growth interface is contained *within* the diode space charge region, *in situ* grown *pn* junctions provide a stringent test of the device quality [58]. Mesa-isolated diodes have been fabricated in epitaxial layers doped using diborane and arsine source gases during growth. Doping levels were 7×10^{17} and 4×20^{16} cm^{-3} for the *p*- and *n*-type layers respectively. Typical forward and reverse diode characteristics are shown in Fig. 12. Forward current ideality factors of 1.01 ± 0.003 are observed over more than 7 decades of current. Very low reverse current densities (less than 3.5 nA/cm^2 at -5 V) and high breakdown slopes of 26–30 dec/V have been achieved [58]. From the reverse bias characteristics, the generation lifetime is estimated to be about 25 μsec.

III. Growth of Strained Silicon–Germanium Alloys

The growth of layers of different materials in a single structure is referred to as *heteroepitaxy*. Rersearch on Si$_{1-x}$Ge$_x$ began with the development of silicon–germanium MBE in the 1960s. Early work was hampered by material quality issues. The 4% lattice mismatch between silicon and germanium plays a key role in determining epitaxial growth and processing conditions. This mismatch produces a driving force for bond breaking and misfit dislocation formation at the Si$_{1-x}$Ge$_x$/Si interface. Thin Si$_{1-x}$Ge$_x$ layers can be grown on Si in a manner such that the bonds in the Si$_{1-x}$Ge$_x$ are distorted, and the layer remains fully *strained*. Section III D of this chapter discusses misfit dislocation formation and strain relaxation during

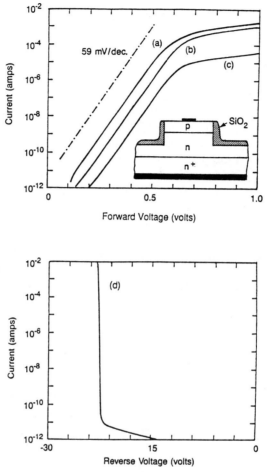

FIGURE 12. Room temperature forward and reverse bias current–voltage characteristics for LRP-grown *pn* diodes. The diode areas are (a) 4.0×10^{-3} cm^{-3}, (b) 4.0×10^{-4} cm^{-3}, (c) 2.5×10^{-5} cm^{-3}, and (d) 1.0×10^{-4} cm^{-3}. The inset shows the diode cross section. The dot-dashed line indicates the ideal forward-bias slope at the measurement temperature. From King *et al.* [58]. Copyright © 1988 IEEE.

growth and processing in more detail. In the 1980s there was significant progress in the area of strained $Si_{1-x}Ge_x$ growth by MBE, UHVCVD, and by rapid thermal techniques. This was accompanied by an increase in research on device fabrication and analysis, particularly in the area of heterojunction bipolar transistors (HBTs). The first CVD-grown HBTs were fabricated by King *et al.* using heteroepitaxial layers grown by LRP [59].

A. Device Applications

Silicon is the dominant material in the integrated circuit industry. High manufacturability, combined with the excellent properties of the surface oxide that forms on silicon, is the key reason for the use of silicon integrated circuits rather than other semiconductor materials such as gallium arsenide (GaAs). In the laboratory, however, scientists have been able to achieve record device operating speeds by designing transistors containing various compound semiconductor materials using heteroepitaxial growth. A number of analogous performance advantages may be achieved by combining layers of silicon with layers of $Si_{1-x}Ge_x$ alloys. Using such techniques, performance limits of existing silicon technologies can be extended, and completely new and useful device structures may be realized. Research in the areas of heterojunction bipolar transistors, modulation doped field effect transistors, various quantum effect devices, photodetectors, and light emitting diodes indicate a number of potential applications in electronics and optoelectronics. References 29 and 60 contain information on such applications. Integrating $Si_{1-x}Ge_x/Si$ heterostructures into Si integrated circuits remains a challenge because of several issues that impact material quality, and hence device fabrication and design.

B. Growth Basics

The difference in lattice constants between Si and $Si_{1-x}Ge_x$ increases the tendency for growth in clumps or islands, rather than as planar films. Experimentally, gross islanding is observed by a characteristic milky appearance under white light. Such films are unusable for device applications. Islanding is a function of surface coverage, growth temperature, and Ge fraction [61]. For MBE growth, it has been shown that islanding can be suppressed at all germanium compositions by growing the $Si_{1-x}Ge_x$ at sufficiently low temperatures (550 °C) [61]. The temperature and germanium composition at which islanding occurs for LRP growth of strained $Si_{1-x}Ge_x$ on Si appears to be qualitatively similar to that reported for MBE growth [4].

Because of such considerations, CVD growth of $Si_{1-x}Ge_x$ is usually performed at temperatures between 550 and 700 °C. The growth rate of epitaxial silicon layers from a dichlorosilane source is low in this temperature range (see Figs. 2 and 8). However, during CVD of $Si_{1-x}Ge_x$ significant enhancements in growth rate are associated with the addition of germane. Enhancements are observed for both silane [62] and dichlorosilane sources, and at various growth pressures [4, 63, 64]. In addition to enhancing the

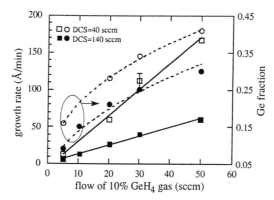

FIGURE 13. $Si_{1-x}Ge_x$ growth rate (solid lines) and germanium fraction (dashed lines) as a function of germane flow rate in an LRP reactor. Total growth pressure is 10 torr. The hydrogen flow rate is 10 slm.

hydrogen desorption rate at low temperatures, germane may act as a catalyst for the decomposition of the silicon source. As shown by the data in Fig. 13, the $Si_{1-x}Ge_x$ growth rate is roughly linearly proportional to the germane flow at 625 °C. The Ge fraction in the alloy increases sublinearly with germane flow.

Impurities such as oxygen incorporate more readily during the growth of $Si_{1-x}Ge_x$ than the growth of Si [57]. Oxygen in the $Si_{1-x}Ge_x$ reduces the minority carrier lifetime [65, 66], but increases the thermal stability of the $Si_{1-x}Ge_x$ against strain relaxation [67]. The lowest oxygen concentrations in the $Si_{1-x}Ge_x$ have been obtained when detailed attention was paid to the reactor ambient. In CVD, the hydrogen is purified relatively easily at the point of use using commercially available equipment [68]. This ultrapure gas is then mixed with very small quantities of high concentration source gases, to dilute impurities from the source gases themselves. Point-of-use purifiers for the various hydride gases are under development. For RTP-based epitaxy, the silicon source is typically dichlorosilane, which minimizes wall deposition and promotes selective epitaxial growth, and the germanium source is usually germane. Arsine, phosphine, and diborane are typical dopant sources.

In the initial work on RTP-based epitaxial growth of $Si_{1-x}Ge_x$, various groups reported high levels of oxygen incorporation in the $Si_{1-x}Ge_x$ films [59, 69]. Prior to the addition of a load-lock chamber, $Si_{1-x}Ge_x$ films grown by LRP at temperatures below 800 °C contained bulk oxygen concentrations on the order of 10^{20} cm^{-3} [59]. This is in contrast to Si epitaxial layers, which were grown in the same reactor without the load-lock, with oxygen concentrations below measurement detection limits (10^{17} cm^{-3}). Addition of a load-lock to the growth chamber has enabled the growth of $Si_{1-x}Ge_x$ with oxygen concentrations below the SIMS detection limits (see Fig. 11). In the load-locked system the base pressure is roughly 10^{-7} torr. While load-locking can certainly be beneficial, it does not solely determine layer contamination, because oxygen incorporation in $Si_{1-x}Ge_x$ is a function of the purity of the source gases, details of the gas injection, and reactor geometry.

C. Heterojunction Bipolar Transistors

In a Si homojunction transistor the barrier for carrier injection from the emitter into the base is controlled by the relative emitter and base doping levels. Higher base doping produces a larger barrier. The addition of germanium to the base of a silicon bipolar transistor reduces the band gap in the base and hence the emitter–base barrier. In this case higher base doping may be used without reducing the current gain. Higher base doping reduces the base sheet resistance, which is an important parameter determining switching speed in both analog and digital applications [70]. For an HBT with a heavily doped base and lightly doped emitter in which the material composition and doping change abruptly, the ratio of the collector current in the HBT to that in a Si device is given by [71, 72]

$$\frac{I_{C(HBT)}}{I_{C(Si)}} = \frac{(N_B W_B / D_B)_{Si}}{(N_B W_B / D_B)_{HBT}} \exp\left(\frac{\Delta E_v}{kT}\right) \quad (3)$$

where N_B, W_B, and D_B are the base doping, width, and minority carrier diffusivity, respectively, and ΔE_v is the valence band discontinuity between Si and $Si_{1-x}Ge_x$. If the germanium fraction is graded across the emitter–base depletion region to eliminate conduction band spikes, the total bandgap difference ΔE_g replaces ΔE_v in the above expression. For strained $Si_{1-x}Ge_x$ on Si, ΔE_v is approximately equal to ΔE_g. It is clear from Eq. 3 that an increase in the base doping in the HBT can easily be compensated by the exponential dependence on the bandgap difference. In addition, the dependence on the thermal voltage kT improves the low-temperature performance. The magnitude of the $Si/Si_{1-x}Ge_x$ bandgap difference as a function of the germanium fraction in the alloy is illustrated in Fig. 14. The symbols represent values extracted from analysis of heterojunction device characteristics. The curves labeled "strained-calc." indicate the published calculations for strained $Si_{1-x}Ge_x$ [72]. The lower curve shows published measurements for unstrained $Si_{1-x}Ge_x$ [73]. The strain in the $Si_{1-x}Ge_x$ splits the valence band degeneracy and accounts for a significant bandgap reduction for a given amount of germanium in the alloy.

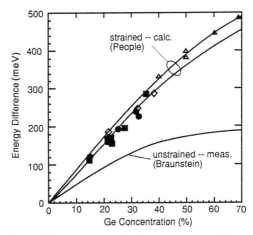

FIGURE 14. Bandgap difference versus Ge fraction extracted from analysis of devices such as ● HBTs and ■ diodes [76], ◇ BICFETs [77], and △ photodiodes [78]. Published calculations for ΔE_g in *strained* $Si_{1-x}Ge_x$ [73], and measurements of ΔE_g for *unstrained* $Si_{1-x}Ge_x$ [74] are also shown.

Using a relatively simple mesa isolated process, HBTs with excellent current characteristics have been fabricated using RTP-based epitaxial growth [75, 66]. In these devices, the epitaxial collector, base, and emitter are grown sequentially in the same chamber, at temperatures of approximately 1000, 650, and 850 °C respectively. Ohmic contacts to the emitter and base are formed by ion implantation and RTA at 850 °C for 10 sec. Individual devices are isolated by plasma etching down to the lightly doped Si collector layer. Oxide formed by low-pressure chemical vapor deposition at 400 °C is used for sidewall passivation. Plots of current gain versus collector current are shown in Fig. 15 for HBTs (solid lines) and a Si control device (dashed line). Although the HBT base doping is about 15 times larger than that in the Si device, a high gain is maintained over many decades of collector current [75]. Fine geometry devices with effective emitter areas as small as $2 \times 1 \mu m \times 10 \mu m$ have been fabricated in LRP-grown material [79]. Using this relatively simple mesa isolation scheme, the measured unity gain cut off frequency f_T is in the range of 30 GHz, and the maximum frequency of oscillation f_{max} is approximately 35 GHz [79]. Very low base resistances (<1 kΩ/☐) have been achieved in thin (~ 250 Å) $Si_{0.69}Ge_{0.31}$ layers. Analysis indicates that the speed is limited by parasitic resistances and collector junction capacitance, rather than the intrinsic portion of the device [79].

Bandgap grading in the neutral portion of the base (by grading the Ge fraction across the base) can further reduce the switching time [80]. In a

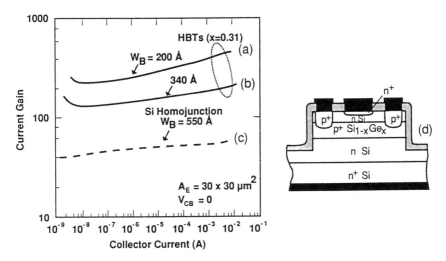

FIGURE 15. Current gain versus collector current for HBTs with a base doping of 7×10^{18} cm^{-3} (a) and (b), and a Si control device (c) with base doping of 5×10^{17} cm^{-3} [76]. The neutral base widths for the devices represented by (a) and (c) are roughly equal. HBT cross section is shown in (d). Adapted from C. King [75].

polysilicon emitter process, IBM has demonstrated HBTs with unity gain cut off frequencies, f_T of 75 GHz [80], and ECL ring oscillator delays of 28 psec [81] using epitaxial Si$_{1-x}$Ge$_x$ bases grown by UHVCVD.

D. MISFIT DISLOCATIONS

Because of the lattice mismatch between silicon and germanium, line defects called misfit dislocations tend to form at the Si$_{1-x}$Ge$_x$/Si interface during growth and processing. A plan-view transmission electron micrograph of an array of interfacial misfit dislocations between a Si$_{1-x}$Ge$_x$ epitaxial layer and a Si substrate is illustrated in Fig. 16. The average spacing between misfit dislocations in this sample is roughly 0.5 µm. Equilibrium theory predicts the thickness, at a given Si$_{1-x}$Ge$_x$ composition, above which it is energetically favorable for misfit dislocations to exist. However, in the Si$_{1-x}$Ge$_x$/Si system, it is generally believed that there is a kinetic barrier to the formation of misfit dislocations [82, 83]. A number of investigations have shown that at low temperatures films with low misfit dislocation density (i.e., large misfit dislocation spacing) can be grown to thicknesses many times the theoretical equilibrium critical thickness [83, 84]. These films are metastable, and strain relaxation will occur upon annealing.

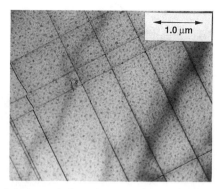

FIGURE 16. Plan view transmission electron micrograph showing misfit dislocations at the interface between an epitaxial $Si_{1-x}Ge_x$ layer and a ⟨100⟩ Si substrate. Photo from Noble [86].

The onset of misfit dislocation formation during LRP growth of strained $Si_{1-x}Ge_x$ on Si at 625 °C has been found to be comparable to that reported for films grown by MBE—4 or 5 times the theoretical, equilibrium critical thickness [85]. However, when such films are heated misfit dislocations will form. The measured misfit dislocation spacing versus $Si_{1-x}Ge_x$ film thickness for layers with 20% Ge grown by LRP at 625 °C is shown in Fig. 17 [85, 86]. The solid line indicates the predictions of the equilibrium

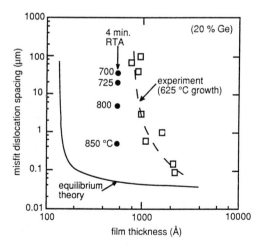

FIGURE 17. Measured dislocation spacing versus $Si_{1-x}Ge_x$ thickness, after LRP growth at 625 °C, with $x \simeq 20\%$ (open squares). The closed symbols show the spacing after 4-min RTA at various temperatures. The solid line indicates the predictions of the equilibrium theory. After Noble [86].

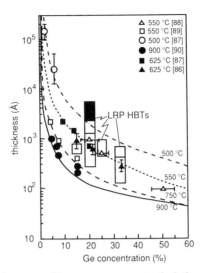

FIGURE 18. HBT design space. The curves represent calculations for the onset of misfit dislocation formation by Houghton [87]. The open symbols are experimental data from HBT-grown material, while the solid symbols refer to material grown by RTP-based techniques. Adapted from Houghton [87]. The shaded rectangular regions pertain to HBTs, and are explained in the text.

theory. Average misfit dislocation spacings were determined by a combination of plan-view transmission electron microscopy and X-ray topographic imaging. Also shown is the evolution of the misfit dislocation spacing upon 4-min rapid thermal annealing at various temperatures (solid symbols). Note that in this annealing experiment misfit dislocations were not observed prior to annealing. The misfit dislocation spacing drops rapidly as the anneal temperature is increased.

Theories that take into account the kinetics of misfit dislocation formation have recently been formulated and compared to experimental determinations of the onset of misfit dislocation formation [87]. The results of such theoretical calacualtions are compared to experimental measurements in Fig. 18. Such information helps to define a "design space" for $Si/Si_{1-x}Ge_x$ HBTs. The lines represent the results of kinetic modeling by Houghton [87], the open symbols represent data from MBE-grown material, and the solid symbols are data from RTP-based epitaxy. Each curve is labeled with the corresponding growth temperature. As expected, the theory predicts that thicker layers can be grown free of misfit dislocations at lower temperatures. The figure also shows rectangular regions that represent areas of the design space in which HBTs have been fabricated by LRP. In the unshaded rectangular regions, no obvious effects of misfit dislocations on the HBT current characteristics were observed. In the hatched regions, nonideal base leakage currents were found. In the solid rectangular regions, strain relaxation was measurable by X-ray measurements of the $Si_{1-x}Ge_x$ lattice constant and by a corresponding drop in the HBT collector current [75].

Because practical devices will probably require some postgrowth processing, methods for misfit dislocation reduction during growth and annealing are being investigated. One method is the use of rapid thermal processing to limit the temperature-time exposure. Other techniques for reducing misfit dislocation formation include minimizing nucleation sources (dust, epitaxial defects, etc.) [91], the use of epitaxial Si capping layers [92], oxygen solution strengthening [67], and limited area growth [93]. The applications of interest will most likely dictate the requirements with respect to misfit dislocation formation. For the HBT, an epitaxial Si cap occurs naturally as the emitter. Oxygen doping is an effective technique for reducing misfit dislocation formation during annealing, but the reduced minority carrier lifetime may limit its application. Limited area growth (e.g., by selective area deposition on oxide patterned wafers) offers the potential to limit misfit dislocation formation during growth, and has the added benefit of reducing parasitic capacitances that can limit device switching speeds [79].

IV. Summary

There are several driving forces for the development of new epitaxial growth techniques, including lower cost, improved capabilities such as the ability to grow thinner layers with tighter doping profile control, and the introduction of features that enable completely new types of electronic and optical devices to be fabricated. Single-wafer epitaxy becomes increasingly attractive as wafer diameters are scaled above 6 in. Epitaxial growth based upon rapid thermal processing offers enhanced processing flexibility. This type of equipment has already provided a wealth of capabilities in the research environment. Temperature measurement and control remain as critical issues inhibiting commercial applications (see Chapter 9).

The suitability of new epitaxial growth processes for silicon device applications may be assessed by a variety of characterization techniques and test devices. Results to date for RTP-based epitaxial growth processes are very promising. Heteroepitaxial growth of $Si_{1-x}Ge_x$ layers on Si is an exciting research area in which RTP-based epitaxy has played a significant role. The wide range of potential applications for such materials has yet to be fully explored.

References

1. J. F. Gibbons, C. M. Gronet, and K. E. Williams, *Appl. Phys. Lett.* **47**, 721 (1986).
2. C. M. Gronet, J. C. Sturm, K. E. Williams, J. F. Gibbons, and S. D. Wilson, *Appl. Phys. Lett.* **48**, 1012 (1986).
3. J. C. Sturm, C. M. Gronet, and J. F. Gibbons, *IEEE Elec. Dev. Lett.* **EDL-7**, 577 (1986).
4. J. L. Hoyt, C. A. King, D. B. Noble, C. M. Gronet, J. F. Gibbons, M. P. Scott, S. S. Laderman, S. J. Rosner, K. Nauka, J. Turner, and T. I. Kamins, *Thin Solid Films* **184**, 93 (1990).
5. S. Reynolds, D. W. Vook, and J. F. Gibbons, *Appl. Phys. Lett.* **49**, 1720 (1986).
6. D. W. Vook and J. F. Gibbons, *J. Appl. Phys.* **67**(4), 2100 (1990).
7. M. L. Green, D. Brasen, H. Luftman, and V. C. Kannan, *J. Appl. Phys.* **65**, 2558 (1989).
8. J. C. Sturm, P. V. Schwartz, E. J. Prinz, and H. Manoharan, *J. Vac. Sci. Technol. B* **9**, 2011 (1991).
9. G. P. Burns and J. G. Wilkes, *Semicond. Sci. Technol.* **3**, 442 (1988).
10. S. K. Lee, Y. H. Ku, and D. L. Kwong, *Appl. Phys. Lett.* **54**, 1775 (1989).
11. M. M. Moslehi and K. C. Saraswat, *IEEE Elec. Dev. Lett.* **8**, 421 (1987).
12. C. W. Pearce, in *VLSI Technology* (S. M. Sze, ed.), pp. 55-94, McGraw-Hill, New York (1988).
13. H. C. Theuerer, J. J. Kleimack, H. H. Loar, and M. Christensen, *Proc. IRE* **48**, 1642 (1960).
14. D. S. Yaney and C. W. Pearce, *IDEM Tech. Dig.*, p. 236 (1981).
15. J. O. Borland, in *Proc. 10th International Conf. on Chemical Vapor Deposition*, Honolulu, 1987 (G. W. Cullen, ed.) Electrochem Soc. 87-8, pp. 307-316, Electrochem. Soc., Pennington, NJ (1987).
16. A. S. Grove, *Physics and Technology of Semiconductor Devices*, pp. 7-18, Wiley, New York (1967).
17. F. C. Eversteyn, *Philips Res. Rep.* **29**, 45 (1974).
18. C. I. Drowley and J. E. Turner, in *Proc. 10th International Conf. on Chemical Vapor Deposition*, Honolulu, 1987 (G. W. Cullen, ed.) Electrochem. Soc. 87-8, pp. 243-252 (1987).
19. C. M. Gronet, *Ph.D. thesis*, Stanford University (Oct. 1988).
20. K. V. Ravi, *Imperfections and Impurities in Semiconductor Silicon*, p. 198, Wiley, New York (1981).
21. E. Zinner, *Scanning* **3**, 57-78 (1980).
22. C. Evans and Assoc., specialists in materials characterization, Redwood City, CA.
23. D. K. Schroder, *Semiconductor Material and Device Characterization*, pp. 298-358, Wiley, New York (1990).
24. G. R. Srinivasan, *J. Electrochem. Soc.* **127**, 1334 (1980).
25. K. V. Ravi, *Imperfections and Impurities in Semiconductor Silicon*, p. 161, Wiley, New York (1981).
26. M. Ogirima and R. Takahashi, in *Proc. 10th International Conf. on Chemical Vapor Deposition*, Honolulu, 1987 (G. W. Cullen, ed,) Electrochem. Soc. 87-8, pp. 204-213, Electrochem. Soc., Pennington, NJ (1987).
27. For example, the Epsilon One reactor by ASM Epitaxy, Tempe, AZ.
28. *Silicon Molecular Beam Epitaxy* (E. Kasper and J. C. Bean, eds.) CRC Press, Boca Raton, FL (1988).
29. See various papers in *Silicon Molecular Beam Epitaxy*, Proc. of the Mat. Res. Soc. Symposium, Vol. 220, Spring, 1991, Anaheim, CA (J. C. Bean, S. S. Iyer, and K. L. Wang, eds.), Mat. Res. Soc., Pittsburgh, PA (1991).

30. G. S. Higashi, J. C. Bean, C. Buescher, R. Yadvish, and H. Temkin, *Appl. Phys. Lett.* **56**, 2560 (1990).
31. H. Hirayama, T. Tatsumi, and N. Aizaki, *Appl. Phys. Lett.* **52**, 1484 (1988).
32. B. S. Meyerson, *Appl. Phys. Lett.* **48**, 797 (1986).
33. D. W. Greve and M. Rancanelli, *J. Vac. Sci. Technol.* B **8**, 511 (1990).
34. D. J. Robbins and I. M. Young, *Appl. Phys. Lett.* **50**, 1575 (1987).
35. G. L. Patton, J. M. C. Stork, J. H. Comfort, E. F. Crabbe, B. S. Meyerson, D. L. Harame, and J. Y.-C. Sun, in *IEDM Tech. Dig.* (Dec. 1990).
36. D. Lubben, R. Tsu, T. R. Bramblett, and J. E. Greene, *J. Vac. Sci. Technol.* A **9**, 3003 (1991).
37. G. N. Parsons, *Appl. Phys. Lett.* **59**, 2546 (1991).
38. See various papers in *Applied Surface Science* **36**, (1989).
39. M. Razeghi, *The MOCVD Challenge*, Adam Hilger, Bristol, UK (1989).
40. P. J. Roksnoer, J. W. F. M. Maes, A. T. Vink, C. J. Vriezema, and P. C. Zalm, *Appl. Phys. Lett.* **58**, 711 (1991).
41. F. Roozeboom and N. Parekh, *J. Vac. Sci. Technol.* B **8**, 1249 (1990).
42. See references in *Rapid Thermal and Integrated Processing Symp. Proc.*, Spring, 1991, Anaheim, CA (J. C. Gelpey, M. L. Green, J. Wortman, and R. Singh, eds.), Mat. Res. Soc., Pittsburgh, PA (1991).
43. J. C. Sturm, P. V. Schwartz, and P. M. Garone, *Appl. Phys. Lett.* **56**, 961 (1990).
44. J. C. Sturm, P. V. Schwartz, and P. M. Garone, *J. Appl. Phys.* **69**, 542 (1991).
45. J. C. Sturm, C. M. Gronet, and J. F. Gibbons, *J. Appl. Phys.* **59**, 4180 (1986).
46. D. Mathiot, J. L. Regolini, and D. Dutartre, *J. Appl. Phys.* **69**, 358 (1991).
47. See references in *Proc. of the Mat. Res. Soc. Symposium*, Vol. 146, Spring, 1989, San Diego, CA (D. Hodul, J. C. Gelpey, M. L. Green, and T. E. Seidel, eds.), Mat. Res. Soc., Pittsburgh, PA (1989).
48. J. Mercier, J. L. Regolini, D. Bensahel, and E. Scheid, *J. Crystal Growth* **94**, 885 (1989).
49. J. L. Regolini, D. Bensahel, J. Mercier, and E. Scheid, *J. Crystal Growth* **96**, 505–512 (1989).
50. J. L. Regolini, D. Bensahel, E. Scheid, and J. Mercier, *Appl. Phys. Lett.* **54**, 658–659 (1989).
51. M. Liehr, M. Greenlief, S. R. Kasi, and M. Offenberg, *Appl. Phys. Lett.* **56**, 629–631 (1990).
52. M. D. Giles, J. L. Hoyt, and J. F. Gibbons, in *Mat. Res. Soc. Symp. Proc.* **69**, pp. 323–328, Mat. Res. Soc., Pittsburgh, PA (1986).
53. J. E. Turner, J. Amano, C. M. Gronet, and J. F. Gibbons, *Appl. Phys. Lett.* **50**, 1601 (1987).
54. T. O. Sedgwick, P. D. Agnello, D. Nguyen Ngoc, T. S. Kuan, and G. Scilla, *Appl. Phys. Lett.* **58**, 1896–1898 (1991).
55. J.-E. Sundgren, J. Knall, W.-X. Ni, M.-A. Hasan, L. C. Markert, and J. E. Green, *Thin Solid Films* **183**, 281–297 (1989).
56. T. Y. Hsieh, K. H. Jung, Y. M. Kim, and D. L. Kwong, *Appl. Phys. Lett.* **58**, 80–82 (1991).
57. J. L. Hoyut, D. B. Noble, T. Ghani, C. A. King, J. F. Gibbons, M. P. Scott, S. S. Laderman, K. Nauka, J. E. Turner, S. J. Rosner, and T. I. Kamins, in *Proceedings of the Second International Conference on Electronic Materials*, p. 551 (R. Chang, T. Sugano, and V. Nguyen, eds.) Mat. Res. Soc., Pittsburgh, PA (1991).
58. C. A. King, C. M. Gronet, and J. F. Gibbons, *IEEE Elec. Dev. Lett.* **9**, 229 (1988).
59. C. A. King, J. L. Hoyt, C. M. Gronet, J. F. Gibbons, M. P. Scott, and J. Turner, *IEEE Elec. Dev. Lett.* **10**, 52 (1989).

60. S. C. Jain and W. Hayes, *Semicond. Sci. Technol.* **6**, 547–576 (1991).
61. J. C. Bean, T. T. Sheng, L. C. Feldman, A. T. Fiory, and R. T. Lynch, *Appl. Phys. Lett.* **44**, 102 (1983).
62. B. S. Meyerson, K. J. Uram, and J. K. LeGoues, *Appl. Phys. Lett.* **53**, 2555–2557 (1988).
63. P. M. Garone, J. C. Sturm, P. V. Schwartz, S. A. Schwarz, and B. J. Wilkens, *Appl. Phys. Lett.* **56**, 1275–1277 (1990).
64. T. I. Kamins and D. J. Meyer, *Appl. Phys. Lett.* **59**, 178–180 (1991).
65. T. Ghani, J. L. Hoyt, D. B. Noble, J. F. Gibbons, J. E. Turner, and T. I. Kamins, *Appl. Phys. Lett.* **58**, 1317 (1991).
66. J. C. Sturm, E. J. Prinz, and C. W. Magee, *IEEE Elec. Dev. Lett.* **12**, 303–305 (1991).
67. D. B. Noble, J. L. Hoyt, W. D. Nix, J. F. Gibbons, S. S. Laderman, J. E. Turner, and M. P. Scott, *Appl. Phys. Lett.* **58**, 1536 (1991).
68. Point-of-use purifiers manufactured by Semi-Gas Systems, San Jose, CA, and Matheson, Inc., Newark, CA.
69. M. L. Green, D. Brasen, M. Geva, W. Reents, F. Stevie, and H. Temkin, *J. Elect. Mat.* **19**, 1015–1019 (1990).
70. J. M. C. Stork, *IEDM Tech. Dig.* p. 550 (Dec. 1988).
71. H. Kromer, *Proc. IEEE* **70**, 13 (1982).
72. A. Chatterjee and A. H. Marshak, *Solid State Elec.* **26**, 59 (1983).
73. R. People, *Phys. Rev. B* **32**, 1405 (1985).
74. R. Braunstein, A. R. Moore, and F. Herman, *Phys. Rev.* **109**, 695 (1958).
75. C. A. King, Ph.D. thesis, Stanford University (1989).
76. C. A. King, J. L. Hoyt, and J. F. Gibbons, *IEEE Trans. Elec. Dev.* **36**, 2093 (1989).
77. R. C. Taft, J. D. Plummer, and S. S. Iyer, *IEDM Tech. Dig.* p. 570 (Dec. 1988).
78. D. V. Lang, R. People, J. C. Bean, and A. M. Sergent, *Appl. Phys. Lett.* **47**, 1333 (1985).
79. T. Kamins, K. Nauka, J. Kruger, L. Camnitz, M. Scott, J. Turner, S. Rosner, J. Hoyt, C. King, D. Noble, and J. Gibbons, *IEDM Tech. Dig.* p. 647 (1989).
80. G. L. Patton et al., *IEEE Elec. Dev. Lett.* **11**, 171 (1990).
81. J. D. Cressler, J. H. Comfort, E. F. Crabbe, G. L. Patton, W. Lee, J. Y.-C. Sun, J. M. C. Stork, and B. S. Meyerson, *IEEE Elec. Dev. Lett.* **12**, 166 (1991).
82. J. W. Matthews and A. E. Blakeslee, *J. Cryst. Growth* **27**, 181 (1974).
83. J. H. van der Merwe, *J. Appl. Phys.* **34**, 123 (1963).
84. Y. Kohama, Y. Fukuda, and M. Seki, *Appl. Phys. Lett.* **52**, 380 (1988).
85. M. P. Scott, S. S. Laderman, T. I. Kamins, S. J. Rosner, K. Nauka, D. B. Noble, J. L. Hoyt, C. A. King, C. M. Gronet, and J. F. Gibbons, in *Thin Films: Stresses and Mechanical Properties* (Mat. Res. Soc. Proc.), Pittsburgh, PA (1989).
86. D. B. Noble, Ph.D. thesis, Stanford University (August, 1991).
87. D. C. Houghton, *J. Appl. Phys.* **70**, 2136–2151 (1991).
88. J. C. Bean, L. C. Feldman, A. T. Fiory, S. Nakahara, and I. K. Robinson, *J. Vac. Sci. Technol. A* **2**, 436 (1986).
89. E. Kasper and H.-J. Herzog, *Thin Solid Films* **44**, 357 (1975).
90. M. L. Green, B. E. Weir, D. Brasen, Y. F. Hseih, G. Higashi, A. Feygenson, L. C. Feldman, and R. L. Headrick, *J. Appl. Phys.* **69**, 745 (1991).
91. C. G. Tuppen, C. J. Gibbings, S. T. Davey, M. H. Lyons, M. Hockly, and M. A. G. Halliwell, in *Proc. 2nd Intl. Symp. on Si MBE*, p. 26 (Electrochem. Soc.), Pennington, NJ (1988).
92. D. B. Noble, J. L. Hoyt, J. F. Gibbons, M. P. Scott, S. S. Laderman, S. J. Rosner, and T. I. Kamins, *Appl. Phys. Lett.* **55**, 1978 (1989).
93. D. B. Noble, J. L. Hoyt, J. F. Gibbons, T. I. Kamins, and M. P. Scott, *Appl. Phys. Lett.* **56**, 51 (1990).

3 Rapid Thermal Growth and Processing of Dielectrics

Hisham Z. Massoud
Semiconductor Research Laboratory
Department of Electrical Engineering
Duke University, Durham, North Carolina

I. Equipment Issues in Rapid Thermal Oxidation	47
A. Heating Sources	48
B. Temperature Measurement and Control	49
C. Manufacturability Issues in RTO	50
II. Rapid Thermal Oxidation Growth Kinetics	51
A. Experimental Observations	51
B. Modeling RTO Growth Kinetics	55
III. Rapid Thermal Processing of Oxides	58
A. Rapid Thermal Annealing	58
B. Rapid Thermal Nitridation	60
C. Rapid Thermal Reoxidation of Nitrided Oxides	61
IV. Electrical Properties of Rapid Thermal Oxidation/Rapid Thermal Processing Oxides	62
A. Interface Traps and Fixed Charges	62
B. Current-Voltage Characteristics	66
C. Dielectric Breakdown	67
D. Carrier Trapping Properties	69
E. Device Performance	70
F. Technology—Current Practice	71
V. Conclusions	72
Acknowledgments	72
References	73

Rapid thermal processing (RTP) has been introduced to meet the thermal budget limitations in the processing of ultrashallow junctions in order to achieve complete dopant activation and defect removal with minimum dopant diffusion. With the continuing trend toward using larger diameter wafers and the increasing emphasis on single-wafer processing, a technology was needed to produce high-quality thin gate dielectrics that is compatible with other RTP applications such as dopant activation, ion-implantation damage annealing, metallization, silicide formation, polysilicon deposition, and oxygen donor annihilation [1]. This technology encompasses many RTP applications in the processing of thin gate dielectrics. To date, these applications include

1. The preoxidation *in situ* rapid thermal wafer cleaning (RTC),
2. The rapid thermal growth of oxides on silicon in dry oxygen, pyrogenic steam, or other ambients such as N_2O,
3. The rapid thermal growth of nitrides and oxynitrides on silicon in N_2 and NH_3,
4. The rapid thermal nitridation (RTN) of oxides and the reoxidation of nitrided oxides,
5. The rapid thermal chemical vapor deposition (RTCVD) of oxides, nitrides, oxynitrides, and doped glasses,
6. The reflow of phosphosilicate (PSG) and borophosphosilicate (BPSG) glasses,
7. The postoxidation high-temperature annealing (RTA) of charges and traps at the $Si-SiO_2$ interface,
8. The optimization of the electrical properties of dielectrics, such as leakage currents, carrier injection and trapping, radiation damage, and dielectric breakdown, and
9. The rapid thermal bonding of wafers for silicon-on-insulator (SOI) applications.

The use of rapid thermal oxidation (RTO) in the growth of thin SiO_2 layers on single-crystal silicon using an incoherent light source was first introduced in 1984 [2]. This first study reported a different activation energy and inferior electrical characteristics for the RTO process when compared with conventional furnace oxidation [2]. Subsequent studies reported RTO oxides grown in a variety of rapid thermal processing systems equipped with different heating sources resulting in electrical properties comparable and in some cases superior to conventional furnace oxides [3-26]. In addition to the formation of SiO_2 layers on single-crystal silicon, RTO has also been used in applications such as the n^+ doping of GaAs by the rapid thermal oxidation of a silicon cap [27], the rapid thermal oxidation of amorphous silicon films obtained by low-pressure chemical vapor deposition [28], the rapid thermal oxidation of SiGe strained layers [29], the formation of titanium dioxide dielectric films [30], the rapid thermal oxidation of polysilicon films [31], the rapid thermal oxidation of silicon monoxide [32], and the improvement of the electrical characteristics of CVD SiO_2 gate dielectrics [33].

This chapter starts with a discussion of RTO equipment issues, pointing out the major challenges in the optimization of the processing of high-quality thin gate dielectrics. This discussion is followed by a review of experimental and modeling studies of RTO growth kinetics, including an evaluation of the different ways of measuring and controlling wafer temperature and its uniformity in RTO systems and the influence of the heating source on oxidation kinetics. The dependence of the electrical properties of thin gate dielectrics formed in RTO systems on processing conditions is then discussed. This chapter concludes with an examination of the future of RTO systems and processes.

I. Equipment Issues in Rapid Thermal Oxidation

The formation and processing of high-quality dielectrics by RTO take place in a rapid thermal processing system capable of switching to an oxidizing ambient during the processing cycle. Thus, an ideal RTO system is also an ideal RTP system that meets the additional needs of processing dielectrics. An optimal RTO system should have the following features [34–42]:

1. The system must be compatible with the use of the reactive and corrosive gases used in the processes of RTC, RTO, RTA, and RTN of high-quality dielectrics such as dry oxygen, pyrogenic steam, anhydrous hydrogen chloride (HCl), bubbled chlorinated compounds, ammonia (NH_3), and nitrous oxide (N_2O).
2. The system must provide ultraclean conditions with minimum levels of contamination and maximum particulate control in the process chamber and the wafer-handling system.
3. The gas handling system must have a vacuum capability and should be able to quickly switch from one processing ambient to another.
4. The process chamber must have a low thermal mass and minimum thermal memory. The wafer holder should conduct the smallest possible amount of heat in order not to affect temperature uniformity.
5. The heating source must have a radiation spectrum that, when combined with the optimal properties of the sample, would result in a uniform temperature distribution in both static and dynamic conditions. This requirement is needed to guarantee the uniformity of the dielectric properties and the suppression of slip.
6. The temperature measurement and control system should be accurate, noncontact, sensitive to the presence and growth of surface films on the wafer back surface or patterns on its front surface, and have a fast response time to track large heating and cooling temperature cycles.
7. Finally, an ideal RTO system must be equipped with computer control and be interfaced with other factory automated equipment or other processing modules in cluster tools.

Equipment issues in RTP systems are discussed in detail in Chapter 9. Those issues that are common with RTO such as the basic system design and characteristics, contamination control, ultraclean processing, process control, and wafer handling and transport will be discussed in that chapter. In the next few sections, however, some equipment issues that are specific to RTO will be briefly examined. These issues include heating sources, temperature measurement and control, and RTO manufacturability issues.

A. Heating Sources

In commercial RTP systems wafers are heated by a variety of heating sources such as tungsten–halogen or tungsten–argon lamp arrays, single water-walled or sealed argon or xenon arc lamps, and continuous resistive heat sources [44]. Heating lamp arrays are usually made of linear lamps that irradiate the wafer either from one side or from both sides. Linear lamp arrays used in heating wafers from both sides are usually configured in a cross-lamp arrangement [44]. Independently controlled tungsten–halogen lamps configured in three concentric rings have been recently shown to result in more flexible temperature control and significant improvement in temperature uniformity [45-47].

The resulting uniformity in the properties of dielectrics processed by RTO is affected by the chamber design and dimensions, the design and optical properties of the heating source, the optical properties of the chamber, the wafer holder, ring, and rotation, and the design of the heat reflectors and selective mirrors. The resulting uniformity is a concern common to RTP and RTO systems. However, differences in the spectral distributions of typical heating sources are of additional interest in RTO because energy absorption occurs predominantly by intrinsic free carriers. Therefore, the wafer heating rate is not only dependent on temperature but also on the dopant concentration in the substrate [48]. In the characterization of the process of silicon oxidation, it has long been established that the oxide growth rate exhibits a strong dependence on dopant concentration [49], and includes a nonthermal photon-induced component observed in the UV, VUV, and visible range [50-67].

The design of the heating system consisting of the heating sources, the heat reflectors, and the wafer holder contributes to the temperature distribution in a wafer during rapid thermal oxidation. In addition to the important issue of temperature uniformity during heating and cooling cycles and the related problem of slip generation, it has been shown that a nonuniform temperature across the wafer results in a thermally induced stress that affects the oxidation kinetics in RTO [68], as will be discussed later.

In conclusion, nonuniformity in oxide thickness and electrical properties of rapid thermal grown oxides are dependent on the heating source and the thermal design of the chamber. Mechanisms resulting in variations in oxide properties include the spectral distribution of the heating sources that affects the nonthermal photon-related component of oxidation growth rate, and nonuniform temperature distributions that affect the thermal component of the oxide growth rate and induce a stress that in turn affects the growth process.

B. Temperature Measurement and Control

The two most widely used methods of RTP temperature measurement are based on the use of thermocouples and pyrometers. These techniques have some limitations in their use in RTP and RTO systems. For a discussion of the proper use of thermocouples and the problems encountered in using single- and multiple-wavelength pyrometry, the reader is referred to Chapter 9 and many recent studies [44, 69, 70]. For microcontamination considerations, the use of thermocouples in RTP systems is limited to calibration purposes [44]. Problems arising from the transparency of silicon wafers below 600 °C have been solved either by using a pyrometer with a bandpass wavelength range outside that of the heating source, by the removal of the interfering radiation from the heating source before reaching the pyrometer, or by special design of the optical properties and thickness of the pyrometer window [44].

Novel techniques of temperature and process control focus on the non-contract measurement of temperature by monitoring its influence on one of the properties of the wafer such as the inelastic scattering of photons as in the Raman effect [70], the optical properties of the wafer as measured either by laser interferometry [71] or *in situ* ellipsometry [72-74], or by photoacoustic wave transmission thermometry [75]. Temperature measurement and process control by *in situ* ellipsometry is particularly well suited to RTO because of its fast response time, which allows for a substantial improvement in temperature control in dynamic situations. The use of *in situ* ellipsometry has also been shown to yield simultaneously both the wafer temperature and the thickness of the growing oxide [74]. This allows for a considerable improvement in RTO process control, as the ellipsometric parameters of the end-of-process target values can be easily used to control the process without relying on detailed and accurate knowledge of RTO growth kinetics and the classical approach to process control based on the accurate control of process time and temperature. *In situ* ellipsometry is also well suited to temperature mapping across the wafer and, therefore, would be ideally suited to be used in RTP systems equipped with independent control of heating elements.

C. Manufacturability Issues in RTO

All issues that are important in the use of RTP in manufacturing are also vital to RTO and its manufacturability. The majority of round-robin-type studies of uniformity issues have concentrated on the temperature uniformity and the influence of wafer patterns consisting of oxides and other films [76]. It was concluded that the manufacturing limits of RTO are dependent on the chamber design, radiation source, and temperature measurement and calibration.

In a recent extensive evaluation of some of the manufacturability issues of RTO based on optimal design of experiments and statistical analysis, the dependence of oxide thickness and oxide thickness variation within a wafer and wafer to wafer on process variables were studied in several RTP systems [77, 78]. Screening experiments based on orthogonal arrays (also called Taguchi or fractional factorial matrices) were carried out to identify the most significant variables. The experimental parameters investigated in the screening experiments were the oxidation temperature, oxidation time, temperature ramp-up and -down rates, percent oxygen and percent HCl in the oxidizing ambient, oxygen flow rate, rapid thermal cleaning, and purge duration [77]. The results of this study indicated that oxidation time (t_{ox}), oxidation temperature (T_{ox}), and time–temperature interactions have the largest effects on oxide thickness and oxide thickness variation. It was also found that faster ramp rates slightly improved the oxide thickness variation and that significant temperature nonuniformities occur in the steady-state period of the time–temperature cycle in comparison with portions of the cycle where the temperature is ramping up or down.

The RTO process latitude improved with progress and advancements in RTP equipment. Improved temperature uniformity and gas handling resulted in smaller oxide thickness variation within a wafer. Greater radiative heat losses at the edge of a wafer than at its center produce a radial temperature gradient that induces a thermal stress known to cause slip generation. This thermally induced stress was also shown to affect RTO kinetics especially at low temperatures and for short oxidations [68]. However, slip suppression with improved slip ring design or by the use of a thin graphite susceptor effectively extending the radius of the wafer was also shown to significantly improve radial temperature uniformity and minimize oxide thickness variation resulting from the thermally induced stress. An additional source of within-a-wafer and wafer-to-wafer oxide thickness variation is the nonthermal photonic component of RTO growth kinetics as discussed earlier.

Chamber design and lamp configuration determine oxide thickness variations. This was most dramatically observed in some RTO systems where the reflector design could be easily discerned in oxide thickness distributions [77, 78].

The standard deviation of oxide thickness within a wafer was less than 10% for all oxidations conditions and all systems [77]. With appropriate optimization of the reactor design, this manufacturing criterion could be substantially improved to within acceptable limits. Wafer-to-wafer oxide thickness variations, however, were not within aceptable limits. The trends observed in all systems studied were for improved repeatability with increasing oxidation temperature. Issues of differences in heating sources, wafer handling, and residual heating between systems should be addressed to improve system-to-system variations. It was concluded, however, that with optimization, RTO is a manufacturable process [77].

II. Rapid Thermal Oxidation Growth Kinetics

In this section, the rapid thermal growth of SiO_2 on silicon is discussed. First, the dependence of RTO growth on experimental conditions is presented. The effects of thermally induced stress on RTO growth kinetics are then discussed. Finally, empirical and physically based modes of oxide kinetics in RTO are introduced. The validity of these models and their limitations are then discussed.

A. EXPERIMENTAL OBSERVATIONS

RTO systems are run either in a temperature-control mode or in intensity-control mode. In the temperature-control mode, a pyrometer is used to measure the temperature and subsequently used to control the power delivered to the heating sources. In the intensity-control mode, the power delivered to the heating source is set to a specified percentage of the maximum available power, and the temperature is not regulated [14]. RTO grown at 1150 °C on ⟨100⟩ silicon using a two-step process showed small differences in RTO growth kinetics between the two modes of operation [14]. The most common use of RTO systems is in the pyrometer-control mode.

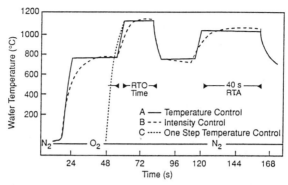

FIGURE 1. An illustrative example of a two-step RTO cycle showing wafer temperature and ambient for RTO followed by RTA. From Nulman et al. [9].

Oxidation cycles in RTO systems are either one-step or two-step time-temperature cycles. An example of a two-step RTO process is illustrated in Fig. 1. It was observed that the oxide thickness grown in two-step cycles is slightly thicker than those grown in one-step cycles at the same processing conditions of time and temperature [20].

RTO growth kinetics was investigated in several studies at temperatures ranging from 800 to 1250 °C. A representative set of one-step RTO on ⟨100⟩ silicon in dry oxygen between 950 and 1200 °C is shown in Fig. 2. It can be seen that the general behavior of oxide growth in RTO is qualitatively

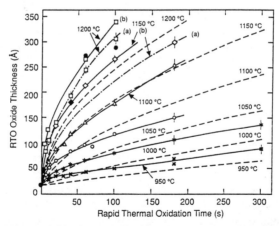

FIGURE 2. RTO growth kinetics on ⟨100⟩ silicon in dry O_2 in the 950–1200 °C temperature range. The RTO oxidation temperatures are designated by × at 950 °C, ✶ at 1000 °C, ○ at 1050 °C, △ at 1100 °C, ◇ at 1150 °C, □ at 1200 °C, ● for 1150 °C, 40 sec and 100 sec RTO each followed by 30 sec Ar annealing, and ■ for two-cycle (20 + 40 sec) RTO at 1200 °C. From Moslehi [4].

similar to that in conventional furnace oxidation in the ultrathin regime [79]. Here, as in thermal oxidation, the rapid thermal oxide thickness increases with time and temperature and the growth rate decreases with time and thickness. Some investigations of RTO growth, however, have reported different or limited kinetics than in thermal oxidation. For example, the data of Sacho and Kiuchi [12] indicate linear kinetics at 1050, 1100, and 1150 °C. The results of Chan Tung et al. [24] also indicate linear growth between 1000 and 1250 °C.

The influence of wafer orientation on RTO growth was reported. It was observed that RTO growth is fastest on ⟨110⟩ and slowest on ⟨100⟩ [6]. The difference on oxide thickness between the different orientations decreased with increasing oxidation temperature, such that at 1200 °C, the oxide thickness grown by RTO on ⟨100⟩, ⟨110⟩, and ⟨111⟩ were similar. The extent of RTO investigations did not indicate a trend towards crossover of the oxidation kinetics on ⟨110⟩ and ⟨111⟩ as was observed in thermal oxidation in the thin regime [80]. RTO growth dependence on wafer orientation is illustrated in Fig. 3. This could be due either to a difference in mechanisms in the growth process or simply to RTO times not being long enough to observe the crossover.

The oxidation ambient during rapid thermal oxidation and its influence on RTO growth kinetics were the subject of many studies. The influence of HCl on RTO growth kinetics was reported by Nulman [20], who compared RTO kinetics without HCl and with 4% HCl at 1100 and 1150 °C. The presence of HCl in the oxidizing ambient enhanced RTO growth much in the same manner observed in conventional thermal oxidation [81]. The effect of diluting oxygen with argon was studied by Gelpey et al. [7], who observed a decrease in the RTO growth rate with increasing the percentage of argon in the ambient. This observation is similar to conventional thermal oxidation where the oxidation rate is influenced by the partial pressure of oxygen in the oxidizing ambient [82]. The influence of oxygen partial pressure on RTO growth is illustrated in Fig. 4.

RTO growth kinetics was compared in systems where the wafers were heated either by tungsten–halogen lamps or water-cooled arc lamps [83]. It was observed that in the initial 20 to 30 sec, the growth rate was linear, becoming nonlinear beyond the first 20 sec. These two growth regimes yielded different activation energies for the growth process. The durations of the initial regime was found to depend on the heating lamp. The activation energy for the linear growth rate was 1.44 and 1.71 eV for growth in tungsten–halogen lamp and arc lamp systems, respectively [83]. By comparison, the activation energy of the linear rate constant in conventional furnace oxidation is 1.76, 1.74, and 2.10 eV for ⟨100⟩, ⟨111⟩, and ⟨110⟩, respectively [79].

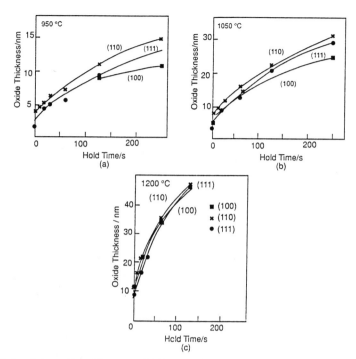

FIGURE 3. Orientation effects on RTO growth kinetics at (a) 950, (b) 1050, and (c) 1200 °C. From Hodge et al. [6]. Copyright © 1985 (work undertaken in DRA Malvern, formerly known as RSRE).

The influence of substrate doping on RTO kinetics was reported in two studies [84, 85]. It was observed that for highly phosphorus-doped ⟨100⟩ silicon wafers with surface carrier concentration of 3 to 4×10^{20} cm^{-3}, the linear oxidation rate constant was enhanced by a factor of 3.6 over lightly doped wafers [84]. This observation is in agreement with heavy doping effects on oxide growth kinetics in conventional oxidation.

Radiative losses from the edge of silicon wafers heated by uniform irradiation during rapid thermal processing result in a radial temperature gradient. This temperature gradient induces a compressive stress at the center and a tensile stress towards the edge of the wafer. This thermally induced stress affects oxidation kinetics during RTO of silicon [68]. Typical oxide distributions are illustrated in Fig. 5. The resolved stress and its effect on oxidation are largest along specific crystal directions, namely the slip directions on slip planes. The stress effect on oxidation is largest at low temperatures and for short times [68].

Caution must be exercised when using RTO oxide thickness as a measurement of temperature uniformity in RTP systems. Because the temperature

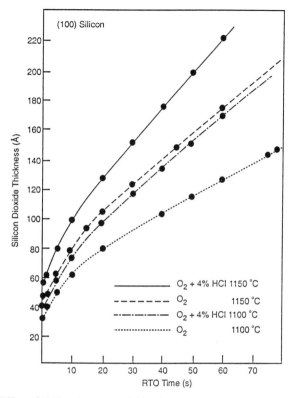

FIGURE 4. Effect of HCl on RTO growth kinetics at 1100 and 1150 °C. From Nulman [20].

gradient and induced stress are functions of wafer diameter, the effect would be larger for larger diameter wafers. In addition, because the thermally induced stress effect is maximum along specific crystal directions, it degrades the oxide uniformity across the wafer. For longer times, the effects of the intrinsic stress at the Si–SiO$_2$ interface dominate the thermally induced stress effect. This RTO mechanism has not been reported previously.

B. Modeling RTO Growth Kinetics

Several RTO growth kinetics modeling studies have been attempted, similar to the modeling of conventional thermal oxidation [86–93]. In the early modeling efforts, attempts were made to identify the activation energies of the growth rate. Several studies divided the RTO growth data into multiple regimes and assigned an activation energy for each regime. For example, Nulman [20] reported activation energies of 0.66 and 0.60 eV for one-step

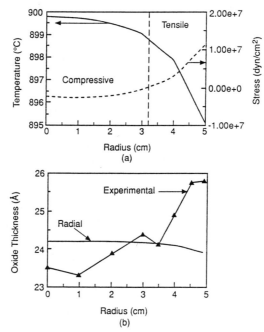

FIGURE 5. Effect of thermally induced stress on the rapid thermal oxidation of silicon. From Deaton and Massoud [68].

and two-step oxidations above 1050 °C, respectively. He also reported linear growth beyond the initial ramp-up phase of two-step oxidations with activation energies of 0.875 and 2.0 eV below and above 1050 °C, respectively. Chan Tung and Caratini [16], however, obtained activation energies of 0.92 and 1.42 eV for the early phase and linear growth phase, respectively. Lassig and Crowley [92] fitted RTO growth data at 1.0 and 0.1 atm to two growth laws, namely, a linear–parabolic and a parabolic relationship. They reported an activation energy of 2.17 eV for the parabolic rate constant, which is in agreement with that of oxygen diffusion in silica [92]. Lassig et al. [93] reported activation energies of 1.31 and 1.20 eV for the linear rate constant of RTO growth in dry and wet oxygen, respectively.

Two detailed models were developed for RTO growth kinetics by Paz de Araujo et al. [89] and Parkhutik et al. [94]. These models are mentioned here because they developed Deal–Grove-like oxidation models that were successfully used to predict selected RTO growth data. At this point, it should be pointed out that RTO growth data are system dependent. This issue was raised by Dilhac [95], who cautioned against seeking a universal RTO growth model, much as the Deal–Grove model and its refinements are used in modeling conventional thermal furnace oxidation.

To illustrate this point, RTO growth kinetics data in dry oxygen on ⟨100⟩ silicon at 950 °C from a number of reports in the literature are plotted in Fig. 6 [2–33, 87–95]. The analytical growth relationship of Massoud and Plummer [79] for conventional thermal oxidation in the ultrathin regime is also plotted at 950 and 1000 °C for comparison. If the scatter in the RTO data is attributed to errors in temperature measurement, then it can be easily seen that the scatter in the RTO data can be attributed to as much as 50 °C uncertainty in the temperature.

The scatter in the data renders the development of a growth kinetics model rather futile. However, the trend in the data is also a strong indication of the constant improvement and progress in RTO equipment over the years in the area of temperature measurement, calibration, and control. It is therefore recommended that a physically based RTO growth kinetics model incorporate the refinements and progress in temperature measurement and control, a better understanding of the contribution of the spectral distribution of the heating sources on the oxidation process, and a better thermal design of the wafer handling system in order to eliminate thermally induced stresses that result from nonuniform radial temperature distributions and are shown to affect the RTO growth kinetics. In the meantime, however, empirical RTO growth laws can be used, provided

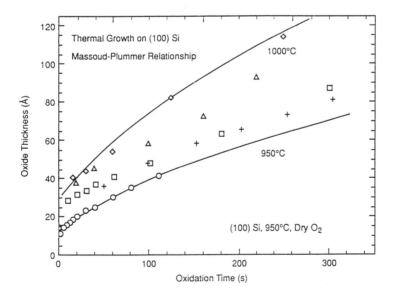

FIGURE 6. Discrepancy in RTO growth kinetics on ⟨100⟩ silicon in dry O_2 at 950 °C. The solid lines are the analytical thermal oxidation growth kinetics relationship of Massoud and Plummer. From Massoud and Plummer [79].

that each system would be independently characterized and modeled. It should be noted that the development of *in situ* process monitoring and control using ellipsometry would provide an alternative approach to process control where the accurate knowledge of the growth rate as a function of temperature is not necessary.

III. Rapid Thermal Processing of Oxides

Rapid thermal processing applications related to dielectrics can be divided into two major areas, namely, the formation of the dielectric films either by thermal growth or chemical vapor deposition, and the postformation treatment of dielectrics. Examples of dielectric formation processes include rapid thermal oxidation, rapid thermal nitridation, rapid thermal oxynitridation, and rapid thermal chemical vapor deposition of oxides, nitrides, and oxynitrides. Examples of postformation treatments include rapid thermal annealing in an inert ambient, rapid thermal annealing in a nitridizing ambient such as in nitrogen or ammonia, rapid thermal fluorination in NF_3, and rapid thermal reoxidation of nitrided thermal or CVD oxides. This section discusses some of these applications, their advantages and limitations.

A. Rapid Thermal Annealing

Thermally grown and chemical vapor deposited oxides formed either in conventional furnaces/reactors or in rapid thermal processing systems have been used in many rapid thermal annealing studies. In these studies, the ambient during RTA was either nitrogen or argon. RTA in nitrogen and argon have the thermal cycle in common but differ in the low-level nitridation that takes place when using N_2. It was later found that RTA in NH_3 resulted in oxides with better properties than those annealed in N_2. This will be discussed in the following two sections. The early studies of rapid thermal annealing investigated the influence of RTA conditions on interface charges, interface trap distributions, and dielectric breakdown.

In studies of the RTA of 100 Å oxides in nitrogen, Nulman *et al.* [96] reported a reduction in the flat-band voltage or in the oxide fixed charge density during the first 20 sec of RTA in N_2 at 1150 °C. They also observed a reduction in the density of midgap interface traps to below $10^{10} \text{eV}^{-1} \text{cm}^{-2}$. An RTA time of 40 sec was found to be optimal in the reduction of interface traps. Nulman *et al.* [96] also reported a considerable increase in the charge-to-breakdown Q_{BD} from 20 C/cm^2 for conventional furnace oxides to 80 C/cm^2 for RTO/RTA oxides of the same oxide thickness of

100 Å. As Q_{BD} is directly correlated to electron trapping in SiO$_2$ [97], it was concluded that electron trapping occurs at a much smaller rate in RTO/RTA oxides than in conventional furnace oxides [98]. Weinberg *et al.* also reported a reduction in electron trapping on water-related centers by RTA for 10 sec in the 600–800 °C range in Ar or N$_2$ ambients [98]. By comparison, hole trapping was optimally reduced by RTA in O$_2$ for a longer time of 100 sec at a higher temperature of 1000 °C.

The kinetics of rapid thermal annealing of interface traps at low temperatures in the 200–400 °C range were studied by Reed and Plummer [99]. The RTA temperature was found to be the most dominant experimental condition affecting the annealing kinetics. They obtained a power-law dependence of D_{it} on annealing time in the form $D_{it} \sim 1/\sqrt{t}$ [99]. The effect of high-temperature rapid thermal annealing on interface trap and oxide defect densities was studied by Vasudev *et al.* [100]. They monitored the change in interface trap density using electron paramagnetic resonance and found that traps could be annealed or generated by RTA depending on the annealing temperature and the initial preannealing concentration of P_b centers. For RTA above 1100 °C, the final concentration of defects were independent of their initial value [100].

The effect of RTA on dielectric breakdown was also studied [96, 101, 102]. Finn and Goe [101] reported no degradation in dielectric breakdown with RTA. Fukada *et al.* [102] reported that RTP oxides with superior electric properties and dielectric reliability to furnace oxides. They correlated the improved breakdown characteristics to the superior Si–SiO$_2$ interface roughness as revealed by oscillations in the Fowler–Nordheim tunneling plots [102].

Rapid thermal processing was also used in the annealing of low-pressure (LP) CVD oxide films [13, 103] and epitaxial dielectrics such as CaF$_2$ on silicon [104], and in the reflow of passivating glasses [105, 106]. In these applications, high-performance gate dielectrics were obtained by RTA of LPCVD oxides, the epitaxial quality of CaF$_2$ grown on silicon was improved, and PSG films were successfully reflowed at temperatures much lower than the 1050–1100 °C range used in furnace PSG reflow.

From the previous discussion, it is concluded that oxides annealed in a rapid thermal processing system generally possess electrical and dielectric properties that are comparable to or better than those of furnace oxides. This conclusion is based, however, on a limited number of studies, as all aspects of dielectric integrity and performance were not fully investigated. This is generally the result of further innovations in postoxidation processing including rapid thermal nitridation and reoxidation, which quickly became the focus of subsequent investigations. These postoxidation treatments are the subject of the following sections.

B. Rapid Thermal Nitridation

The incorporation of nitrogen in the SiO_2 network of an oxide film by rapid thermal nitridation results in a nitrogen-rich surface layer, a bulk oxide layer with small concentrations of nitrogen, and an interface region consisting of an interfacial nitride layer and an oxygen-rich layer. Rapid thermal nitridation results in oxides with several advantages such as an increased barrier to dopant diffusion [107, 108], an increased resistance to interface trap generation under electrical stress [109], a reduced sensitivity to radiation [110], an increased endurance to hot-electron injection [111, 112], and a dielectric constant larger than that of pure oxides [113]. Rapid thermal nitridation is usually done in pure ammonia gas and is well suited for the accurate control of the amount of nitrogen incorporated in the oxide, which in turn determines the range of control for the optimization of the above mentioned advantages.

Following the rapid thermal nitridation of SiO_2, nitrogen depth profiles obtained by Auger electron spectroscopy indicate that the surface nitrogen concentration increases with RTN temperature and time, and saturates at 1.7×10^{22} cm^{-3} [114]. In contrast, the interfacial nitrogen concentration does not saturate and increases with RTN time and temperature. This observation suggests that the concentration of the nitriding species at the interface is not rate limiting in the process. Joshi *et al* [114] concluded that surface nitridation is essentially a reaction rate limited process with the nitrogen concentration depending on RTN temperature and independent of oxide thickness, and that the nitridation process in the bulk of the oxide is slow and dependent on the oxide thickness.

The increase in nitrogen concentration at the surface is accompanied by a decrease in the concentration of oxygen. The nitridation reaction was proposed to be a displacement reaction between nitrogen and oxygen, as defined by the reaction [115]

$$4NH_3 + 3SiO_2 \rightarrow Si_3N_4 + 6H_2O. \qquad (1)$$

This reaction is important because it indicates the formation of water as a reaction byproduct. Water present in SiO_2 layers has long been shown to influence their electric properties, as will be discussed later. Joshi *et al.* [114] proposed that the nitridation in the bulk of the oxide is suppressed because of the presence of water and its failure to escape from that region of the oxide. Under the same RTN conditions, thinner oxides had higher surface nitrogen concentrations than those in thicker oxides [114]. This observation and the assumption of a thin interfacial strained layer in the oxide support the increased probability of the formation of Si—N bonds in a strained Si—O network as proposed by Vasquez and Madhukar [116].

Rapid thermal nitridation proceeds by NH_3 quickly permeating through the oxide. During this diffusion, nitridation takes place throughout the oxide with increased rate at the surface and the interface regions due to the easier removal of water generated as a reaction byproduct [114, 117]. The presence of water at the interface also results in the formation of an oxygen-rich layer [118]. The physical, optical, and electrical properties discussed later are related to the presence and behavior of the nitridation reaction byproducts in the oxide.

Despite many improved properties of nitrided oxides, the rapid thermal nitridation of SiO_2 introduces a large concentration of electron traps [119], results in fixed-charge buildup [120], and causes leakage current to increase [121]. The degradation in oxide properties with RTN was found to depend on RTN conditions, which in turn determine the concentration of nitrogen and hydrogen incorporated in the oxide. Postnitridation anneals in an inert or oxidizing ambient were found to suppress or reduce some of these disadvantages while preserving the advantages of nitridation. Post-RTN annealing in oxygen is especially advantageous and will be discussed next.

C. RAPID THERMAL REOXIDATION OF NITRIDED OXIDES

The rapid thermal reoxidation of nitrided oxides was observed to reduce hot-carrier degradation of N-channel metal oxide semiconductor (NMOS) field-effect transistors (FETs) by several orders of magnitude because of the suppression of interface and bulk electron trap generation by hot-hole injection [122–124]. Reoxidation was also shown to reduce the enhancement in gate-induced drain leakage caused by hot-electron stress in comparison with control oxides and nitrided oxides [125].

Hori *et al.* [126] proposed a model to explain the effects of rapid thermal reoxidation of nitrided oxides on the electrical properties of these dielectrics. This model is based on previous studies that established the role of hydrogen-containing species such as —H and —OH bonds in electron trapping [127]. They proposed that —H and —OH bonds are created during RTN as a result of the dissociation of NH_3 [116] and are responsible for the large number of electron traps in nitrided oxides. Reoxidation of nitrided oxides eliminates these hydrogen-containing species via the reactions [126]

$$2(\equiv Si-H) + O \rightarrow \; \equiv Si-O-Si\equiv \; + H_2 \quad (2)$$

and

$$\equiv Si-H \; + \; \equiv Si-OH + O \rightarrow \; \equiv Si-O-Si\equiv \; + H_2O, \quad (3)$$

which result in the reduction of electron traps in nitrided oxides while maintaining all other advantages of RTN. The findings of this model were supported by the experimental observation of the simultaneous monotonical decrease in the flat-band voltage shift induced by high-field stressing and the concentration of hydrogen measured by SIMS [126].

Hori et al. [126] concluded that reoxidized nitrided oxides prepared by rapid thermal processing should be considered one of the most promising candidates to replace thermally grown convential oxides for future ultra-large-scale integration (ULSI) devices. The electrical properties of these oxides are discussed next.

IV. Electrical Properties of Rapid Thermal Oxidation/Rapid Thermal Processing Oxides

The electrical properties of oxides prepared by rapid thermal processing were measured and compared to those of furnace-grown oxides in a large number of studies. As indicated earlier, these properties were found to be comparable to or better than those of furnace-grown oxides. The electrical properties of interest here are the interface traps and fixed charges, the current–voltage and tunneling characteristics, the hot-electron trapping properties and degradation sources, and the dielectric breakdown distributions. The dependence of these properties on the experimental conditions during rapid thermal oxidation, nitridation, reoxidation, and annealing is discussed next.

A. Interface Traps and Fixed Charges

Interface trap distributions in the silicon band gap and oxide fixed charges at the Si–SiO$_2$ interface are obtained from high-frequency and quasi-static C–V measurements. The interface trap distributions of furnace-grown and rapid thermally grown oxides are compared in Fig. 7 [128]. The furnace-grown oxides (FO) were grown at 800 °C in a 2:1 N$_2$:O$_2$ dry mixture, at 850 °C in a 1:1 H$_2$:O$_2$ wet mixture, and in the 950–1000 °C range in dry O$_2$ to a thickness of 98 Å. The rapid thermally grown oxides were grown in dry O$_2$ in the 1000–1200 °C range. Diluted and wet FO oxide samples have nearly U-shaped D_{it} distributions, and RTO samples have similarly shaped D_{it} distribution with smaller concentrations than FO samples [128]. It is noted that the D_{it} distributions of FO and RTO samples become nearly equal as the RTO temperature is decreased.

Rapid thermal oxidation followed by rapid thermal nitridation in ammonia has been shown to improve many properties of RTO oxides, such as increasing the barrier to dopant and contaminant diffusion, improving the dielectric strength, and reducing damage induced by radiation and

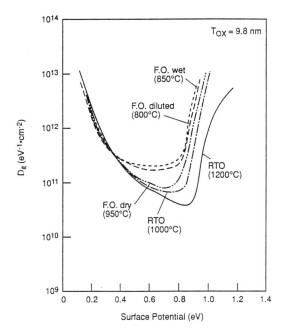

FIGURE 7. Comparison of interface trap distributions in furnace oxides (FO) and rapid thermal oxides (RTO). From Fukuda et al. [128]. Copyright © 1992 IEEE.

high-field stress. Rapid thermal nitridation has the advantage of limited dopant diffusion in comparison with long thermal nitridations done in conventional resistively heated furnaces [113, 129]. Dielectrics prepared this way are designated as RTO/RTN oxides.

In thin nitrided oxides, however, it was observed that interface traps and fixed charges vary strongly with the RTN time and temperature both in thermal and rapid thermal nitridation [107, 111–113, 129–140]. Rapid thermal nitridation affects the C-V characteristics of MOS capacitors (MOSCAPs) in several ways. The MOSCAP capacitance in accumulation was observed to increase with RTN time and temperature [114, 120, 139, 140]. Increases of as much as 20% in C_{ox} were observed after RTN at 1150 °C [114]. Hori et al. [120, 139] concluded from Auger electron spectroscopy measurements that the thickness of the nitrided oxide did not change with RTN and that the increase in the accumulation capacitance was due to an increase in the dielectric constant of the nitrided oxide. The dielectric constant was reported to increase with RTN temperature, and increase then saturate with RTN time [120]. For an 80-Å-thick oxide rapid-thermally nitrided at 1150 °C for 300 sec, the dielectric constant of the oxide increased from 3.9 to 4.6 upon nitridation [120].

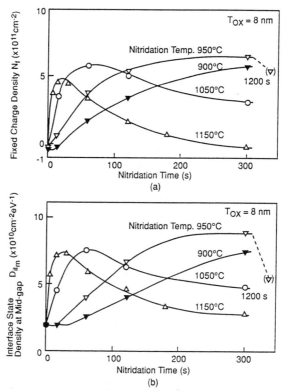

FIGURE 8. Changes in oxide fixed charge density N_f and midgap interface trap density $D_{it,m}$ with rapid thermal nitration time and temperature. From [120, 139]. Copyright © 1992, 1986, IEEE.

As rapid thermal nitridation proceeds, a shift ΔV_{fb} in the flat-band voltage is observed in the high-frequency C-V measurements. This shift is used to calculate a shift ΔN_f in the fixed charge density N_f at the Si-SiO$_2$ interface. When combined with the shifts in the quasi-static measurements, a shift $\Delta D_{it,m}$ in the midgap density D_{it} of interface traps is obtained. An example of the changes in the oxide fixed charge and interface trap density at midgap is shown in Fig. 8 [120, 138]. Both N_f and $D_{it,m}$ first increase with a rate depending on the RTN temperature, reach a maximum, and then turn around and decrease monotonically with a rate depending on the RTN temperature [114, 120, 139].

The similarity in the behavior of N_f and $D_{it,m}$ suggests that the physical and chemical mechanisms underlying these changes are the same in both cases. Models explaining the role of RTN in the trends in N_f and $D_{it,m}$ based on mechanisms involving nitrogen, hydrogen, and Si—O bonds were evaluated [120]. Hori *et al.* [120] proposed a two-step model consistent with the dependence of ΔN_f and $\Delta D_{it,m}$ on RTN time and temperature that would predict the turnaround observations. The first step is a defect formation step resulting from the incorporation of nitrogen and the second step is the reduction of these defects in an annealing-type process. An alternative explanation is to consider the thermal and chemical changes at the interface and their influence on the partial charge transfer in interface bonds [141]. This alternative explanation was successful in predicting the turnaround effect in conventional thermal annealing of MOS structures in nitrogen [141].

Hori *et al.* [142] studied the dependence of fixed charge densities and interface trap distributions on the experimental conditions during nitridation and reoxidation. In their studies, rapid-thermally grown oxides were subjected to different RTN treatments in NH_3 ranging from 15 sec at 950 °C to 60 sec at 1150 °C. Oxides nitrided at 950 °C for 15 sec in 100% NH_3 were designated as NO_S samples, those nitrided at 950 °C for 60 sec in 100% NH_3 as *NO* samples, and those nitrided at 1150 °C for 60 sec in 100% NH_3 as NO_{HT} samples. Prior to reoxidation, the NO_S, *NO*, and NO_{HT} samples had oxide fixed charges densities N_f of 1.6 ± 10^{11}, 3.8×10^{11}, and 1.8×10^{11} cm^{-2}, respectively. The NO_S and *NO* samples were nitrided in the preturnaround stage where N_f increases with nitridation time, while the NO_{HT} samples were nitrided in the postturnaround stage where N_f decreases with nitridation time. These samples were reoxidized at 950, 1050, and 1150 °C.

The dependence of oxide fixed charge density of the NO_S, *NO*, and NO_{HT} samples on reoxidation time is shown in Fig. 9. It is observed that N_f exhibits a turnaround as reoxidation proceeds at any given temperature, increasing in the early stages of reoxidation, reaching a maximum at a reoxidation time t_{max}, and then gradually decreasing [142]. It was found that t_{max} gets shorter and N_f decreases more rapidly at higher reoxidation temperatures. Finally, N_f reaches a low value of 4×10^{10} cm^{-2}, which is comparable to that of thermally grown oxides [142]. The conditions of rapid thermal nitridation were also to affect changes in N_f with reoxidation. The turnaround and the decrease of N_f in the final reoxidation stage takes place earlier in NO_S samples than in *NO* samples, and the increase in N_f is significantly larger in heavily nitrided samples than in lightly nitrided ones [142].

To explain the turnaround in N_f with reoxidation, Hori *et al.* [142] extended their two-step model that explained the turnaround behavior of N_f with nitridation to reoxidation. This was achieved by extending the second step of the model to include two annealing-type processes instead of one. These two processes consist of one that involves hydrogen and one that

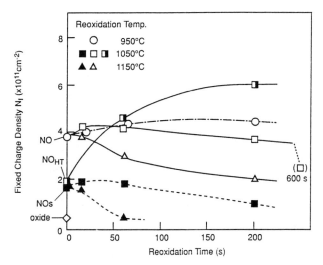

FIGURE 9. Dependence of the oxide fixed charge density N_f of reoxidized nitrided oxides on reoxidation time. From Hori et al. [142]. The symbols NO_S, NO, and NO_{HT} are described in the text. Copyright © 1989 IEEE.

does not [142]. When the postnitridation ambient is switched to one with no hydrogen such as N_2 or O_2, hydrogen diffuses out of the oxide and activates hydrogen-passivated sites, which are then reduced by the annealing-type process that does not involve hydrogen.

In all samples studied, the midgap interface trap density $D_{it,m}$ ranged from 1.5×10^{10} to 2.5×10^{10} eV^{-1} cm^{-2}. This indicates that these interface traps were annealed out at the lowest processing temperature [142].

B. Current–Voltage Characteristics

The dc I–V characteristics of oxides grown in conventional furnaces and rapid thermal processing equipment were compared by Moslehi et al. [140]. The conventional furnace oxides were 99 Å thick and grown at 850 °C, while the RTO oxides were 104 Å thick and grown at 1150 °C. The leakage current was smaller than 1 pA for samples biased in accumulation with up to 7.5 V. For voltages larger than 7.5 V, the I–V characteristics were in the Fowler–Nordheim tunneling regime with two distinct slopes [140]. The I–V characteristics of FO and RTO oxides were similar except that the RTO samples exhibited significantly less voltage stress effects than the FO samples.

The effects of rapid thermal nitridation on the I–V characteristics of RTO oxides were studied in several investigations [114, 140]. The dependence of

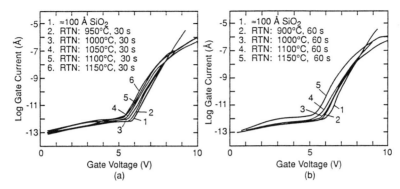

FIGURE 10. Current-voltage $I-V$ characteristics of nitrided oxides showing their dependence on nitridation conditions. From Moslehi [140].

the $I-V$ characteristics on RTN temperature and time is shown in Fig. 10 [140]. It is observed that the current passing through the oxide increases for electric fields in the 6–8 MV/cm range, with the current increase becoming larger in samples nitrided at higher temperatures. At fields larger than 8 MV/cm, the current through the oxide decreases. This behavior was attributed by Moslehi *et al.* [140] to the role of nitrogen affecting the tunneling barrier for electrons and holes, the role of the changing interfacial fixed charge density, and the lower conductivity of RTN oxides and its weaker field dependence. Joshi *et al.* [114] made similar observations and noted the absence of changes in the $I-V$ characteristics in samples thermally nitrided in conventional furnaces.

C. Dielectric Breakdown

Early investigations of the dielectric breakdown properties of RTO oxides indicated that the breakdown characteristics of RTO and FO oxides were similar with an average breakdown field of 13.5 MV/cm [140]. Later studies indicated that RTO samples exhibited breakdown characteristics superior to those of FO samples, with excellent breakdown uniformity, an average breakdown field of 15.0 MV/cm, and an average breakdown charge of 50 C/cm^2 at a stress current density of 1 A/cm^2 [143].

The time-zero dielectric breakdown (TZDB) histograms of FO and RTO samples grown to the same thickness are compared in Fig. 11. In TZDB tests, a negative gate voltage is applied and the breakdown voltage is defined as the voltage across the oxide when a current of 1 μA through it is measured [128]. Compared in Fig. 11 are the TZDB characteristics of a furnace oxide grown at 950 °C in dry O$_2$ and a rapid thermal oxide grown

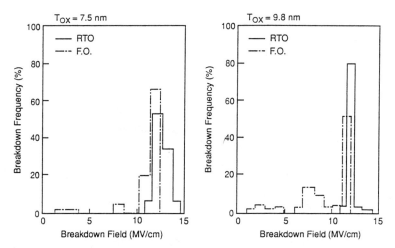

FIGURE 11. Time-zero dielectric breakdown (TZDB) histograms of FO and RTO MOSCAPs with 75 and 98 Å. From Fukuda [128]. Copyright © 1992 IEEE.

at 1000 °C in dry O_2 at two thicknesses of 75 and 98 Å. It can be seen that the TZDB of the RTO oxide is better than that of the FO oxide in two ways. First, low-field (<1 MV/cm) and medium-field (2–8 MV/cm) breakdown rarely occur in RTO oxides. Second, the maximum breakdown frequency occurs at higher fields in RTO oxides [128].

The time to breakdown (t_{BD}) was measured on FO and RTO oxides by applying a negative gate voltage with a constant field of 13 MV/cm or a constant current density in the 10^{-2}–10 A/cm² range [128]. The results for 98-Å-thick FO and RTO oxides are compared in Fig. 12. It was found that the dependence of t_{BD} on the current density in the oxide J_{ox} was in the form $t_{BD} \sim J_{ox}^{-1.4}$, that t_{BD} was consistently longer in RTO samples than FO samples, and that t_{BD} increased with RTO temperature [128]. Fukuda *et al.* [128] explained the longer t_{BD} of RTO oxides by the reduction in strained Si—O—Si and Si dangling bonds at the Si–SiO$_2$ interface when the RTO oxide is grown at a higher temperature.

The *in situ* removal of the native oxide and metallic contaminants from the silicon surface in hydrogen or 1% HCl/Ar ambients prior to rapid thermal oxidation by rapid thermal cleaning was found to depend strongly on the processing conditions [96, 128]. Fukada *et al.* [128] recommend RTC times of 20–60 sec and temperatures in the 700–900 °C range to avoid etch pits on Si and make the time to 50% cumulative failure longer.

It was observed that rapid thermal nitridation increased the average breakdown field of RTO samples by up to 10% [140]. This improvement in dielectric properties was attributed to the formation of the nitrogen-rich

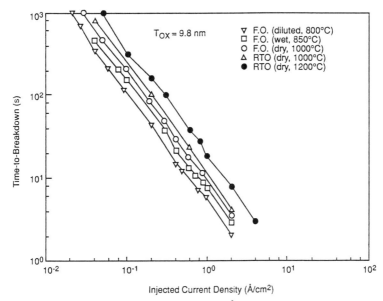

FIGURE 12. Mean time to breakdown (t_{BD}) of 98 Å thick FO and RTO oxides. From Fukuda [128]. Copyright © 1992 IEEE.

layer at the surface of the oxide and at the Si–SiO$_2$ interface, and to bulk electron trapping [140]. In recent studies of the time-dependent dielectric breakdown of reoxidized nitrided oxides, Liu et al. [144, 145] reported comparable impact ionization coefficients and activation energies, though the charge to breakdown was somewhat improved. These observations are related to the charge-trapping properties of oxides formed by RTP as will be discussed in the next section.

D. Carrier Trapping Properties

Constant-current stressing is widely used in the characterization of hot-carrier trapping in dielectrics. In this test, current is maintained at a constant value through the oxide by continuously changing the gate voltage of an MOS capacitor structure. At different time intervals, high-frequency and quasi-static C-V measurements are done to monitor changes in the flat-band voltage ΔV_{fb} and thus in the oxide fixed charge density ΔN_f, and in the interface trap density at midgap $\Delta D_{it,m}$. The change in the oxide fixed charge density ΔN_f is a monitor of the amount of charge trapped in the oxide, and the change in the interface trap density $\Delta D_{it,m}$ at midgap is a monitor of the influence of the passing of hot carriers through the

Si–SiO$_2$ interface on interface trap generation. These parameters are routinely plotted as functions of the injection time t_{inj} or the charge injected through the oxide Q_{inj}. These two parameters are related by $Q_{inj} = J_{inj} t_{inj}$, where J_{inj} is the electric current density passing through the oxide, and is sometimes designated by J_{ox}.

When a control oxide grown in a conventional furnace on n-type silicon to form an MOS capacitor structure is electrically stressed such that the surface is in accumulation, Moslehi et al. [140] reported that the voltage needed to maintain a constant current decreased with stress time t_{inj} indicating a higher probability of tunneling resulting from positive charge trapping. Nitrided oxides of the same thickness exhibited a voltage that increased with t_{inj} and a C-V curve with a larger parallel flat-band voltage shift in the positive direction indicating electron trapping [140]. In the process of current stressing, i.e., high-field carrier injection, new interface traps are generated at the Si–SiO$_2$ interface. The buildup of interface traps with stress time t_{inj} was observed in control and nitrided oxides, with the RTN oxides having generally smaller values of $D_{it,m}$ [140]. RTN oxides are thus more resistant to interface trap generation by high-field electron current stressing. Joshi et al. [114] observed that electron trapping increases with RTN temperature and time, except at 900 °C where the samples nitrided for 60 sec showed lower electron trapping than those nitrided for 30 sec.

The effects of reoxidation on electron trapping were reported in many studies [114, 142, 144, 145]. Hori et al. [142] observed improved resistance against electron injection in optimally reoxidized nitrided oxides, which can have ΔV_{fb} and $\Delta D_{it,m}$ smaller by more than two orders of magnitude than those of a thermal oxide. They suggested that to minimize charge trapping by reoxidation of a nitrided oxide, electron trap elimination should be of greater concern than minimizing the generation of interface traps [142]. They also found that, in most cases, little differences exist between post-nitridation treatments in N$_2$ and O$_2$.

E. DEVICE PERFORMANCE

Many studies have reported the use of RTP dielectrics in NMOS, PMOS, and complementary MOS (CMOS) process integration. These studies report a list of advantages of the application of RTP in ultralarge-scale integration devices and circuits. Hot-carrier immunity was improved, with lifetimes for reaching 30 mV threshold voltage shift and 10% transconductance degradation being improved by 3 and 1.5 orders of magnitudes, respectively

[146]. The electron mobility could be increased over that of devices with thermal oxides by light nitridation [147]. Rapid thermal processing is indispensable in light nitridation treatments and in compatibility with cluster tools. Excellent immunity to gate-induced drain leakage (GIDL) from hot-electron stress especially at low drain voltages was demonstrated [125].

The use of reoxidized nitrided oxides was shown to significantly increase circuit lifetime over conventional oxides [124]. The recovery of channel hot-carrier damage in reoxidizes nitrided oxide n- and p-channel MOSFETs is substantially greater than in conventional furnace oxides, especially in p-MOSFETs [148]. A CMOS gate delay time of 55 psec/stage was demonstrated on the basis of an improved saturation transconductance in n-MOSFETs and a small degradation in that of p-MOSFETs [149]. Hori et al. [149] obtained device lifetime improvement of over 100 times to exceed 10 years with respect to both on- and off-state hot-carrier reliability for n-MOSFETs and unchanged hot-carrier reliability for p-MOSFETs. The above performance was obtained by light nitridation followed by reoxidation in O_2 or reannealing in N_2. On the basis of these results, it was concluded that deep-submicron CMOS with reannealed or reoxidized nitrided oxide gate dielectrics satisfy the requirements of CMOS operation with a 3.3 V power supply. Chapter 8 presents RTP device applications and process integration issues in further detail.

F. TECHNOLOGY—CURRENT PRACTICE

It is projected that the manufacturing of integrated circuits in the ULSI era for devices with feature dimensions of less than 0.5 μm will be carried out in multichamber and multiprocess cluster tools consisting mainly of rapid thermal and plasma unit processes [150]. These cluster tools will meet the process technology needs of 16–256 MB DRAMS and logic integrated circuits of equivalent complexity by providing automated connection between processes, a controlled interprocess environment, cleaner wafer handling between processes, and greater manufacturing flexibility [150]. The rapid thermal growth and processing of high-quality gate dielectrics will become an important unit in these cluster tools rather than being a high throughput stand alone system.

A recent evaluation of the present status of RTO and RTP of dielectrics concluded that rapid thermal processing is a proven, necessary technique for interconnect technology using titanium silicide and nitrides, and that its wide use for oxidation, glass reflow, and contact alloying is unclear and controversial at the present time [151]. But, with the clear hot-electron reliability advantages of reoxidized lightly nitrided oxides over thermal oxides, and the integration requirements into future cluster tools driven by cost reduction, RTO and RTP of dielectrics must become a manufacturing process. Before that is achieved in a stand-alone system or a cluster tool, advances must be made to improve temperature measurement accuracy and control, temperature uniformity, and an increased throughput [152, 153]. With the constant improvement in RTP equipment and its needed optimization for oxidation process, it is expected that the rapid thermal growth and processing of dielectrics in state-of-the-art devices will become a standard technology.

V. Conclusions

In this chapter the rapid thermal oxidation and dielectric processing in silicon technology were reviewed. A discussion of the equipment issues related to process optimization indicates that RTO offers definite advantages in processing technology but offers some challenges in optimizing the equipment for use in RTO/RTP applications. Temperature uniformity, temperature measurement and control, reactor design, and radiation sources are important in determining the performance of RTP systems in dielectric applications. Because of issues related to temperature measurement resulting from emissivity changes and differences in compensation schemes, it was demonstrated that rapid thermal oxidation modeling of the oxide growth kinetics is hampered by the uncertainty of the temperature during the process. The properties of oxides grown in RTO systems were presented and compared with those of furnace oxides and those of nitrided and reoxidized/nitrided oxides. It was shown that the rapid thermal processing of dielectrics yields oxides with superior properties and unique advantages, especially in the area of hot-electron reliability. The potential of RTO in a manufacturing setting was evaluated, and it is concluded that RTO is a necessary process in future cluster tools.

Acknowledgments

The author would like to thank R. Deaton, R. Sampson, Y. Kim, K. Hunt, R. Fair, J. Wortman, M. Öztürk, J. Hauser, N. Masnari, E. Irene, M. Moslehi, J. Gelpey, J. Kuehne, D. Hodul, A. Gat, and J. Nulman for many helpful discussions. This study was supported by the NSF/ERC on Advanced Electronic Materials Processing at North Carolina State University.

References

1. For a review and general RTP bibliography, see for example R. Singh, *J. Appl. Phys.* **63**, R59-R114 (1988).
2. G. J. Grant, G. Brown, J. Shu, E. Lee, and J. Reynolds, *J. Electrochem. Soc.* **131**, 469C (1984).
3. J. Nulman, J. P. Krusius, and A. Gat, *IEEE Electron Device Lett.* **6**, 205-207 (1985).
4. M. M. Moslehi, S. C. Shatas, and K. C. Saraswat, *Appl. Phys. Lett.* **47**, 1353-1355 (1985).
5. C. M. Gronet, J. C. Sturm, K. E. Williams, and J. F. Gibbons, *Mat. Res. Soc. Symp. Proc.* **52**, 305-312 (1985).
6. A. M. Hodge, C. Pickering, A. J. Pidduck, and R. W. Hardeman, *Mat. Res. Soc. Symp. Proc.* **52**, 313-319 (1985).
7. J. C. Gelpey, P. O. Stump, and R. A. Capodilupo, *Mat. Res. Soc. Symp. Proc.* **52**, 321-325 (1985).
8. Z. A. Weinberg, T. N. Nguyen, S. A. Cohen, and R. Kalish, *Mat. Res. Soc. Symp. Proc.* **52**, 327-332 (1985).
9. J. Nulman, J. P. Krusius, and P. Renteln, *Mat. Res. Soc. Symp. Proc.* **52**, 341-348 (1985).
10. A. Gat and J. Nulman, *Semicond. Int.* **8**, 120-124 (1985).
11. J. P. Ponpon, J. J. Grob, A. Grob, and R. Stuck, *J. Appl. Phys.* **59**, 3921-3923 (1986).
12. Y. Sato and K. Kiuchi, *J. Electrochem. Soc.* **133**, 652-654 (1986).
13. S. T. Ang and J. J. Wortman, *J. Electrochem. Soc.* **133**, 2361-2362 (1986).
14. J. Nulman, J. P. Krusius, N. Shah, A. Gat, and A. Baldwin, *J. Vac. Sci. Tech. A* **4**, 1005-1008 (1986).
15. J. F. Gibbons, C. M. Gronet, J. C. Sturm, C. King, and K. Williams, *Mat. Res. Soc. Symp. Proc.* **74**, 629-639 (1986).
16. N. Chan Tung and Y. Caratini, *Electron. Lett.* **22**, 694-696 (1986).
17. A. Maury, S. C. Kim, A. Manocha, K. H. Oh, D. Kostelnick, and S. Shive, in *Technical Digest of the International Electron Devices Meeting*, Cat. No. 86CH238-12, pp. 676-679, IEEE, New York (1986).
18. M. Moslehi, *Mat. Res. Soc. Symp. Proc.* **92**, 73-87 (1987).
19. J. T. Fitch and G. Lucovsky, *Mat. Res. Soc. Symp. Proc.* **92**, 89-94 (1987).
20. J. Nulman, *Mat. Res. Soc. Symp. Proc.* **92**, 141-146 (1987).
21. S. Mehta, D. T. Hodul, and C. J. Russo, *Nucl. Instrum. Meth. Phys. Res. B* **21**, 629-632 (1987).
22. T. Fujii and F. Ohara, *Oyo Buturi* **56**, 937-941 (1987).
23. M. M. Moslehi, M. Wong, K. C. Saraswat, and S. C. Shatas, in *Digest of Technical Papers, 1987 Symposium on VLSI Technology*, pp. 21-22, Business Center Academic Societies of Japan, Tokyo, Japan (1987).
24. N. C. Tung, Y. Caratini, C. D'Anterroches, and J. L. Buevoz, *Appl. Phys. A* **47**, 237-247 (1988).
25. Y. Miyai, K. Yoneda, H. Oishi, H. Uchida, and M. Inoue, *J. Electrochem. Soc.* **135**, 150-155 (1988).
26. P. B. Moynagh, P. J. Rosser, and R. B. Calligarro, in *UK IT 88 Conference Publication*, pp. 540-543, Inf. Eng. Directorate, London, UK (1988).
27. D. K. Sadana, J. P. de Souza, and F. Cardone, *Appl. Phys. Lett.* **57**, 1681-1683 (1990).
28. F. Gualandris and M. Gregori, *J. Vac. Sci. Tech. B* **8**, 10-15 (1990).
29. D. Nayak, K. Kamjoo, J. C. S. Woo, J. S. Park, and K. L. Wang, *Appl. Phys. Lett.* **56**, 66-68 (1990).
30. G. P. Burns, *J. Appl. Phys.* **65**, 2095-2097 (1989).

31. J. Nulman, *Mat. Res. Soc. Symp. Proc.* **74**, 641-646 (1987).
32. E. Fogarassy, A. Slaoui, C. Fuchs, and J. L. Regolini, *Appl. Phys. Lett.* **51**, 337-339 (1987).
33. W. Ting, P. C. Li, G. Q. Lo, D. L. Kwong, and N. S. Alvi, *J. Appl. Phys.* **66**, 5641-5643 (1989).
34. S. Mehta and D. Hodul, *Mat. Res. Soc. Conf. Proc.* **92**, 95-101 (1987).
35. S. Leavitt, *Semicond. Int.* **10(4)**, 64-70 (1987).
36. J. C. Gelpey, P. O. Stump, J. Blake, A. Michel, and W. Rausch, *Nucl. Instrum. Meth. Phys. Res.* **B 21**, 612-617 (1987).
37. D. Aitken, S. Mehta, N. Parisi, C. J. Russo, and V. Schwartz, *Nucl. Instrum. Meth. Phys. Res.* **B 21**, 621-626 (1987).
38. D. Hodul and S. Mehta, *Solid State Tech.* **31(5)**, 209-211 (1988).
39. R. Kakoschke, *Nucl. Instrum. Meth. Phys. Res.* **B 37-38**, 753-759 (1989),
40. D. Hodul and S. Mehta, *Nucl. Instrum Meth. Phys. Res.* **B 37-38**, 818-822 (1989).
41. F. Y. Sorrell, C. P. Eakes, M. C. Öztürk, and J. J. Wortman, *Proc. SPIE* **1189**, 55-63 (1990).
42. W. A. Keenan, W. H. Johnson, D. Hodul, and D. Mordo, *Nucl. Instrum. Meth. Phys. Res. B* **55**, 269-274 (1991).
43. R. Kakoschke, *Mat. Res. Soc. Symp. Proc.* **224**, 159-170 (1991).
44. See, for example, F. Roozeboom and N. Parekh, *J. Vac. Sci. Tech. B* **8**, 1249-1259 (1990), and references therein.
45. P. P. Apte, S. Wood, L. Booth, K. C. Saraswat, and M. M. Moslehi, *Mat. Res. Soc. Symp. Proc.* **224**, 209-214 (1991).
46. M. M. Moslehi, J. Kuehne, R. Yeakley, L. Velo, H. Najm, B. Dostalik, D. Yin, and C. J. Davis, *Mat. Res. Soc. Symp. Proc.* **224**, 143-157 (1991).
47. M. M. Moslehi, H. Najm, L. Velo, R. Yeakley, J. Kuehne, B. Dostalik, D. Yin, and C. J. Davis, in *ULSI Science and Technology 1991* (J. M. Andrews and G. K. Celler, eds.), Proc. Vol. 91-11, pp. 503-527, Electrochemical Society, Pennington, NJ (1991).
48. T. Sato, *Jap. J. Appl. Phys.* **6**, 339-347 (1967).
49. C. P. Ho and J. D. Plummer, *J. Electrochem. Soc.* **126**, 1516-1522 and 1523-1530 (1979).
50. R. Oren and S. K. Ghandhi, *J. Appl. Phys.* **42**, 752-756 (1971).
51. S. A. Schafer and S. A. Lyon, *J. Vac. Sci. Technol.* **19**, 494-497 (1981).
52. S. A. Schafer and S. A. Lyon, *J. Vac. Sci. Technol.* **21**, 422-425 (1982).
53. I. W. Boyd, *Appl. Phys. Lett.* **42**, 728-730 (1983).
54. R. A. B. Devine and G. Auvert, *Appl. Phys. Lett.* **49**, 1605-1607 (1986).
55. F. Micheli and I. W. Boyd, *Electron. Lett.* **23**, 298-300 (1987).
56. E. M. Young and W. A. Tiller, *Appl. Phys. Lett.* **50**, 46-48 (1987).
57. E. M. Young and W. A. Tiller, *Appl. Phys. Lett.* **50**, 80-82 (1987).
58. F. Micheli and I. W. Boyd, *Appl. Phys. Lett.* **51**, 1149-1151 (1987).
59. E. M. Young and W. A. Tiller, *J. Appl. Phys.* **62**, 2086-2094 (1987).
60. I. W. Boyd and F. Micheli, in *Emerging Technologies for In Situ Processing*, Proc. of the Advanced Nato Research Workshop (D. J. Ehrlich and V. T. Nguyen, eds.), pp. 171-178, Martinus Nijhoff, Dordrecht, The Netherlands (1988).
61. F. Micheli and I. W. Boyd, *Appl. Phys. A* **47**, 249-253 (1988).
62. I. W. Boyd, *Mat. Res. Soc. Symp. Proc.* **105**, 23-34 (1988).
63. I. W. Boyd, *Mat. Res. Soc. Symp. Proc.* **129**, 421-433 (1989).
64. V. Nayar and I. W. Boyd, *Chemtronics* **4**, 101-103 (1989).
65. V. Nayar, P. Patel, and I. W. Boyd, *Electron. Lett.* **26**, 205-206 (1990).
66. P. Patel and I. W. Boyd, *Apppl. Surf. Sci.* **46**, 352-356 (1990).

67. A. Kazor and I. W. Boyd, *Electron. Lett.* **27**, 909–911 (1991).
68. R. Deaton and H. Z. Massoud, *J. Appl. Phys.* **70**, 3588–3592 (1991).
69. P. J. Rosser, P. B. Moynagh, and K. B. Affolter, *Proc. SPIE* **1393**, 49–66 (1990).
70. D. Peyton, H. Kinoshita, G. Q. Lo, and D. L. Kwong, *Proc. SPIE* **1393**, 295–308 (1990).
71. V. M. Donnelly and J. A. McCaulley, *J. Vac. Sci. Technol. A* **8**, 84–92 (1990).
72. H. Z. Massoud, R. K. Sampson, K. A. Conrad, Y.-Z. Hu, and E. A. Irene, in *ULSI Science and Technology 1991* (J. M. Andrews and G. C. Celler, eds.), Proc. Vol. 91-11, pp. 541–550, Electrochemical Society, Pennington, NJ (1991).
73. H. Z. Massoud, R. K. Sampson, K. A. Conrad, Y.-Z. Hu, and E. A. Irene, *Mat. Res. Soc. Symp. Proc.* **224**, 17–22 (1991).
74. R. K. Sampson and H. Z. Massoud, in *ULSI Science and Technology 1991* (J. M. Andrews and G. C. Celler, eds.), Proc. Vol. 91-11, pp. 574–581, Electrochemical Society, Pennington, NJ (1991).
75. Y. J. Lee, C. H. Chou, B. T. Khuri-Yakub, and K. C. Saraswat, in *Digest of Technical Papers, 1990 Symposium on VLSI Technology*, IEEE Cat. No. 90-CH2874-6, pp. 105–106, IEEE, New York (1990).
78. R. Deaton and H. Z. Massoud, *Mat. Res. Soc. Symp. Proc.* **224**, 373–378 (1991).
79. H. Z. Massoud and J. D. Plummer, *J. Appl. Phys.* **62**, 3416–3423 (1987).
80. E. A. Irene, H. Z. Massoud, and E. Tierney, *J. Electrochem. Soc.* **133**, 1253–1256 (1986).
81. D. W. Hess and B. E. Deal, *J. Electrochem. Soc.* **124**, 735–739 (1977).
82. T. Nakayama and F. C. Collins, *J. Electrochem. Soc.* **113**, 706–713 (1966).
83. C. A. Paz de Araujo, J. C. Gelpey, Y. P. Huang, and R. Kwor, *Mat. Res. Soc. Symp. Proc.* **92**, 133–140 (1987).
84. A. Yehia-Messaoud, G. Sarrabayrouse, A. Claveri, A. Martinez, E. Scheid, E. Campo, and M. Faye, *Mat. Res. Soc. Symp. Proc.* **224**, 391–396 (1991).
85. A. Slaoui, B. Hartiti, M. C. Busch, J. C. Muller, and P. Siffert, *Mat. Res. Soc. Symp. Proc.* **224**, 409–414 (1991).
86. P. J. Rosser, P. B. Moynagh, and C. N. Duckworth, *Mat. Res. Soc. Symp. Proc.* **74**, 611–616 (1986).
87. S. P. Murarka, in *Proceedings of the First International Symposium on Ultra Large Integration Science and Technology* (S. Broydo and C. M. Osburn, eds.), Proc. Vol. 87-11, pp. 87–100, Electrochemical Society, Pennington, NJ (1987).
88. R. Singh, N. E. McGruer, K. Rajkanan, and J. H. Weiss, *J. Vac. Sci. Technol. A* **6**, 1480–1483 (1988).
89. C. A. Paz de Araujo, R. W. Gallegos, and Y. P. Huang, *J. Electrochem. Soc.* **136**, 2673–2676 (1989).
90. V. Murali and S. P. Murarka, in *Proceedings of the First International Symposium on Ultra Large Scale Integration Science and Technology* (S. Broydo and C. M. Osburn, eds.), Proc. Vol. 87-11, pp. 133–140, Electrochemical Society, Pennington, NJ (1987).
91. Y. L. Chiou, C. H. Sow, G. Li, and K. A. Ports, *Appl. Phys. Lett.* **57**, 881–883 (1990).
92. S. E. Lassig and J. L. Crowley, *Mat. Res. Soc. Symp. Proc.* **146**, 307–312 (1989).
93. S. E. Lassig, T. J. DeBolske, and J. L. Crowley, *Mat. Res. Soc. Symp. Proc.* **92**, 103–108 (1987).
94. V. P. Parkhutik, V. A. Labunov, and G. G. Chigir, *Phys. Stat. Sol. (a)* **96**, 11–18 (1986).
95. J.-M. Dilhac, *Mat. Res. Soc. Symp. Proc.* **146**, 333–338 (1989).
96. J. Nulman, J. Scarpulla, T. Mele, and J. P. Krusius, in *Technical Digest of the International Electron Devices Meeting*, Cat. No. 85CH2252-5, pp. 376–379, IEEE, New York (1985).
97. M. S. Liang and C. Hu, in *Technical Digest of the International Electron Devices Meeting*, Cat. No. 81CH1708-7, pp. 396–399, IEEE, New York (1981).

98. Z. A. Weinberg, D. R. Young, J. A. Calise, S. A. Cohen, J. C. DeLuca, and V. R. Deline, *Appl. Phys. Lett.* **45**, 1204-1206 (1984).
99. M. L. Reed and J. D. Plummer, *Mat. Res. Soc. Symp. Proc.* **52**, 333-340 (1986).
100. P. K. Vasudev, R. C. Henderson, P. J. Kaplan, and E. H. Poindexter, Abs. No. 566 in *ECS Extended Abstracts* **86-2**, 849 (1986).
101. M. A. Finn and M. E. Goe, in *Proceedings of the Symposium on Reduced Temperature Processing for VLSI* (R. Reif and G. R. Srinivascan, eds.), Vol. 86-5, pp. 85-94, Electrochemical Society, Pennington, NJ (1986).
102. H. Fukuda, T. Iwabuchi, and S. Ohno, *Jap. J. Appl. Phys.* **27**, L2164-L2167 (1988).
103. J. Lee, I. C. Chen, and C. Hu, *IEEE Elec. Dev. Lett.* **7**, 506-509 (1986).
104. L. Pfeiffer, J. M. Phillips, T. P. Smith III, W. M. Augustyniak, and K. W. West, *Appl. Phys. Lett.* **46**, 947-949 (1985).
105. D. F. Downey, C. J. Russo, and J. T. White, *Solid State Technol.* **25**(9), 87-93 (1982).
106. J. S. Mercier, L. D. Madsen, and I. Calder, *Mat. Res. Soc. Symp. Proc.* **52**, 251-258 (1986).
107. T. Ito, T. Nakamura, and H. Ishikawa, *J. Electrochem. Soc.* **129**, 184-188 (1982).
108. I. Kato, T. Ito, S. Inoue, T. Nakamura, and H. Ishiwaka, *Jap. J. Appl. Phys.* **21**, 153-160 (1982).
109. M. A. Schmidt, F. L. Terry Jr., B. P. Mathur, and S. D. Senturia, *IEEE Trans. Elec. Dev.* **35**, 1627-1632 (1988).
110. Z. Celik-Butler and T. Y. Hsiang, *IEEE Trans. Elec. Dev.* **35**, 1651-1655 (1988).
111. M. M. Moslehi, K. C. Saraswat, and S. C. Shatas, *Appl. Phys. Lett.* **47**, 1113-1115 (1985).
112. T. Ito, T. Nakamura, and H. Ishikawa, *IEEE Trans. Elec. Dev.* **29**, 498-502 (1982).
113. J. Nulman and J. P. Krusius, *Appl. Phys. Lett.* **47**, 148-150 (1985).
114. A. B. Joshi, G. Q. Lo, D. K. Shih, and D. L. Kwong, *Proc. SPIE* **1393**, 122-149 (1990).
115. Y. Hayafuji and K. Kajiwara, *J. Electrochem. Soc.* **129**, 2102-2108 (1982).
116. R. P. Vasquez and A. Madhukar, *J. Appl. Phys.* **60**, 234-242 (1986).
117. A. E. T. Kuiper, M. F. C. Willemsen, A. M. L. Theunissen, W. M. van de Wijgert, F. H. P. M. Habraken, R. G. H. Tijhaar, W. F. van der Weg, and J. T. Chen, *J. Appl. Phys.* **59**, 2765-2772 (1986).
118. R. P. Vasquez, A. Madhukar, F. J. Grunthaner, and M. L. Naiman, *Appl. Phys. Lett.* **46**, 361-363 (1985).
119. S. T. Chang, N. M. Johnson, and S. A. Lyon, *Appl. Phys. Lett.* **44**, 316-318 (1984).
120. T. Hori, H. Iwasaki, Y. Naito, and H. Esaki, *IEEE Trans. Elec. Dev.* **34**, 2238-2244 (1987).
121. E. Suzuki, D. K. Schroder, and Y. Hayashi, *J. Appl. Phys.* **60**, 3616-3621 (1986).
122. G. J. Dunn and S. A. Scott, *IEEE Trans. Elec. Dev.* **37**, 1719-1726 (1990).
123. B. S. Doyle and G. J. Dunn, *IEEE Elec. Dev. Lett.* **12**, 63-65 (1991).
124. G. J. Dunn and J. T. Krick, *IEEE Trans. Elec. Dev.* **38**, 901-906 (1991).
125. A. B. Joshi and D. L. Kwong, *IEEE Elec. Dev. Lett.* **13**, 47-49 (1992).
126. T. Hori, H. Iwasaki, Y. Yoshioka, and M. Sato, *Appl. Phys. Lett.* **52**, 736-738 (1988).
127. A. Hartstein and D. R. Young, *Appl. Phys. Lett.* **38**, 631-633 (1981).
128. H. Fukuda, T. Arakawa, and S. Ohno, *IEEE Trans. Elec. Dev.* **39**, 127-133 (1992).
129. J. Nulman, J. P. Krusius, and L. Rathbun, in *Technical Digest of the IEEE International Electron Devices Meeting 1984*, IEEE Cat. No. 84CH2099-0, pp. 169-172, IEEE, New York (1984).
130. M. M. Moslehi and K. C. Saraswat, *IEEE Trans. Elec. Dev.* **32**, 106-123 (1985).
131. F. L. Terry Jr., R. J. Aucion, M. L. Naiman, P. W. Wyatt, and S. D. Senturia, *IEEE Elec. Dev. Lett.* **4**, 191-193 (1983).

132. S. K. Lai, J. Lee, and V. K. Dham, in *Technical Digest of the IEEE International Electron Devices Meeting 1983*, IEEE Cat. No. 83CH1973-7, p. 190-193, IEEE, New York (1983).
133. C. C. Chang, A. Kamgar, and D. Kahng, *IEEE Elec. Dev. Lett.* **6**, 476-478 (1985).
134. M. Severi, L. Dori, and M. Impronta, *IEEE Elec. Dev. Lett.* **6**, 3-5 (1985).
135. P. Pan and C. Paquette, *Appl. Phys. Lett.* **47**, 473-475 (1985).
136. M. M. Moslehi, S. C. Shatas, and K. C. Saraswat, in *Proceedings of the Fifth International Symposium on Silicon Materials Science and Technology: Semiconductor Silicon 1986* (H. R. Huff, T. Abe, and B. Kolbesen, eds.), pp. 379-397, Electrochemical Society, Pennington, NJ (1986).
137. C.-T. Chen, F.-C. Tseng, C.-Y. Chang, and M.-K. Lee, *J. Electrochem. Soc.* **131**, 875-877 (1984).
138. G. Ruggles and J. R. Monkowski, *J. Electrochem. Soc.* **133**, 787-793 (1986).
139. T. Hori, Y. Naito, H. Iwasaki, and H. Esaki, *IEEE Elec. Dev. Lett.* **7**, 669-671 (1986).
140. M. M. Moslehi, K. C. Saraswat, and S. C. Shatas, *Proc. SPIE* **623**, 92-114 (1986).
141. H. Z. Massoud, *J. Appl. Phys.* **63**, 2000-2005 (1988).
142. T. Hori, H. Iwasaki, and K. Tsuji, *IEEE Trans. Elec. Dev.* **36**, 340-350 (1989).
143. M. M. Moslehi, S. C. Shatas, K. C. Saraswat, and J. D. Meindl, *IEEE Trans. Elec. Dev.* **34**, 1407-1410 (1987).
144. Z. H. Liu, P. Nee, P. K. Ko, C. Hu, C. G. Sodini, B. J. Gross, T.-P. Ma, and Y. C. Cheng, *IEEE Elec. Dev. Lett.* **13**, 41-43 (1992).
145. Z. H. Liu, P. Nee, P. K. Ko, C. Hu, C. G. Sodini, B. J. Gross, T. P. Ma, and Y. C. Cheng, in *Extended Abstracts of the 1991 International Conference on Solid State Devices and Materials*, pp. 26-28, Bus. Center Acad. Soc. Japan, Tokyo, Japan. (1991).
146. T. Hori and H. Iwasaki, *IEEE Elec. Dev. Lett.* **10**, 64-66 (1989).
147. T. Hori, *IEEE Trans. Elec. Dev.* **37**, 2058-2069 (1990).
148. B. S. Doyle and G. J. Dunn, *IEEE Elec. Dev. Lett.* **13**, 38-40 (1992).
149. T. Hori, S. Akamatsu, and Y. Odake, *IEEE Trans. Elec. Dev.* **39**, 118-126 (1992).
150. P. Burggraaf, *Semicond. Int.* **14(11)**, 66-70 (1991).
151. L. Peters, *Semicond. Int.* **14(10)**, 72-74 (1991).
152. D. Mordo, Y. Wasserman, and A. Gat, *Semicond. Int.* **14(11)**, 86-90 (1991).
153. M. M. Moslehi, R. A. Chapman, M. Wong, A. Paranjpe, H. N. Najm, J. Kuehne, R. L. Yeakley, and J. C. Davies, *IEEE Trans. Elec. Dev.* **39**, 4-32 (1992).

4 Thin-Film Deposition

Mehmet C. Öztürk
North Carolina State University
Department of Electrical and Computer Engineering
Raleigh, North Carolina

I. Equipment	81
II. Thin-Film Deposition Processes	83
A. Silicon Dioxide	84
B. Silicon Nitride	90
C. Polycrystalline Silicon	93
D. Polycrystalline Si–Ge	97
E. Epitaxy	103
F. Conductors	103
III. *In Situ* Processing—Applications	106
IV. Equipment Issues	111
A. Temperature Uniformity	111
B. Temperature Measurement	114
C. Deposition of Films on Windows and Chamber Walls	116
V. Summary	118
References	118

Thin-film deposition is one of the key processes used in microelectronics manufacturing. Fabrication of modern integrated circuits requires the ability to deposit semiconductors, metals, and insulators. Thin films are deposited using a large variety of methods. In silicon processing, physical vapor deposition (PVD) and chemical vapor deposition (CVD) are the two most commonly used techniques. The use of more elaborate techniques such as molecular beam epitaxy (MBE) has so far been largely restricted to research laboratories. Commonly used metals such as aluminum and titanium are deposited by PVD techniques (e.g., evaporation or sputtering) whereas CVD is generally preferred for silicon, silicon dioxide, and silicon nitride. However, new CVD processes such as tungsten and selective epitaxy are also finding new and exciting applications in microelectronics manufacturing.

This chapter concerns the art of CVD in lamp-heated rapid thermal processing (RTP) systems. In the following sections we will review the results from recent publications on rapid thermal CVD processes and equipment design issues. In doing so we will assume that the reader is familiar with the basic principles of CVD. Those readers who do not have sufficient background on the technique should refer to one of the introductory books on semiconductor processing [1-3].

Chemical vapor deposition can be performed in a wide pressure range from atmosphere to ultra high vacuum. A CVD process is usually identified with its preferred pressure regime: atmospheric pressure chemical vapor deposition, low-pressure chemical vapor deposition (LPCVD), and recently, ultrahigh vacuum chemical vapor deposition (UHV-CVD) are three different forms of thermal CVD that rely on pyrolysis of source gases at elevated temperatures. These different pressure regimes come with different sets of equipment specifications. At present, LPCVD is the preferred technique in silicon processing for many thin-film deposition processes because of the high deposition rates and excellent uniformity it can provide. It is very likely that UHV-CVD will be limited to low-temperature growth of high-quality epitaxial films because of the cost involved in its machinery [4, 5]. Recent results on epitaxial growth of Si and Si_xGe_{1-x} at atmospheric pressure in a clean process environment are also very encouraging [6] (see also Chapter 2).

Application of RTP to chemical vapor deposition was accomplished after implant annealing, oxidation, and silicide formation. The technique was first referred to as limited reaction processing (LRP) by the originators at Stanford University. In their early papers, the Stanford group demonstrated polysilicon LPCVD, silicon epitaxy and Si/Ge epitaxy and used these thin films in *in situ* processing of several device structures [7-13]. Over the years, the name rapid thermal chemical vapor deposition (RTCVD— similar to RTO for rapid thermal oxidation and RTA for rapid thermal annealing) became more commonly used by workers in the field.

To date, the majority of the work on RTCVD has been carried out in reactors designed to operate at reduced pressures. In principle, low-pressure RTCVD can be viewed as a single-wafer variant of conventional LPCVD. The main difference between RTCVD and LPCVD is that in an RTCVD system, temperature is used in place of gas flows to initiate and terminate chemical reactions on wafer surfaces. The wafer, which is inserted into the chamber at room temperature, can be heated to the process temperature in a matter of seconds, and the deposition time can be computer controlled to a very high degree of precision. As a result, the technique lends itself to deposition of ultrathin films once only attainable by techniques such as MBE.

At present, RTCVD is being studied as one of the potential techniques for thin film deposition in future single-wafer cluster tools. For this application, the technique is currently holding a great deal of promise because short-time processing—forte of RTP—is essential to maximize the throughput in single-wafer manufacturing tools.

I. Equipment

Rapid thermal processing systems designed for thin-film deposition do not differ greatly from the other RTP equipment used for annealing or oxidation. The heart of the reactor is the heat source that is either a single, high-energy arc lamp or a bank of tungsten halogen lamps. Arc lamps emit radiation in a wavelength band of ~ 0.2–$1.5\,\mu$m, whereas tungsten halogen lamps continue to emit radiation from $\sim 0.3\,\mu$m up to $\sim 4.0\,\mu$m. The reaction chamber is constructed using either quartz or stainless steel. A quartz chamber is cooled by flowing air around it. Since quartz does not absorb light efficiently within the wavelength band of the lamps, it can be maintained at a low temperature using this modest cooling. A stainless steel chamber, on the other hand, must be water cooled. In this case, a sealed window is used to pass the light energy onto the wafer. Regardless of the material used for chamber construction, the wafer rests on a quartz sample holder consisting of a pedestal and 3 or 4 very sharp quartz pins. The cross-sectional view of a RAPRO™ RTP chamber constructed for thin film deposition is shown in Fig. 1. The chamber design fits the description given above for the stainless-steel-walled reactor very closely. The chamber has all the necessary seals to enable low-pressure depositions. To maximize the deposition uniformly on the wafer, the chamber has additional, non-standard features such as wafer rotation and individual lamp power control. These features naturally change from vendor to vendor.

Even though the majority of the RTP equipment manufacturers have so far preferred to build reactors for low-pressure CVD, the basic principles of RTCVD can be equally applied to reactors that deposition thin films at atmospheric pressure. The difference between conventional CVD and RTCVD is the sequence of the necessary tasks. These two different sequences are shown in Fig. 1 for a generic low-pressure thin-film deposition process. The basic difference is the mechanism that determines the duration of the deposition cycle. In conventional LPCVD, this is accomplished by switching the gas flows. In RTCVD, temperature is the only switch used to start and stop chemical reactions on the wafer surface. As in other RTP processes, the wafer temperature can reach the process temperature in a few seconds while the reactant gases are flowing. One of

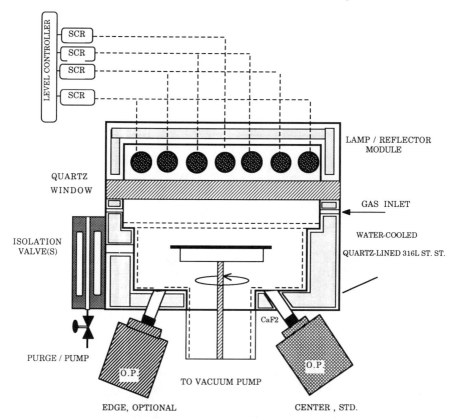

FIGURE 1. Cross-sectional view of the RAPRO™ RTCVD chamber.

the advantages of RTCVD is apparent in Fig. 2. That is, in furnace processing, long ramp-up and ramp-down times can increase the total thermal budget by a large amount.

Another important difference between LPCVD and RTCVD lies in the temperature of the chamber walls when the reactant gases are flowing. In LPCVD, the chamber walls are hot and in equilibrium with the rest of the system. Hence, deposition on chamber walls is inevitable. On the other hand, RTCVD chambers are cold-walled (especially if water-cooled stainless chambers are used). Water cooling practically eliminates deposition on chamber walls. The only region that can potentially get hot in such a reactor is the quartz window. Deposition on this window can be minimized by using high temperature/short time processing.

Equipment issues specifically related to RTCVD are discussed in detail in Section 5 and in Chapter 9.

Thin-Film Deposition

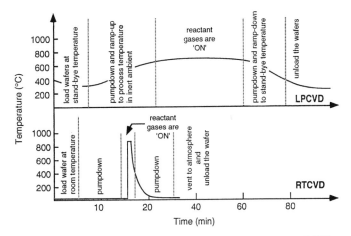

FIGURE 2. Process sequence for conventional LPCVD and RTCVD. In LPCVD, gas flows determine the duration of the deposition cycle. In RTCVD, temperature is used to start and stop the deposition process. To deposit the same amount of material in a short time, RTCVD is performed at a higher temperature.

II. Thin-Film Deposition Processes

As we have discussed in the introduction, interest in RTCVD stems from two reasons. First, the technique provides a new means of depositing thin films in a controllable manner. Heterojunction devices that rely on thin alternating layers of epitaxial films can greatly benefit from this property. This subject is treated in detail in Chapter 2. Second, RTCVD is presently being considered as a potential technique for use in single-wafer cluster tools. This requires RTCVD of materials such as silicon dioxide, silicon nitride, polysilicon, and metals commonly used in silicon processing. An appreciable number of these processes have already been demonstrated in RTCVD reactors. In this section, we will review the information currently available on these processes.

Since RTCVD is a single-wafer technique, throughput is an important concern. In general, processes that do not require a deposition time of more than 1–2 min are preferred. However, conventional LPCVD processes can deliver deposition rates around 100 Å/min, which is definitely too slow for single-wafer manufacturing. Fortunately, in some cases standard chemical recipes can be carried out at slightly higher temperatures, providing much higher deposition rates. This becomes possible for many known CVD processes, the temperature dependence of the deposition rate is exponential and follows the Arrhenius relation

$$R = Ae^{-E_a/kT} \tag{4.1}$$

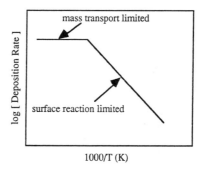

FIGURE 3. Temperature dependence of the deposition rate for a typical CVD process.

where R is the deposition rate, E_a is the activation energy in eV, T is the absolute temperature in K, A is the frequency factor, k is Boltzmann's constant (1.381×10^{-23} J/K) and q is the electronic charge (1.602×10^{-19} C). Equation 4.1 is valid only if the deposition rate is determined by the rate of surface reactions. Such a process is said to occur in a *surface reaction limited regime*. However, above a certain threshold temperature the reaction rate becomes faster than the rate at which unreacted molecules are delivered to the surface. When this occurs, the deposition rate no longer increases with temperature and the deposition is said to occur in a *mass transport limited regime*. In this regime, depositions are dependent on reactant concentration, reactor geometry, and gas flows. These two regimes are demonstrated in Fig. 3, which shows the temperature dependence of the deposition rate for a typical CVD process. Therefore, in a true mass transport limited regime, the deposition rate is completely independent of temperature. Unfortunately, some CVD processes reach this point before they can deliver a deposition rate acceptable for single-wafer processing. In such cases, new chemical recipes must be developed that can meet the throughput requirements.

Other incompatibility problems can exist such as excessive particulate generation by some reactive gases in cold-walled chambers and gas pre-heating outside the chamber. In such cases, new processes are proposed and studied to overcome the problems posed by the existing processes. In this section, we shall review results from recent publications on RTCVD of thin films currently used or studied for use in microelectronics manufacturing.

A. SILICON DIOXIDE

The success of the silicon-based microelectronics technologies relies on the existence of high-quality insulators and insulator/silicon interfaces. Silicon

Thin-Film Deposition

dioxide grown on silicon by thermal oxidation is an excellent insulator that can meet the stringent demands of advanced device structures. The metal-oxide semiconductor (MOS) transistor technology practically relies on the high-quality oxides and oxide/silicon interfaces that can be achieved by thermal oxidation of silicon. Thermally grown oxides are also successfully used for device isolation in silicon-based processing technologies. Silicon dioxide films deposited by conventional LPCVD techniques are used wherever there is a need for a low-temperature process or when silicon consumption that occurs during oxide growth is not tolerable.

In the following subsections, rapid thermal SiO_2 deposition processes are evaluated for thin gate dielctric and thick insulator applications.

1. Thin Deposited Oxide as a Gate Dielectric

One of the most promising applications of single-wafer cluster tools is the *in situ* formation of MOS gate structures. We can envision a multichamber cluster tool with chambers constructed to clean and prepare the wafer surface, to form the thin gate oxide, and finally to deposit the polysilicon gate electrode.

The presently available and proven gate oxide process is the thermal oxidation of silicon. This process, which has also been successfully demonstrated in RTP systems (rapid thermal oxidation), is treated in detail in Chapter 3. In cluster tools, low-pressure processes will most likely be preferred over atmospheric-pressure processes. Unfortunately, RTO is a high-pressure process. Thus, it would be very inconvenient to combine RTO with other low-pressure processes in the cluster tool.

Another disadvantage of RTO is that it is a very high-temperature process (>950 °C), which requires process times on the order of 0.5–2 min. Therefore significant dopant diffusion can occur in the substrate during RTO. This may present a problem if the channel region has established doping profiles to enhance the performance of the device. Nonuniform profiles are necessary to control punchthrough and bring the threshold voltage of a MOS transistor to its desired value. At present, the commonly used process for channel doping is ion implantation through the gate oxide. However, if the gate stack is formed *in situ*, deposition of the polysilicon gate electrode will be performed immediately after the formation of the gate oxide unless ion implantation can be carried out in the cluster tool in a separate chamber. Hence, alteration of the doping profile under the gate must be accomplished prior to the formation of the gate stack. Therefore, in this application a low-temperature, non–Si-consuming dielectric process would be highly preferable to RTO.

Even though a deposited low-temperature oxide process has been desired and researched for long, it has not yet been possible to match the quality of the thermal oxides. However, recent results on deposited oxides are very promising [14]. It should be emphasized that today, like any other process, oxide deposition will benefit from vastly improved equipment, as well as ultraclean gases and chemicals. Hence, major advances in the field can be anticipated.

A potential gate quality, low-pressure oxide deposition process makes use of SiH_2Cl_2 and N_2O as the two reactive gases. The deposition takes place at temperatures around 90 °C providing deposition rates less than 100 Å/min [15]. A potential problem of this process is chlorine contamination that can degrade the electrical properties. The process has been used by Ting *et al.* at the University of Texas to fabricate oxide/nitride/oxide stacked gate insulator structures by rapid thermal processing [16, 17]. The process is a relatively high-temperature process, although significant savings are obtained in thermal budget compared to RTO. Unfortunately, there is very little information in the literature on the electrical quality of these oxides.

The lower temperature oxide deposition process commonly used in LPCVD furnaces makes use of the reactive gases SiH_4 and O_2. The reaction proceeds according to [2]

$$SiH_4 + O_2 \rightarrow SiO_2 + 2H_2 \qquad (4.2)$$

resulting in SiO_2 films in the 300–450 °C range. Using this process to deposit the gate oxide, Ahn *et al.* successfully fabricated MOS transistors and compared these devices with standard MOS transistors with thermally grown oxides [14]. However, this is probably not the right process to use in RTCVD reactors because of excessive particulate generation and cagedboat requirements.

A new process that is currently under study at North Carolina State University makes use of SiH_4 and N_2O [18]. This process is advantageous over the SiH_2Cl_2 process because SiH_4 dissociates at lower temperatures than SiH_2Cl_2, providing higher deposition rates at lower temperatures. The same reaction has been previously demonstrated in conventional furnaces [3]. It has been proposed that the reaction between SiH_4 and N_2O proceeds according to

$$SiH_4 + 2N_2O \rightarrow SiO_2 + 2N_2 + 2H_2O \qquad (4.3)$$

Xu *et al.* used a low-pressure RTCVD reactor with a base pressure of $\sim 10^{-8}$ torr to deposit 80–200-Å thick oxides. The silicon-to-oxygen atomic ratio as a function of the flow ratio of the reactive gases SiH_4 and N_2O is shown in Fig. 4. Argon was premixed with SiH_4 in the bottle (10% SiH_4 and 90% Ar) to reduce the safety risks associated with the pyrophic

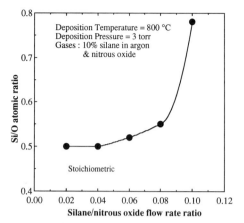

FIGURE 4. Si:O atomic ratio of oxides deposited at 800 °C using SiH_4 and N_2O as a function of the flow ratio of the two gases. From Xu et al. [18].

nature of SiH_4. The films were deposited at 800 °C. The data was obtained from 200-Å thick films using Auger electron spectroscopy and Rutherford backscattering spectroscopy. As shown, for $SiH_4:N_2O$ ratios less than approximately 4%, stoichiometric films are obtained. The temperature dependence of the deposition rate is shown in Fig. 5 for two $SiH_4:N_2O$ ratios. A higher $SiH_4:N_2O$ ratio results in a higher deposition rate at all temperatures studied. The deposition rates clearly follow the Arrhenius behavior. The activation energies of the processes with $SiH_4:N_2O$ ratios of 2 and 4% are 40.5 and 35.5 kcal/mol (1.76 and 1.54 eV) respectively. The films deposited at temperatures between 800 and 900 °C were also found to

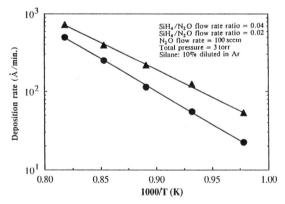

FIGURE 5. Temperature dependence of the oxide deposition rate using SiH_4 and N_2O as the reactive gases. From Xu et al. [18].

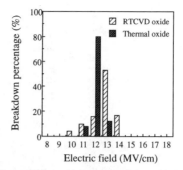

FIGURE 6. Catastrophic breakdown histograms of thermal dry oxide and RTCVD oxide. The oxide was deposited using SiH_4 and N_2O at 800 °C and $SiH_4 : N_2O$ flow ratio of 2%. From Xu et al. [18].

be stoichiometric SiO_2 when the gas flow ratio was limited to 2%. As shown in Fig. 5, the deposition rate at 800 °C is ~55 Å/min, which is nearly two orders of magnitude higher than the thermal dry oxidation rate at the same temperature and at atmospheric pressure. Xu et al. also studied the electrical properties of these oxides as a function of the deposition parameters. It was found that nonstoichiometric, silicon-rich films deposited on Si yielded high densities of interface traps (D_{it}). However, the D_{it} of stoichiometric silicon dioxide films deposited at temperatures between 765 and 950 °C were all within the smallest detectable levels. A study of the catastropic electrical breakdown of the deposited oxides yielded an average breakdown field of 13 MV/cm, which is comparable to that of the best thermally grown oxides. This is demonstrated in Fig. 6.

In summary, the new process described above appears to be a potential candidate for RTCVD of gate quality SiO_2 films in cluster tools. The advantages of such a process over rapid thermal oxidation are significant savings in process thermal budget and a process compatible with other chambers in a cluster tool. Obviously, much more work needs to be done on a new oxide process before it can replace the mature and reliable thermal oxidation technology.

2. Thick Deposited Oxide for Isolation

Deposited oxides are also needed to isolate different conducting layers in integrated circuits. These applications require oxides that are often several thousands of angstroms in thickness, for which throughput becomes a major concern in single-wafer systems. Practical RTCVD processes developed for these applications must therefore aim for deposition rates in excess of ~1000 Å/min.

An important property of an oxide deposition process is its step coverage. Step coverage is a measure of the ability of a thin film deposition process to coat horizontal and vertical regions of a wafer in an equal manner. Some applications require excellent step coverage. A good example is the sidewall spacer oxide surrounding the polysilicon gate in a self-aligned MOS process. On the other hand, when the oxide is deposited for planarization, good step coverage may not be desirable. The step coverage of a CVD process can be characterized by a single parameter called the "sticking coefficient" [19]. A small sticking coefficient implies a greater tendency for the adsorbed atoms to diffuse on the surface or become reemitted. Both diffusion and reemission are known to improve the step coverage. However, according to Cheng et al., reemission is the dominant mechanism. It has been shown that a CVD process with a smaller sticking coefficient provides better step coverage and conformality. Table I gives sticking coefficients of four silicon sources used in oxide CVD processes [19]. On the basis of these data, TEOS (tetraethoxysilane or tetraethylorthosilicate) and TMCTS (tetramethylcyclotrasiloxane) appear as the better choices for achieving good step coverage.

RTCVD of SiO_2 by pyrolysis of TEOS has been demonstrated [20, 21]. The deposition occurs according to the following proposed chemical reaction [22]:

$$Si(OC_2H_5)_4 \rightarrow SiO_2 + 2H_2O + 4C_2H_4. \quad (4.4)$$

The temperature dependence of the deposition rate of silicon dioxide obtained from pyrolysis of TEOS is shown in Fig. 7 at three different pressures in a lamp-heated RTCVD reactor [20]. Below 800 °C, SiO_2 deposition appears to be controlled by surface reactions with an activation energy of 76 kcal/mol (~ 3.3 eV). This activation energy is higher than those reported in the literature for the TEOS process [22–24]. Above 800 °C a sharp decrease in the activation energy can be observed, which is suggestive of a transition into the mass transport limited regime. As shown, the deposition rate is approaching 1000 Å/min around the transition

Table I Sticking coefficients of oxide LPCVD processes [19].

Oxide Deposition Process	Sticking Coefficient
Silane	0.26
Diethylsilane	0.1
TEOS	0.045
TMCTS	0.04

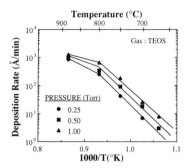

FIGURE 7. Temperature dependence of the silicon dioxide deposition rate at different pressures using tetrathylorthosilicate (TEOS) in an RTCVD reactor. From Miller et al. [20].

temperature. Therefore, the process can indeed provide high throughput with a relatively low thermal budget process (~ 800 °C/min) for applications that can tolerate such thermal budgets and require good step coverage. A suitable application for the process can be the formation of the sidewall spacer.

The etch rates of the deposited oxides shown in Fig. 7 were found to be greater than that of the thermal oxide in a 1.0% HF solution. This is suggestive of a less dense oxide. However, the deposited oxide can be densified by annealing either in a furnace or RTA system at elevated temperatures as shown in Fig. 8. Both anneals can provide etch rates comparable to the etch rate of the thermal oxide when the annealing temperature is sufficiently high. After a postdeposition anneal, the oxide can yield acceptable electrical properties for passivation purposes. However, these properties are not good enough to make the process a candidate for gate applications.

The TEOS process described above has an obvious disadvantage: the deposition temperature can not be made low enough for use in multi level metallization processes that make use of aluminum. For this purpose, a process that can provide an acceptable throughput below approximately 450 °C is needed. Such an RTCVD process has not been found yet. For this application, instead of pure RTCVD, it may be necessary to consider hybrid processes that combine RTCVD with remote plasma or UV-assisted processing.

B. SILICON NITRIDE

Silicon nitride (Si_3N_4) has many important applications in silicon processing. Conventional applications include device isolation via selective oxidation

Thin-Film Deposition

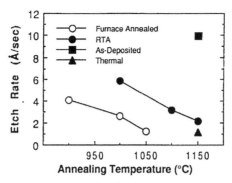

FIGURE 8. TEOS–SiO$_2$ etch rate in 1.0% HF after furnace and rapid thermal postdeposition anneals. Also shown for comparison are the etch rates of thermal oxide and as-deposited TEOS oxide. From Miller et al. [20].

of silicon (LOCOS) and gate dielectrics in metal-nitride-oxide-silicon transistor memory structures. Silicon nitride is also used for passivation purposes because of its good barrier properties to water and sodium. In a new application, deposited Si$_3$N$_4$ is being considered as an interlevel dielectric in the fabrication of oxide/nitride/oxide stacked gate structures for future MOS devices [16, 17].

Silicon nitride has been deposited in conventional systems by reacting ammonia (NH$_3$) with either silane (SiH$_4$) at atmospheric pressure or dichlorosilane (SiH$_2$Cl$_2$) at reduced pressure. Because the reduced pressure technique outperforms the atmospheric counterpart in deposition uniformity and wafer throughput, it has become the standard technique used in industry.

The reaction between NH$_3$ and SiH$_2$Cl$_2$ can be described by the following equation:

$$3SiH_2Cl_2 + 10NH_3 \rightarrow Si_3N_4 + 6NH_4Cl + 6H_2 \quad (4.5)$$

The reaction byproduct, ammonium chloride (NH$_4$Cl), has a vapor pressure of approximately 1 torr at 160 °C and has long been a problem in the standard LPCVD furnaces. The problem is ammonium chloride deposits in the form of a fine powder in the cooler areas of the system that eventually clogs the exhaust flanges and pipes of standard hot-wall LPCVD furnaces. In a cold-wall RTCVD chamber, this becomes a major particulate problem because NH$_4$Cl deposits essentially everywhere including the quartz windows and the chamber walls. Therefore, the standard silicon nitride process can not and should not be used in cold-wall RTCVD reactors.

This particulate problem created the need for the development of a new reduced-pressure process. RTCVD of Si$_3$N$_4$ at reduced pressure using SiH$_4$ and NH$_3$ was studied and successful results were obtained [25].

The same process has also been used for the *in situ* formation of an oxide/nitride/polysilicon stack by RTCVD [16, 17]. The process can be described with the following equation:

$$3SiH_4 + 4NH_3 \rightarrow Si_3N_4 + 12H_2 \qquad (4.6)$$

The use of silane in conventional tube furnaces at low pressures resulted in radial nonuniformities due to mass transfer effects [26]. The nonuniformity was in the form of a decreased deposition in the center of the wafer. This type of nonuniformity is often called the bullseye effect and is dependent on the wafer-to-wafer spacing. For this reason, the silane process was not preferred at reduced pressures. However, RTCVD systems are single-wafer reactors. Therefore, depletion effects observed in tube-type furnaces may not necessarily apply. Indeed, when silane was used for Si_3N_4 deposition at reduced pressures the uniformity problems which normally occur in tube furnaces were not experienced [25].

Using the SiH_4 reaction given in Eq. 4.6, Johnson *et al.* deposited Si_3N_4 using SiH_4 premixed with Ar (10% SiH_4) [25]. The deposition rate obtained at 785 °C using $SiH_4 : N_2O$ flow ratios ranging from 10 to 200 is shown in Fig. 9. The deposition rate was found to increase with decreasing $NH_3 : SiH_4$ ratio, which was also observed in conventional LPCVD furnaces [26]. Auger electron spectroscopy of these samples indicated that stoichiometric nitride resulted for depositions using $NH_3 : SiH_4$ greater than 120:1. The films were found to be Si rich below a ratio of 80:1. The temperature dependence of the nitride deposition rate is shown in Fig. 10 for a total pressure of 8 torr and $NH_3 : SiH_4$ flow ratio of 120:1 [25]. The activation energy of the process is 47.2 kcal/mol (~2.1 eV). It is clear from Fig. 10 that the process is suitable for the deposition of thin films. This may be

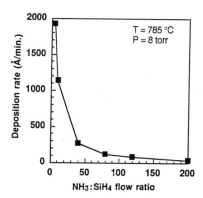

FIGURE 9. Si_3N_4 deposition rate versus reactant gas ratio. The films were deposited using NH_3 and 10% SiH_4 in Ar at 785 °C and a total pressure of 8 torr. From Johnson *et al.* [25].

Thin-Film Deposition

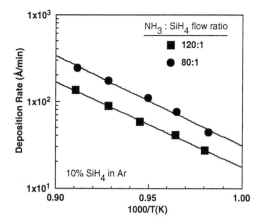

FIGURE 10. Temperature dependence of the silicon nitride deposition rate using NH_3 and 10% SiH_4 in Ar. The total pressure is 8 torr. From Johnson et al. [25].

acceptable for new applications of Si_3N_4 where only a few nanometers of the material must be deposited [16, 17]. However, new processes are needed if RTCVD nitride is to be used for more conventional applications. Similar to the low-temperature LPCVD of thick oxides, CVD of nitride may require hybrid processes in order to meet the demands of single-wafer manufacturing.

C. POLYCRYSTALLINE SILICON

Polysilicon has many important applications in the fabrication of silicon devices. With the advent of self-aligned MOS transistor technology, polysilicon has become and continues to be the preferred gate electrode of state-of-the-art MOS transistors. It also is an ideal interconnect when used with a metal such as titanium to form a low-resistivity silicide. In the bipolar area, the high performance of *npn* transistors is largely due to the success of their polysilicon emitters. Another application of the material is that undoped or lightly doped polysilicon can serve as a high value resistor. Finally, polysilicon is now being considered as the active region material for thin film transistors. In summary, it is clear that polysilicon will preserve its popularity in silicon-based microelectronics technologies. Possibly because of its wide range of applications, polysilicon was one of the first films studied in RTCVD reactors [11, 21, 27–34].

Polysilicon is deposited in conventional LPCVD reactors, usually by pyrolysis of pure silane according to the following chemical reaction [35]:

$$SiH_4(gas) \rightarrow SiH_4(adsorbed) \rightarrow Si + 2H_2. \qquad (4.7)$$

Deposition is carried out at reduced pressures (0.2 to 1.0 torr) and at temperatures around 600 °C. Typical deposition rates are within 100–200 Å/min. These rates are certainly too slow for use in single-wafer reactors. Therefore, either a new process that can deliver higher rates must be developed or the existing process must be improved with some modifications.

An important property of polysilicon deposited on SiO_2 is its surface morphology. Very smooth surfaces are needed to meet the demands of decreasing dimensions of advanced MOS and bipolar devices. The dependence of polysilicon surface roughness on deposition conditions has been well documented. It has been shown that deposition temperature plays a significant role in determining the surface roughness of the films [36–38]. Rougher films are obtained at higher temperatures. Indeed, to optimize the surface morphology it has been suggested that polysilicon should be deposited at lower temperatures in the amorphous state and subsequently annealed to achieve the amorphous-to-polycrystalline transition [37]. It has been shown that the smooth surface obtained by an amorphous deposition is retained during this transformation.

The transition from amorphous-to-polycrystalline deposition occurs around 575 °C [38]. However, at this temperature the deposition rate can be as low as 20 Å/min. LPCVD of amorphous silicon is an extremely slow process that can only be considered in conventional batch processing systems. To meet the throughput requirements of single-wafer manufacturing, much higher deposition rates are obviously needed. For instance, when polysilicon is used as the gate electrode, 2000–3000 Å thick polysilicon is usually needed. Therefore, deposition rates in excess of 2000 Å/min should be attainable in RTCVD reactors.

The primary factors that influence the polysilicon deposition rate are deposition temperature and the partial pressure of the reactants. However, the carrier gas can also play a significant role on the deposition rate. The effect of the carrier gas on the deposition rate of polysilicon can be explained as follows: The pressure dependence of the polysilicon deposition rate from SiH_4 can be expressed as [39]

$$G = \frac{ap_{SiH_4}}{1 + bp_{SiH_4}} = \frac{C_1 C_2 \alpha p_{SiH_4}}{1 + C_2 \alpha p_{SiH_4}} \quad (4.8)$$

where p_{SiH_4} is the partial pressure of SiH_4 and a, b, C_1, C_2 and α are constants. The constant α is given by

$$\alpha = \frac{K_1}{K_1 + p_{H_2}}. \quad (4.9)$$

where K_1 is another constant and p_{H_2} is the partial pressure of hydrogen. Therefore, if hydrogen is used as a carrier gas the hydrogen partial pressure in the reactor can be much more than the hydrogen partial pressure generated as a result of the silane decomposition. As a result, the use of hydrogen can cause an appreciable reduction in the decomposition rate. To avoid this unnecessary loss of the growth rate, Ar or N_2 can be used instead of H_2. These gases also reduce the safety concerns associated with the pyrophoric nature of SiH_4 and explosive nature of H_2.

The easiest way to increase the deposition rate is to increase the deposition temperature. Several studies reported in the literature considered RTCVD of polysilicon at temperatures substantially higher (>700 °C) than the temperatures used in conventional LPCVD furnaces [21, 29, 34]. In this temperature range, acceptable throughput levels for single-wafer manufacturing can be achieved.

The temperature dependence of polysilicon deposited using 10% SiH_4 in Ar is shown in Fig. 11. Dilution of SiH_4 with Ar to levels below ~10% was preferred in order to reduce the safety concerns associated with the pyrophoric nature of SiH_4. According to Fig. 11, deposition rates exceeding 2000 Å/min can be achieved at temperatures above 700 °C. The activation energy is 39.2 kcal/mol (~1.7 eV) which is close to the activation energies reported for LPCVD of polysilicon. Above 780 °C, mass transport effects result in a reduction of the activation energy. However, because a true mass transport limited process should be completely independent of temperature, there could be more than one mechanism playing a role above this temperature.

Even though acceptable throughput levels can be achieved at higher temperatures, there is a concern about the surface morphology of the films.

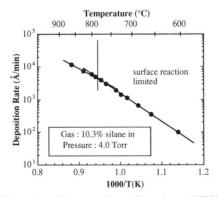

FIGURE 11. Polysilicon deposition rate from silane in an RTCVD reactor. Silane was diluted with argon to a concentration of 10.3%. From Öztürk et al. [21].

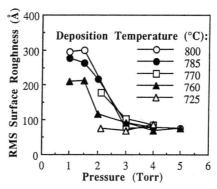

FIGURE 12. Root-mean-square surface roughness of ~2000 Å thick polysilicon deposited on SiO$_2$ using 10% SiH$_4$ in Ar at various temperatures and pressures. The roughness data for the samples deposited at 770 and 725 °C was obtained by cross-sectional TEM analysis. The rest of the data was obtained using ultraviolet surface reflectance measurements. From Ren *et al.* [40].

Currently there is limited information on the surface morphology of RTCVD polysilicon. However, the reports published so far are suggestive of a different behavior. It appears that surprisingly smooth films can be obtained at relatively high temperatures [21, 29, 40]. The root-mean-square (rms) surface roughness of ~2000-Å thick polysilicon films deposited on SiO$_2$ by RTCVD is shown in Fig. 12. The data were obtained using UV surface reflectance and cross-sectional transmission electron microscopy (TEM) measurements. As shown, very smooth films can be obtained at temperatures well above 700 °C. Such low rms roughness values can be obtained in conventional LPCVD furnaces only if an amorphous film is deposited at very low temperatures. The smooth surface morphology can also be observed in Fig. 13, which shows a cross-sectional transmission electron micrograph of a polysilicon film deposited by RTCVD at 750 °C on SiO$_2$ at 4 torr. It can also be seen that the film has a columnar grain structure. Ren *et al.* measured O$_2$ levels in these films and found oxygen levels on the order of a few percent [40]. They hypothesized that this excess oxygen could be responsible for the smooth morphology of polysilicon deposited in these cold-walled RTCVD reactors. Similarly high oxygen concentrations were observed by others as well [41]. Ren *et al.* compared electrical properties of MOS structures with polysilicon gate electrodes formed by RTCVD and conventional LPCVD. The results obtained from 80–200-Å thick gate oxides indicated comparable or better properties with the RTCVD polysilicon. In spite of the relatively high oxygen levels in the films, both boron and arsenic implanted films yielded resistivities comparable to LPCVD polysilicon [34].

Thin-Film Deposition

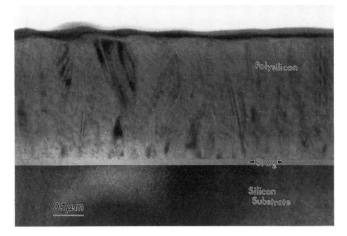

FIGURE 13. Cross-sectional transmission electron micrograph of polycrystalline silicon deposited on SiO_2 at 750 °C and 4 torr using 10% SiH_4 in Ar.

Polysilicon can also be deposited selectively in an RTCVD reactor [42]. Kwong *et al.* showed that by using 2% SiH_2Cl_2 diluted in hydrogen, polysilicon could be deposited selectively at elevated temperatures (~ 850 °C). At this temperature, grains can be as large as 2000 Å. In this work, selectively deposited and *in situ* arsenic doped polysilicon was used as a solid diffusion source to form shallow n^+p junctions in silicon. The disadvantage of the process is the high growth temperature (~ 850 °C) needed for selectivity. At this temperature, the deposited material has a rough surface morphology and dopant diffusion in silicon is not negligible.

In summary, results obtained so far on polysilicon are promising. The material can be deposited at rates that are sufficiently high for single-wafer processing. This is accomplished by depositing polysilicon using SiH_4/Ar at temperatures around 750 °C, which is well above the temperatures used in conventional systems. At these high temperatures, RTCVD can be successfully used to obtain smooth and low-resistivity films.

D. POLYCRYSTALLINE SI–GE

Polycrystalline Si–Ge is currently being investigated as an alternative material to polycrystalline silicon in several applications in microelectronics. These include gate electrode for advanced MOS transistors [43–45], substrate material for thin film transistors [46], and finally diffusion source to form ultrashallow junctions in silicon [44, 47, 48].

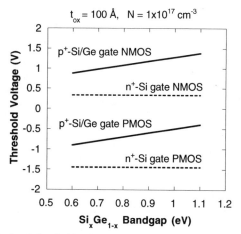

FIGURE 14. Calculated threshold voltage of a two-terminal MOS structure with a p^+ Si-Ge gate electrode as a function of the band gap of the alloy on p- and n-type silicon. Also shown for comparison are the threshold voltages obtained with n^+ polysilicon gates on the same substrates.

Polycrystalline Si-Ge offers some useful properties. When used as a p-type gate electrode, the work function of the material can be tailored to the desired level by carefully adjusting the amount of germanium in the alloy. This is due to the fact that the band gap of the material can be varied from 1.12 eV for pure silicon to 0.57 eV for pure Ge. Since Si and Si-Ge have almost identical electron affinities, any changes in the band gap translate into changes in the work function of the gate electrode, providing another degree of freedom in threshold voltage control of MOS transistors. The calculated threshold voltage for a two-terminal p^+ Si-Ge gated MOS structure is shown in Fig. 14 as a function of band gap for both p-channel and n-channel devices. Also shown for comparison are the threshold voltages obtained using the standard n^+ polysilicon gates on these devices.

It has also been shown that boron activation in pure Ge and Si-Ge alloys takes place at considerably lower temperatures. For instance, in pure Ge, implanted boron can be activated at temperatures as low as 400 °C [49]. Even though this advantage is somewhat lost when Si is added to the alloy, polycrystalline Si-Ge still offers lower activation temperatures for boron compared to polysilicon [43, 44, 49]. This is demonstrated in Figure 15, which shows the sheet resistance of implanted polycrystalline Si-Ge alloys with different compositions as a function of the RTA temperature. The samples were implanted with boron at 10 keV/1×10^{16} cm^{-3} and annealed for 10 sec.

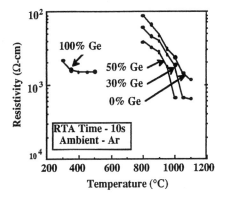

FIGURE 15. Sheet resistivity of boron-implanted polycrystalline Si-Ge after annealing in Ar at different temperatures for 10 sec. From Öztürk et al. [44].

Polycrystalline $Si_{1-x}Ge_x$ alloys can be deposited by mixing GeH_4 with SiH_4, Si_2H_6, or SiH_2Cl_2. The deposition process requires precise control over the deposition temperature and the flow ratio of the Si and Ge source gases used in the process. This is due to the fact that the film composition is a strong function of both of these parameters. This is demonstrated in Fig. 16, which shows the percentage Ge concentration in polycrystalline $Si_{1-x}Ge_x$ deposited by RTCVD on Si at temperatures ranging from 550 to

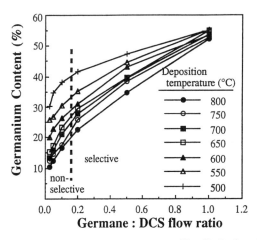

FIGURE 16. Percentage Ge concentration in polycrystalline Si-Ge deposited by RTCVD at different temperatures. As shown, the Ge concentration is a strong function of both the deposition temperature and the GeH_4 (germane) to SiH_2Cl_2 (dichlorosilane, DCS) gas flow ratio. From Zhong et al. [50].

800 °C using different flow ratios of germane (GeH_4) and dichlorosilane (SiH_2Cl_2) at a total pressure of 2.5 torr [50]. The GeH_4 used in this study was premixed with hydrogen to a dilution of 7.8%. The flow rate of GeH_4 was kept constant at 5 sccm and the flow rate of SiH_2Cl_2 was varied from 200 to 5 sccm, corresponding to $GeH_4 : SiH_2Cl_2$ flows ratios from 0.025 to 1. As shown, for gas flow ratios less than 0.2, the Ge content in the alloy is a strong function of the gas flow ratio and rapidly diminishes with increasing SiH_2Cl_2 in the gas stream. As the ratio is increased above 0.1, the germanium content gradually increases with increasing gas flow ratio. For the entire $GeH_4 : SiH_2Cl_2$ gas flow ratios shown in Fig. 16, the Ge content is increasing with decreasing deposition temperature. This is because GeH_4 can readily decompose at low temperatures whereas Si deposition from SiH_2Cl_2 is considerably slower below 600 °C. As the GeH_4 flow in the gas stream approaches that of SiH_2Cl_2, the germanium content of the alloy becomes less sensitive to the deposition temperature.

Below a $SiH_2Cl_2 : GeH_4$ flow ratio of 0.2, it is also indicated in Fig. 16 that deposition on silicon dioxide does not occur. The transition point at which the selective depositions start to occur is expected to be system dependent because of variations in factors such as base pressure and system and gas purity, which are likely to have a strong influence on the selectivity process. Nevertheless, the role of GeH_4 as a selectivity enhancing agent is evident from the data of Fig. 16. The selectivity of the process has been attributed to the formation of highly volatile GeO during deposition through the following reactions [50]:

$$GeH_4 \xrightarrow{heat} Ge + 2H_2$$
$$Ge + SiO_2 \longrightarrow GeO_2 + Si \quad (4.10)$$
$$GeO_2 + Ge \longrightarrow 2GeO$$

While higher Ge concentrations can improve selectivity, increased lattice mismatch can produce a rough surface morphology if the deposition conditions are not optimized. Three dimensional growth is expected to dominate when the interfacial energy between the alloy and the substrate is greater than the sum of the alloy/ambient and substrate/ambient surface energies. In an earlier work, Bean *et al.* have investigated the MBE growth of Si_xGe_{1-x} alloys on Si and have shown that smooth Si_xGe_{1-x} films on Si can be obtained by growing the films under a critical temperature determined by the germanium content in the alloy [51]. It is believed that at lower temperatures the surface mobility of the depositing species is reduced considerably, resulting in very small surface migration lengths. Islanding is thus avoided at lower temperatures. Results obtained with RTCVD are in general agreement with this earlier work [52]. The transition between

Thin-Film Deposition

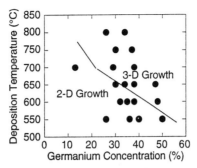

FIGURE 17. Deposition temperature as a function of Ge concentration and regions of two- and three-dimensional growth. From Sanganeria et al. [52]. Reprinted with permission of the publisher, The Electrochemical Society.

two- and three-dimensional growth mechanism as a function of Ge concentration is shown in Fig. 17. The data shown were published in an article by Sanganeria et al. in a different form [52]. As shown, just like MBE, the RTCVD technique requires lower temperatures to grow smooth films on Si with high Ge concentrations.

RTCVD has been also used to deposit *in situ* boron-doped polycrystalline Si–Ge on Si and SiO_2 [45, 53, 54]. It has been demonstrated that the alloys can be heavily doped at relatively low temperatures suitable for gate electrode and diffusion source applications. In these studies, *in situ* doping was achieved by adding B_2H_6 (diborane) to the gas stream. The resistivity of *in situ* boron-doped polycrystalline Si–Ge deposited at 650 °C on SiO_2 is shown in Fig. 18 as a function of the diborane flow. Germanium

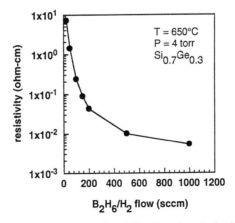

FIGURE 18. Resistivity of *in situ* boron-doped $Si_{0.7}Ge_{0.3}$ deposited using the gases SiH_2Cl_2, GeH_4, B_2H_6, and H_2 as a function of the B_2H_6 flow.

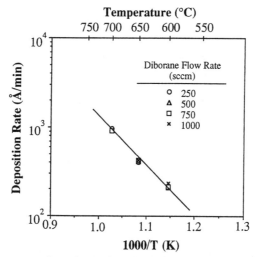

FIGURE 19. Temperature dependence of the deposition rate for *in situ* boron-doped polycrystalline $Si_{0.7}Ge_{0.3}$ deposited using different flows of the dopant gas, diborane (B_2H_6). From Sanganeria *et al.* [53].

concentration in the alloy is 30%. As shown, the sheet resistivity in the alloy can be changed by three orders of magnitude simply by changing the amount of diborane flow. Low-resistivity films can be obtained with resistivities comparable to degenerately doped polysilicon without a subsequent high-temperature annealing step.

The temperature dependence of the deposition rate for 30% Ge is shown in Fig. 19 for four different diborane flows. A typical Arrhenius-type temperature dependence has been obtained with an activation energy of ~ 25 kcal/mol (~ 1.1 eV) [53]. The depositions were carried out using 7.8% GeH_4 and 40 ppm B_2H_6, both diluted with H_2. The total pressure was kept fixed at 4.0 torr. As shown, addition of B_2H_6 to the gas stream has negligible or no detectable influence on the deposition rate. On the other hand, diborane is known to increase the deposition rate of Si from both SiH_4 [55] and Si_2H_6 [56]. An advantage of the process in single-wafer processing is evident from Fig. 19. That is, the alloy can be deposited at higher rates compared to silicon. The deposition rate of Si from SiH_2Cl_2 and H_2 is negligibly small at temperatures where the $GeH_4 : SiH_2Cl_2 : H_2$ system can deliver several hundreds of angstroms per minute. This is an important advantage of polycrystalline Si–Ge over Si.

These heavily doped alloys can still be deposited selectively. Using this unique advantage, Grider *et al.* used *in situ* boron-doped polycrystalline Si–Ge as a diffusion source and formed high-quality ultrashallow p^+n junctions in silicon [54].

Thin-Film Deposition

Polycrystalline Si-Ge is a relatively new material that has certain advantages compared to polysilicon. Its high deposition rate and lower dopant activation temperatures give Si-Ge an edge in low-temperature single-wafer manufacturing. The possibility of selective deposition with an enhanced selectivity compared to polysilicon will also be very useful.

E. Epitaxy

Epitaxy is an area of great interest for many researchers. Epitaxial growth by RTCVD provides a new alternative to techniques such as plasma enhanced CVD, ultra-high vacuum CVD, and molecular beam epitaxy, which are capable of growing ultrathin, high-quality epitaxial layers. By using high-temperature and short-time growth cycles, RTCVD can grow films in modest reactors at much higher deposition rates suitable to a single-wafer manufacturing environment. Precision temperature/time profile control allows growth of films on the order of tens of angstroms with abrupt dopant transition profiles. Epitaxial growth by RTCVD has been demonstrated for silicon [7, 10, 11, 57–70], $Si_{1-x}Ge_x$ [8, 9, 12, 57, 62, 65, 71–85], III–V compounds [57, 86] and silicon carbide [87, 88]. Readers interested in this application of RTP should refer to Chapter 2.

F. Conductors

At present, chemical vapor deposition is not a widely used technique for deposition of metals. Instead, physical vapor deposition techniques such as evaporation and sputtering are generally preferred. However, these conventional techniques can no longer satisfy the growing demands of modern integrated circuits. Specifically, selective deposition techniques are becoming very attractive with the continued scaling of device geometries to deep submicron dimensions. Consequently, there is a growing interest in new metal deposition techniques. Chemical vapor deposition (either conventional CVD or plasma assisted) has been investigated to deposit a number of conductors including tungsten, copper, titanium silicide, and titanium nitride. However, only tungsten (W) and titanium silicide ($TiSi_2$) have so far been attempted in RTCVD reactors.

1. Titanium Silicide ($TiSi_2$)

Titanium (di)silicide ($TiSi_2$) is presently used in microelectronics to form self-aligned contacts to gate electrodes and source/drain junctions of submicrometer MOS transistors. The current process in use today relies on the selective removal of unreacted Ti on SiO_2 following an intermediate, low-temperature anneal [89]. The selective Ti etch leaves a Ti-Si compound on silicon that can be further annealed to obtain stoichiometric $TiSi_2$. The technique, which is referred to as SALICIDE (self-aligned silicide), forms $TiSi_2$ on the source and drain junctions and the polysilicon gate. Even though the SALICIDE process is widely used in industry today, its application to future deep submicrometer technologies is highly questionable. The main disadvantage of the process is the inevitable consumption of the substrate or the gate silicon, which can impose serious limitations on the junction depth because of junction leakage [90] and the gate length due to silicide bowing [91].

An alternative to the SALICIDE process described above is selective deposition of $TiSi_2$. The LPCVD of $TiSi_2$ has been first accomplished in 1985 by Tedrow et al. of the Massachusetts Institute of Technology [92]. The reactor used in this study was essentially an RTCVD reactor with all the standard features (lamp heating, quartz chamber, quartz wafer holder, pyrometer, etc.). However, Tedrow et al. preferred to refer to their homemade system as an LPCVD reactor and did not utilize the rapid temperature switching capability of the system. Ironically, in the same year Gibbons et al. using a very similar system in the RTP mode, reported the technique as limited reaction processing (LRP) [7] and have attracted a lot of interest from many researchers, which led to the technical material covered in two chapters of this book.

To deposit $TiSi_2$ on Si, Tedrow et al. used titanium tetrachloride ($TiCl_4$) and silane (SiH_4) as the two source gases. The base pressure of the reactor used was less than 10^{-7} torr. The investigators were able to deposit stoichiometric $TiSi_2$ films with a minimum resistivity of 22 $\mu\Omega$-cm. In a later study, Ilderem and Reif reported their optimized deposition parameters [93]. Using SiH_4 and $TiCl_4$, the authors reported low resistivity (15-20 $\mu\Omega$-cm) $TiSi_2$ films deposited at 730 °C and 67 mtorr. At this temperature and pressure, the deposition rate (1000-4000 Å/min determined by the flow conditions) is sufficient for a single-wafer process.

Ilderem and Reif propose the following reaction at the start of the deposition process:

$$TiCl_4(g) + 2SiH_4(g) + Si(s) \rightarrow TiSi_2(s) + SiClH_3(g) + 3HCl(g) + H_2(g) \tag{4.11}$$

which is suggestive of silicon (Si(s)) consumption from the substrate. The silicon consumption is closely coupled to the $TiCl_4$ flow rate. This is because $TiCl_4$ decomposes into $TiCl_2$ and HCl through

$$TiCl_4 + H_2 \rightarrow TiCl_2 + 2HCl(g). \tag{4.12}$$

Both $TiCl_2$ and HCl react with Si through a series of reactions [93]. It is therefore proposed that the increased flux of $TiCl_4$ onto the surface leads to an increase in the partial pressures of $TiCl_2$ and HCl and hence higher silicon consumption rates. Once the silicon consumption rate diminishes, it is proposed that the reaction reduces to

$$TiCl_4(g) + 3SiH_4(g) \rightarrow TiSi_2(s) + SiClH_3(g) + 3HCl(g) + 3H_2(g) \tag{4.13}$$

Using the optimized conditions, Ilderem and Reif applied their deposited silicide films to device fabrication, which involved fabrication of *pn*-junctions and MOS transistors, both of which delivered promising results [94]. However, silicon consumption remained a problem.

LPCVD of $TiSi_2$ without Si consumption was first realized in 1989 by Regolini *et al.* from France in a true RTCVD reactor [70, 95, 96]. The group used a lamp-heated, cold-walled reactor in the fast-switching mode and deposited $TiSi_2$ selectivity using $TiCl_4$ and SiH_4. This work, in contrast to work done at MIT, features (1) the use of hydrogen as a dilutant and carrier gas (dilution of the reactive species down to ~1%) and (2) a higher total pressure of 2 torr during deposition. The selectivity is attributed to the production of HCl through the following reactions:

$$TiCl_4 + 2SiH_4 \rightarrow TiSi_2 + 4HCl + 2H_2 \tag{4.14a}$$

$$Si(s) + 2HCl \rightarrow SiCl_2(g) + H_2 \tag{4.14b}$$

However, to achieve selectivity with no substrate consumption, the flow conditions must be optimized. An excess flow of SiH_4 can cause a Si pedestal to grow underneath the deposited $TiSi_2$, while SiH_4 deficiency can cause silicon consumption. The group demonstrated selective deposition of ~1700 Å $TiSi_2$ on Si with no silicon consumption at 800 °C. At this temperature, the deposited films are polycrystalline, single phase, stoichiometric $TiSi_2$ and exhibit the stable C54 structure.

Selective CVD of silicides with no silicon consumption will be very useful when the ultrashallow junctions that are being engineered today become inevitable in deep submicron microelectronics. It will also be useful in eliminating the wet chemistry associated with the conventional SALICIDE process. This will enable cluster tool developers to design machines that will be able to form the entire raised source/drain junctions *in situ* including the self-aligned metal contacts. However, more work is in order to develop a reliable selective process. Issues such as selectivity with respect to reactive ion etching (RIE)-damaged side-wall oxides and possibly other insulators such as Si_3N_4 must be addressed.

2. Tungsten

Tungsten (W) is one of the most widely researched conductors. Even though the main thrust for tungsten has been via filling, tungsten has also been considered as a diffusion barrier for Al on Si and on $TiSi_2$ [97], and also as an alternative gate electrode to polysilicon [98-100].

At Stanford University, Moslehi and Saraswat developed a novel rapid thermal/plasma multiprocessing reactor to deposit tungsten films on Si and SiO_2 [98-100]. Selective tungsten depositions were accomplished using a mixture of WF_6, H_2, and Ar without using the plasma capability of the reactor. Selectivity was found superior to that obtained in hot-wall furnaces. Nonselective depositions on SiO_2 were achieved using the same gases but with the microwave plasma capability generating Ar, Ar + H_2, or WF_6 + Ar plasmas. The nonselective depositions of tungsten on SiO_2 were achieved without the need for a glue layer. Interestingly, this unique flexible concept of combining RTP and plasma processing has not been exploited by equipment manufacturers.

III. *In Situ* Processing—Applications

In situ processing is a technique in which several processes are carried out in sequence without exposing the wafer to air between the process steps. The technique also aims at minimizing the wafer handling in order to reduce the number of particulates. The term *in situ* processing is used rather loosely to describe two different approaches. One approach utilizes a single RTP chamber for a number of *in situ* processes [42]. These processes can be combinations of different RTP processes such as oxidation, annealing, and CVD. This approach utilizes the cold-wall nature of the deposition chamber and assumes that chamber walls are always free of any reactive contaminants. Therefore, by changing gases, one can deposit or grow different layers sequentially. Nevertheless, this requires processes that are compatible with each other. For instance, it is difficult to envision a chamber that can be used for oxidation and epitaxy sequentially. The second approach uses different chambers for different processes. The wafer is transferred from one chamber to the other with a robotic arm. Because of its cold-wall nature and rapid heating and cooling advantages, RTCVD is well suited to both approaches.

The adaptability of RTCVD to *in situ* processing was rightfully exploited even in the very early LRP publications of the Stanford University researchers. In 1986, Sturm *et al.* combined CVD of polysilicon with rapid thermal oxidation (RTO) and rapid thermal annealing (RTA) and fabricated MOS structures [11]. Excellent interface properties were demonstrated.

This was followed by another report published in the same year in which Sturm *et al.* combined selective silicon epitaxy with RTO and doped polysilicon deposition in the same reactor [101]. The authors used the *in situ*-formed structure to fabricate MOS transistors.

One of the most promising applications of *in situ* processing can be found in the fabrication of MOS structures for advanced devices. Many RTP processes including surface cleaning, rapid thermal oxidation, nitridation, and polysilicon deposition can be carried out without wafer handling and exposure to air. It is anticipated that MOS structures with ultrathin gate dielectrics will benefit from such an approach. Electrical properties of thermal oxides (RTO) with and without *in situ*-deposited polysilicon gate electrodes were studied [27, 41]. These studies indicated comparable electrical properties for both cases. However, MOS structures with *in situ*-deposited gate electrodes exhibited either higher breakdown voltages or tighter breakdown voltage distributions, which were attributed to lower defect generation during processing and wafer transport.

With similar motivations, Ting *et al.* combined RTO and RTCVD of SiO_2 and polysilicon in a single RTRP reactor [16, 17, 42]. The structures formed in this report were NO (nitride/oxide) or ONO (oxide/nitride/oxide) structures for dynamic random access memories and nonvolatile memories. In this work, the goal was to combine the excellent properties of the thermal oxide/silicon interface with the higher dielectric constant and good diffusion barrier properties of the silicon nitride. Ting *et al.* also proposed that if an additional thin oxide layer is used on top of the nitride, hole injection from the gate can be significantly reduced, thereby reducing the current conduction through the stacked structure. The process sequence used by the researchers is shown in Fig. 20. As shown, by using the rapid temperature switching capability of RTCVD (ramp rates of several hundred degrees per second), it is possible to use extremely short process times.

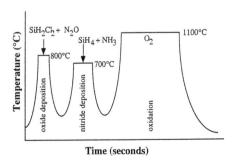

FIGURE 20. Time–temperature profile for ONO stacked layer fabricated by RTP-CVD. From Kwong *et al.* [42].

In situ growth of epitaxial layers with different properties (i.e., doping, band gap, etc.) can also enable fabrication of very complicated device structures. For instance, if the gate stack of a MOSFET is formed *in situ*, one can not perform a standard "threshold adjust" implant through the gate oxide. Instead, one can envison a process in which the channel is grown to the desired doping density followed by the formation of the gate stack. Such a fabrication sequence can be carried out in a cluster tool with chambers for epitaxy, RTO, and RTCVD.

In situ processing using RTP techniques has been proposed for other applications. An application of *in situ* processing in the bipolar area has been proposed by Ruddell *et al.* [28]. In this application, the goal was to form polysilicon emitter contacts with superior characteristics. The researchers used a novel RTCVD reactor that combined a lamp-heated quartz chamber with microwave plasma capability. In this reactor, it was possible to remove the native oxide (by *in situ* CF_4 plasma clean), grow a thin plasma oxide (using N_2O), and deposit polysilicon. Such a structure is shown in Fig. 21. Highest gains were obtained from transistors with interfacial oxides. Later Kermani and Wong proposed a similar process for the same application [33]. Their approach was to grow a thin interfacial oxide by RTO after removing the chemical oxide and to deposit polysilicon *in situ* without exposing the wafer to any source of contamination. Kermani and Wong argued that the approach could yield a stable and atomically smooth oxide/polysilicon interface that did not suffer from the breakup problems of chemical oxides that occur during emitter and base drive-in steps.

One of the most promising applications of *in situ* processing is in the formation of raised source/drain junctions for deep submicron MOS

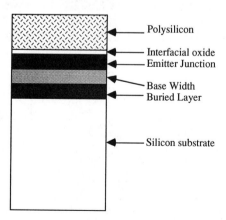

FIGURE 21. Cross section of the polysilicon/interfacial oxide/silicon substrate structure for a bipolar device. From Kermani and Wong [33].

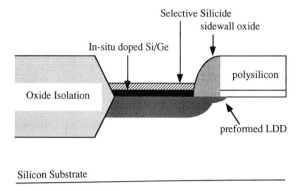

FIGURE 22. Cross-sectional view of a raised MOS transistor junction.

transistors. By definition, a raised junction is a junction raised above the level of the gate oxide/substrate interface by either selective area deposition or selective area processing of several thin films. The basic raised junction is shown in Fig. 22 as part of an MOS transistor. A possible fabrication sequence for the structure is shown in Fig. 23. The structure consists of two selective layers: (1) dopant source (polycrystalline or epitaxial and (2) metal contact (e.g., titanium silicide). In this approach, the selectively deposited dopant source can also serve as a sacrificial layer during the SALICIDE process. If the silicide is formed by selective CVD techniques, the sacrificial layer can be very thin. More complex raised source/drain junction structures have also been proposed. A unique raised structure proposed by Tasch *et al.* aims at reducing the lateral electric field in the channel region without the need for a lightly doped drain (LDD) structure [102].

Parts of the sequence shown in Fig. 23 are currently under investigation. Kwong *et al.* attempted using *in situ* arsenic doped polysilicon as a selective diffusion source to form shallow n^+p junctions in silicon [42]. However, the relatively high deposition temperature (800–850 °C—possibly used to achieve selective growth) resulted in arsenic penetration into the substrate during growth as well as arsenic pile-up at the surface and at the polysilicon/substrate interfaces. This caused arsenic levels in polysilicon to drop to levels on the order of 10^{17} cm^{-3}, which is undesirable because the series resistance introduced by the layer. As discussed in Section III D, in this application an alternative to polysilicon is polycrystalline Si–Ge that can be deposited selectively at low temperatures without sacrificing the high deposition rate required by single-wafer manufacturing. Grider *et al.* exploited this property of polycrystalline Si–Ge to form ultrashallow p^+n junctions in silicon [54]. In this study, polycrystalline Si–Ge was deposited at 650 °C using SiH$_2$Cl$_2$ and GeH$_4$ was heavily boron doped *in situ* using

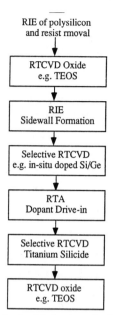

FIGURE 23. Sequence for *in situ* fabrication of raised source/drain structures.

B_2H_6. Because of the low deposition temperature used, no dopant diffusion into the substrate was experienced during growth. The junctions were formed by RTA subsequent to the Si-Ge deposition. The resulting boron profiles obtained by secondary ion mass spectroscopy (SIMS) are shown in Fig. 24.

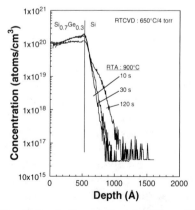

FIGURE 24. SIMS profiles of ultra shallow p^+n junctions formed in silicon using selectively deposited *in situ* boron-doped polycrystalline Si-Ge as a diffusion source.

Thin-Film Deposition

In situ processing is expected to prove useful, especially with demanding geometries of future deep submicron technologies. At those levels, particulate and contamination control will undoubtedly be very critical. Formation of thin native oxides that form at semiconductor and conductor interfaces will have to be eliminated as well. *In situ* processing can be a solution to many of these problems. In future cluster tools, rapid thermal processing is a strong candidate for many of the process steps currently used in microelectronics.

IV. Equipment Issues

During recent years, RTP systems have matured appreciably for applications like oxidation and annealing. However, the concept of CVD in RTP systems is relatively new and not many vendors have yet engaged in construction of RTCVD equipment. In this section, we will briefly address equipment issues that are specifically related to rapid thermal chemical vapor deposition. A more general discussion on RTP equipment can be found in Chapters 9 and 10.

A. Temperature Uniformity

In early RTP reactors, temperature uniformity was a major problem. It was found that radiant heat losses from the edge of the wafer forced the wafer edge to be cooler than the wafer center. This was true even if uniform radiation was incident on the wafer. Theoretical studies showed later that with the dimensions of the current RTP systems and lamp reflectors, flat reflectors used behind lamps could not provide uniform radiation to the wafer [103]. In fact, calculations showed by simple ray tracing that these reflectors provided even more radiation to the center of the wafer, increasing the magnitude of the problem [103, 104]. To address these issues equipment vendors have developed new approaches and some of these proved very successful. At present, steady-state temperature uniformity in commercial reactors appears to be satisfactory for many applications such as RTA and RTO. However, uniformity during heat-up and cool-down still has a lot of room for improvement. Fortunately, temperature uniformity during these cycles may not be very critical for RTCVD. This is due to the following reasons: (1) The majority of the RTCVD processes are low-temperature processes. Therefore, the slip associated with nonuniform cooling of the wafer from high temperatures does not exist. (2) The majority of the RTP processes require deposition times much longer (several minutes) than the ramp-up and cool-down times.

Because most CVD processes have an exponential temperature dependence, negligible film deposition takes place during ramp-up and ramp-down. Therefore, unless very high temperature or very short time processes are used, temperature uniformity during heat-up and cool-down may not be very critical in RTCVD reactors.

Another potential source of temperature nonuniformity is localized absorptivity (i.e., emissivity) variations on a wafer. Here the absorptivity of interest is the average absorptivity $\bar{a}$ obtained by averaging absorptivity $a(\lambda)$ over the entire spectrum of wavelengths. The average absorptivity can thus be determined from

$$\bar{a} = \frac{\int_0^\infty \lambda a(\lambda) \, d\lambda}{\int_0^\infty a(\lambda) \, d\lambda} \tag{4.15}$$

Average absorptivity can vary locally on a silicon surface because of patterns created with films of different optical properties (e.g., SiO_2 and Si). These issues were studied for the first time by Vandenabeele et al. [105]. The researchers used silicon wafers that contained large and small area patterns etched in SiO_2. They showed by implant annealing and oxidation that large temperature variations can occur on a wafer if the pattern size is on the order of the diffusion length for lateral heat diffusion, L_d. The diffusion length is given by

$$L_d = \sqrt{\frac{Kt}{4\varepsilon\sigma T^3}} \tag{4.16}$$

where K is the thermal conduction coefficient of silicon (W/m·K), t is the sample thickness, ε is the total emissivity (front and back side) of the wafer, σ is the Stefan–Boltzmann constant (5.67×10^{-8} W/m^2 K^4), and T is the absolute temperature in Kelvins. Equation 4.16 gives values on the order of a few millimeters for Si (4 mm at 1100 °C). Since L_d is inversely proportional to $T^{3/2}$, at lower RTCVD temperatures larger patterns are needed to create the same temperature nonuniformities. However most RTCVD processes are much more temperature sensitive than the sheet resistance of an implant profile. For instance, Vandenabeele et al. demonstrated a 10% variation in the sheet resistance of an implant due to a temperature difference of 26 °C [105]. As we shall discuss below, with polysilicon such a temperature difference would cause a much larger deposition nonuniformity on the wafer. Even though this appears to be a serious problem, by carefully arranging devices and test sites on a lithography mask one can avoid local variations in wafer absorptivity.

The influence of the deposition system and patterns on thickness uniformity can be further reduced by choosing the correct chemistry. The percentage process uniformity $\%X$ across the wafer can be expressed as

Thin-Film Deposition

a function of the percentage temperature uniformity $\%T$ according to [107]

$$\%X = \left\{1 - \exp\left[-\frac{E_a}{kT_{\max}}\left(\frac{\%T}{100 - \%T}\right)\right]\right\} \times 100 \qquad (4.17)$$

where E_a is the activation energy, $T_{\max}$ is the maximum temperature on the wafer, and k is Boltzmann's constant. Note that when E_a is zero $\%X$ is zero, indicating a perfect uniformity. Equation 4.17 indicates that processes with lower activation energies will result in better process uniformities. Equation 4.17 is plotted in Fig. 25 for three different deposition temperatures using the activation energy of polysilicon deposition (1.6 eV) [106]. As shown, the uniformity of the deposited film can be quite unacceptable even though the temperature nonuniformity is within a few percent. Figure 25 clearly shows that in order to achieve a deposition uniformity within 5%, the temperature nonuniformities should be less than 0.25%. At 700 °C, this corresponds to a center-to-edge temperature difference $T_c - T_e$ of only 1.75 °C. This is a challenging, yet an essential goal to make RTCVD a manufacturing process.

Unfortunately, temperature uniformity of an RTCVD process is further complicated by another physical mechanism unique to lamp-heated RTP. During an RTCVD cycle, the average absorptivity goes through a continuous change as the film is deposited, because the absorptivity is determined not only by what is being deposited but also how thick and on what it is being deposited. Now, suppose there exists a small initial temperature nonuniformity on an oxidized wafer that is going through a polysilicon deposition process. The small temperature nonuniformity will lead to a small initial deposition nonuniformity on the wafer. Unfortunately, this initial nonuniformity will also lead to variations in wafer absorptivity across the wafer. That is, even if the mask design is made in a way to avoid

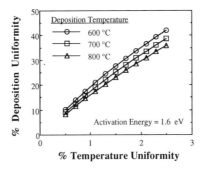

FIGURE 25. Deposition uniformity of polysilicon as a function of the percentage temperature uniformity across the wafer. From Öztürk et al. [106]. Copyright © 1991 IEEE.

absorptivity variations, after some point during the deposition cycle different points will have different light absorption properties if the starting temperature uniformity is not absolutely perfect. It has been shown that this effect can lead to extremely poor uniformity figures [107].

In summary, temperature uniformity in a lamp-heated RTCVD reactor is a complex issue and relies on optimization of several factors. While the equipment manufacturers are working to maximize the temperature uniformity in their systems it is also very important to advance our understanding of the processes and materials. It will be safe to assume that there will always be a small percentage of temperature nonuniformity across a large-area silicon wafer. Hence, the trick will be to choose processes, materials, and structures that are least sensitive to temperature variations on a silicon wafer.

B. Temperature Measurement

Temperature measurement has always been one of the key issues in rapid thermal processing. First-generation rapid thermal processors used a thermocouple embedded in a small silicon chip placed near the wafer in process for temperature measurement and control. Even though this simple technique proved useful for annealing, it is not suitable to RTCVD. This is due to the fact that deposition will also occur on the control wafer, drastically changing its absorptivity. Therefore, a fresh wafer can have an absorptivity quite different than the absorptivity of the thermocouple wafer with a thick layer deposited on it.

Single-wavelength pyrometry is used widely in commercial RTP reactors (see Chapter 9). The technique has achieved a wide acceptance among equipment manufacturers due to its maturity and noncontact nature. Pyrometry makes a single measurement, based on emitted radiation, which is a complex combination of surface temperature and surface emissivity. At high temperatures ($\geq 600\,°C$), with highly doped substrates, and for processes where the emissivity does not change during the process, the single wavelength pyrometer has proved successful [108]. There is some dependence of emissivity on surface roughness, but this is minor relative to the complications that exist in a CVD environment.

The emissivity of a silicon wafer depends strongly on the properties of the films deposited on the wafer. During a deposition process, the surface emissivity continuously changes as the thickness of the deposited film increases. If single-wavelength pyrometry is used, this can cause unacceptably large temperature errors. This is demonstrated in Fig. 26, which shows the

Thin-Film Deposition

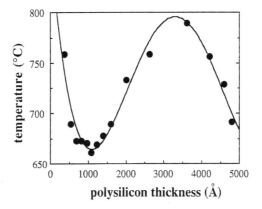

FIGURE 26. Variations in actual deposition temperature during RTCVD of polysilicon on 1000-Å thick SiO_2. The wafer temperature was measured with a 3.8-μm pyrometer and the pyrometer output signal was used in a feedback loop. During the entire deposition, the pyrometer was operated in an endeavor to maintain the temperature constant at the set temperature. From Öztürk et al. [106]. Copyright © 1991 IEEE.

variations in deposition temperature during RTCVD of polysilicon on 1000 Å of SiO_2 [106]. The changes occur as a result of the changing surface emissivity during deposition. As shown, the wafer temperature can vary significantly from the starting temperature. The magnitude of the excursions is actually determined by the thickness of the SiO_2 layer under polysilicon. This is demonstrated in Fig. 27, which shows the calculated wafer emissivity as a function of the polysilicon thickness for three different oxide

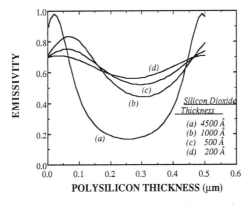

FIGURE 27. Calculated wafer emissivity of the polysilicon a function of polysilicon thickness for different SiO_2 thicknesses. From Öztürk et al. [106]. Copyright © 1991 IEEE.

thicknesses. A detailed description of the simulation procedures is reported elsewhere [103]. The problem still exists when a single layer is deposited on silicon. However, simulations show that if very thin films (<500 Å) are deposited or grown, the changes in the wafer emissivity are much less dramatic, which is not true for multilayers.

It is clear that single-wavelength pyrometry will not be useful in RTCVD reactors. However, *in situ* surface emissivity measurement can be extremely helpful. The alternative is to use a technique that does not rely on the optical properties of the wafer surface to obtain an accurate temperature measurement. Such techniques are currently under investigation by various research groups.

C. Deposition of Films on Windows and Chamber Walls

As LPCVD arises as a new application area for rapid thermal processing, several new chamber design requirements are imposed on equipment manufacturers. In today's RTCVD reactors, the chamber walls are usually made of water cooled stainless steel with a quartz window for passing the lamp radiation. O-rings are usually used for vacuum sealing. A drawback of this approach is that the quartz window must be sufficiently thick to handle the stress that develops from the pressure differential between the inner and outer surfaces. The maximum bending stress, σ, of a circular plate can be calculated from

$$\sigma = k_1 \left(\frac{d}{t}\right)^2 p, \tag{4.18}$$

where d is the window diameter, t is the quartz thickness, p is the pressure, and k_1 is the stress factor. Therefore, to keep the stress at a minimum the window thickness must be increased with the window diameter. Since one of the driving forces for single-wafer manufacturing is to increase wafer size, the window diameters of interest should be greater than 20 cm. Thicker quartz is needed for larger windows, which will require a very effective window cooling. Equipment designers have employed both air and water cooling on the windows outside the chamber.

On the other hand, inside the chamber, the hot silicon wafer still remains as a major source of heat for the window. It is natural to assume that lower process temperatures will lead to cooler windows, because the radiation power $P(\lambda)$ decreases rapidly with decreasing temperature. However, in order to evaluate the contribution of the hot wafer to window heating, we also need to consider the processing time. Using a 3000-Å poly-Si deposition as an example, the total energy E_T absorbed by the window has

Thin-Film Deposition

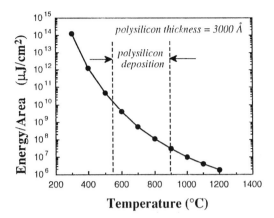

FIGURE 28. Radiant energy absorbed by the window during RTCVD of 3000 Å polysilicon. The calculations were made assuming that the entire energy emitted by the wafer was absorbed by the window. From Öztürk et al. [106]. Copyright © 1991 IEEE.

been calculated. The result is shown in Fig. 28. In these calculations, the energy per unit wavelength $E(\lambda)$ given off by the wafer at a particular wavelength was found by multiplying the emitted power from a gray body at that wavelength with the deposition time. The absorption spectrum of a 1-cm quartz window was used to calculate the amount of absorbed radiation $E_a(\lambda)$. Finally, the total energy E_T absorbed by the window was found by integrating $E_a(\lambda)$ over the entire wavelength range. The activation energy was taken as 1.6 eV. As shown, E_T decreases rapidly as the process temperature is increased. Therefore, the effect of the silicon wafer on the quartz heating will be minimized if higher temperature/shorter time processes are used. It can be speculated that with lower temperature/longer time processes, there would be more time to remove the heat from the window. However, inside the chamber, at pressures typical for LPCVD, the amount of convective cooling should be small with negligible effect on window temperature.

Window heating is a critical issue for RTCVD. Certain gases such as SiH_4 can easily decompose at lower temperatures (above 400 °C), which may result in deposition on the window. If this happens a chain reaction is initiated. The film deposited on the window absorbs more radiation and the window temperature increases and as a result, deposition rate on the window is enhanced. It is essential that deposition on the window be prevented or at least *in situ* chamber cleaning be performed to maintain controllability. To minimize this undesirable effect, chamber designs that allow thinner quartz must be provided. Internal cooling such as spraying an inert gas compatible with the process chemistry may also be considered.

The issues discussed in this section can be resolved by further research and development on equipment design. This is actively pursued by both equipment manufacturers and research laboratories in universities and other institutions.

V. Summary

In this chapter we have reviewed material on thin-film deposition in RTP systems published in technical journals and presented at conferences. The technique has advantages over conventional CVD, including the ability to deposit ultrathin films and the suitability for *in situ* processing and single-wafer cluster tools. Several thin films used in silicon processing have already been demonstrated in RTCVD reactors. These are silicon dioxide, silicon nitride, and polysilicon. New films such as $Si_{1-x}Ge_x$ and $TiSi_2$ are also being studied. The objective is to deposit high-quality films that can be deposited at rates sufficient for single-wafer processing. However, issues such as deposition on chamber walls must also be addressed. Temperature uniformity appears to be less of a problem; however, issues such as mask design must be understood and carefully considered. Temperature measurement for RTCVD is still a challenging task that needs to be resolved. Important advances in this area can be anticipated in the next few years. Overall the preliminary results are promising, but further work is needed to move from research laboratories to a semiconductor manufacturing environment.

References

1. S. K. Ghandi, *VLSI Fabrication Principles, Silicon and Gallium Arsenide*, Wiley, New York (1983).
2. S. M. Sze, ed., *VLSI Technology*, 2nd ed., McGraw-Hill, New York (1988).
3. A. Sherman, *Chemical Vapor Deposition for Microelectronics*, Noyes Publications, Park Ridge, New Jersey (1987).
4. B. S. Meyerson, *Appl. Phys. Lett.* **48**, 797 (1986).
5. B. S. Meyerson, F. K. LeGoues, T. N. Nguyen, and D. L. Harame, *Appl. Phys. Lett.* **50**, 113 (1987).
6. T. O. Sedgwick, V. P. Kesan, P. D. Agnello, D. A. Grützmacher, D. Nguyen-Ngoc, S. S. Iyer, D. J. Meyer, and A. P. Ferro, *IEDM Tech. Dig.* **451**, 000 (1991).
7. J. F. Gibbons, C. M. Gronet, and K. E. Williams, *Appl. Phys. Lett.* **47**, 721 (1985).
8. C. M. Gronet, J. C. Sturm, K. E. Williams, and J. F. Gibbons, *Symposium on Rapid Thermal Processing, MRS Symposia Proceedings*, 305 (1985).
9. C. M. Gronet, C. A. King, and J. F. Gibbons, *Symposium on Materials Issues in Silicon Integrated Circuit Processing, MRS Symposia Proceedings*, 107 (1986).
10. C. M. Gronet, J. C. Sturm, K. E. Williams, and J. F. Gibbons, *Appl. Phys. Lett.* **48**, 1012 (1986).

Thin-Film Deposition

11. J. C. Sturm, C. M. Gronet, and J. F. Gibbons, *IEEE Electron Dev. Lett.* **EDL-7**, 282 (1986).
12. J. C. Sturm, C. M. Gronet, and J. F. Gibbons, *J. Appl. Phys.* **59**, 4180 (1986).
13. J. F. Gibbons, C. M. Gronet, J. C. Sturm, C. King, K. Williams, S. Wilson, S. Reynolds, D. Vook, M. Scott, R. Hull, C. Nauka, J. Turner, S. Laderman, and G. Reid, *Beam-Solid Interactions and Transient Processes Symposium, MRS Symposia Proceedings*, 629 (1986).
14. J. Ahn, W. Ting, and D. L. Kwong, *49th Annual Device Research Conference, DRC Extended Abstracts*, IVB1 (1991).
15. K. Watanabe, T. Tanigaki, and S. Wakayama, *J. Electrochem. Soc.* **128**, 2630 (1981).
16. W. Ting, S. N. Lin, and D. L. Kwong, *Symposium on Rapid Thermal Annealing/Chemical Vapor Deposition and Integrated Processing, MRS Symposia Proceedings*, 351 (1989).
17. W. Ting, S. N. Lin, and D. L. Kwong, *Appl. Phys. Lett.* **55**, 2312 (1989).
18. X. Xu, R. Kuehn, J. J. Wortman, and M. C. Öztürk, to be published (1992).
19. L. Y. Cheng, J. C. Rey, J. P. McVittie, and K. Saraswat, IEEE VMIC Conference, June 12-14, *Extended Abstracts*, 404 (1990).
20. R. Miller, M. C. Öztürk, J. J. Wortman, F. S. Johnson, and D. T. Grider, *Mat. Lett.* **8**, 353 (1989).
21. M. C. Öztürk, J. J. Wortman, Y. Zhong, X. Ren, R. M. Miller, F. S. Johnson, D. T. Grider, and D. A. Abercrombie, *Rapid Thermal Annealing/Chemical Vapor Deposition and Integrated Processing Symposium, MRS Symposia Proceedings*, 109 (1989).
22. F. S. Becker, D. Pawlik, H. Anzinger, and A. Spitzer, *J. Vac. Sci. Tech. B* **5**, 1555 (1987).
23. A. C. Adams and C. D. Capio, *J. Electrochem. Soc.* **126**, 1042 (1979).
24. H. Huppertz and W. L. Engl. *IEEE Trans. Electron Dev.* **ED-26**, 658 (1979).
25. F. S. Johnson, R. M. Miller, M. C. Öztürk, and J. J. Wortman, *Rapid Thermal Annealing/Chemical Vapor Deposition and Integrated Processing Symposium, MRS Symposia Proceedings*, 345 (1989).
26. R. S. Rosler, *Solid State Tech.* **4-77**, 63 (1977).
27. P. Pan, A. Kermani, W. Berry, and J. Liao, *Symposium on Rapid Thermal Annealing/Chemical Vapor Deposition and Integrated Processing, MRS Symposia Proceedings*, 47 (1989).
28. F. Ruddell, C. Parkes, B. M. Armstrong, and H. S. Gamble, *Symposium on Rapid Thermal Annealing/Chemical Vapor Deposition, MRS Symposia Proceedings*, 133 (1989).
29. J. C. Liao, J. L. Crowley, and T. I. Kamins, *Rapid Thermal Annealing/Chemical Vapor Deposition and Integrated Processing Symposium, MRS Symposia Proceedings*, 97 (1989).
30. F. Ruddell, C. Parkes, B. M. Armstrong, and H. S. Gamble, *Semicon. Sci. Tech.* **5**, 765 (1990).
31. T. Y. Hsieh, H. G. Chun, D. L. Kwong, and D. B. Spratt, *Appl. Phys. Lett.* **56**, 1778 (1990).
32. J. C. Liao and T. I. Kamins, *J. Appl. Phys.* **67**, 3848 (1990).
33. A. Kermani and F. Wong, *Solid State Tech.* **33**, 41 (1990).
34. X. Ren, M. C. Öztürk, J. J. Wortman, C. Blat, and E. Niccolian, *J. Elect. Mat.* **20**, 251 (1991).
35. R. F. C. Farrow, *J. Electrochem. Soc.* **121**, 899 (1974).
36. K. L. Chiang, C. J. Dell'Oca, and F. N. Schwettmann, *J. Electrochem. Soc.* **126**, 2267 (1979).
37. G. Harbeke, L. Krausbuaer, E. F. Steigmeier, A. E. Widmer, J. Kappert, and G. Neugebauer, *J. Electrochem. Soc.* **131**, 675 (1984).
38. D. Forster, A. Learn, and T. Kamins, *Solid State Tech.* **29**, 227 (1986).

39. C. H. J. V. D. Brekel and L. J. M. Bollen, *J. Crystal Growth* **54**, 310 (1981).
40. X. Ren, M. C. Öztürk, J. J. Wortman, B. Zhang, and D. Maher, *J. Vac. Sci. Tech.* B **10**, 1081 (1992).
41. V. Murali, A. T. Wu, L. Dass, M. R. Frost, D. B. Fraser, J. Liao, and J. Crowley, *J. Electronic Mat.* **18**, 731 (1989).
42. D. L. Kwong, T. Y. Hsieh, K. H. Jung, W. Ting, and S. K. Lee, *Symposium on Rapid Isothermal Processing, SPIE Proceedings*, 109 (1989).
43. T. J. King, J. R. Pfiester, J. D. Shott, J. P. McVittie, and K. C. Saraswat, *IEDM Tech. Dig.* 253 (1990).
44. M. C. Öztürk, D. T. Grider, S. P. Ashburn, M. Sanganeria, and J. J. Wortman, *Materials Research Society, Symposium Proceedings, Spring 1991* **224**, 223 (1991).
45. D. T. Grider, M. C. Öztürk, M. Sanganeria, S. Ashburn, and J. J. Wortman, *Spring Electronic Materials Conference* (1991).
46. T.-J. King and K. C. Saraswat, *IEDM Tech. Dig.* 567 (1991).
47. M. C. Öztürk, Y. Zhong, D. T. Grider, M. Sanganeria, J. J. Wortman, and M. A. Littlejohn, *Rapid Thermal and Related Processing Techniques, SPIE Proceedings* **1393**, 260 (1991).
48. D. T. Grider, M. C. Öztürk, and J. J. Wortman, *Proceedings of the Third International Symposium on ULSI Science and Technology, Electrochemical Society Extended Abstracts*, 296 (1991).
49. M. C. Öztürk, D. T. Grider, J. J. Wortman, M. A. Littlejohn, Y. Zhong, D. Batchelor, and P. Russell, *J. Electron. Mat.* **19**, 1129 (1990).
50. Y. Zhong, M. C. Öztürk, D. T. Grider, J. J. Wortman, and M. A. Littlejohn, *Appl. Phys. Lett.* **57**, 2092 (1990).
51. J. C. Bean, T. T. Sheng, L. C. Feldman, A. T. Fiory, and R. T. Lynch, *Appl. Phys. Lett.* **44**, 102 (1984).
52. M. Sanganeria, M. C. Öztürk, D. M. Maher, J. J. Wortman, D. Batchelor, B. Zhang, and Y. L. Zhong, *Proceedings of the Third International Symposium on ULSI Science and Technology, Electrochemical Society Extended Abstracts*, 851 (1991).
53. M. Sanganeria, D. T. Grider, M. C. Öztürk, J. J. Wortman, *J. Electron. Mat.* **21**, 61 (1991).
54. D. T. Grider and M. C. Öztürk, to be published (1992).
55. C. M. Maritan, L. P. Berndt, N. G. Tarr, J. M. Bullerwell, and G. M. Jenkins, *J. Electrochem. Soc.* **135**, 1793 (1988).
56. S. Nakayama, I. Kawashima, and J. Murota, *J. Electrochem. Soc.* **133**, 1721 (1986).
57. J. F. Gibbons, S. Reynolds, C. Gronet, D. Vook, and C. King, *Rapid Thermal Processing of Electronic Materials Symposium, MRS Symposia Proceedings*, 281 (1987).
58. C. A. King, C. M. Gronet, J. F. Gibbons, and S. D. Wilson, *IEEE Electron Dev. Lett.* **9**, 229 (1988).
59. J. L. Regolini, D. Bensahel, E. Scheid, A. Perio, and J. Mercier, *Appl. Sur. Sci.* **36**, 673 (1989).
60. J. L. Regolini, D. Bensahel, J. Mercier, and E. Scheid, *J. Crystal Growth* **96**, 505 (1989).
61. F. Wong, C. Y. Chen, and K. Yen-Hui, *Symposium on Rapid Thermal Annealing/Chemical Vapor Deposition and Integrated Processing, MRS Symposia Proceedings*, 27 (1989).
62. M. L. Green, D. Brasen, H. Temkin, C. Kannan, and H. S. Luftman, *Rapid Thermal Annealing/Chemical Vapor Deposition and Integrated Processing Symposium, MRS Symposia Proceedings*, 55 (1989).
63. M. L. Green, D. Brasen, H. Luftman, and V. C. Kannan, *Appl. Phys. Lett.* **65**, 2558 (1989).

64. S. K. Lee, Y. H. Ku, and D. L. Kwong, *Rapid Thermal Annealing/Chemical Vapor Deposition and Integrated Processing Symposium, MRS Symposia Proceedings*, 127 (1989).
65. M. L. Green, H. Temkin, and D. Brasen, *Symposium on Rapid Isothermal Processing, SPIE Proceedings*, 106 (1989).
66. M. L. Green, D. Brasen, M. Geva, W. Reents, F. S. Jr., and H. Temkin, *J. Electron. Mat.* **19,** 1015 (1990).
67. F. Wong and Y. Ku, *Microelectronic Manuf. Test.* **13,** 25 (1990).
68. D. Mathiot, J. L. Regolini, and D. Dutarte, *J. Appl. Phys.* **69,** 358 (1991).
69. T. Y. Hseih, K. H. Jung, D. L. Kwong, and S. K. Lee, *J. Electrochem. Soc.* **138,** 1188 (1991).
70. J. L. Regolini, D. Dutartre, D. Bensahel, and J. Penelon, *Solid State Tech.* **34,** 47 (1991).
71. C. M. Gronet, C. A. King, W. Opyd, J. F. Gibbons, S. D. Wilson, and R. Hull, *J. Appl. Phys.* **61,** 2407 (1987).
72. C. A. King, J. L. Hoyt, C. M. Gronet, J. F. Gibbons, M. P. Scott, S. J. Rosner, G. Reid, S. Laderman, K. Nauka, and T. I. Kamins, *IEEE Trans. Electron Dev.* **35,** 2454 (1988).
73. J. F. Gibbons, C. A. King, J. L. Hoyt, D. B. Noble, C. M. Gronet, M. P. Scott, S. J. Rosner, G. Reid, S. Laderman, K. Nauka, and J. Turner, *IEDM Tech. Dig.* 566 (1988).
74. M. P. Scott, S. S. Laderman, T. I. Kamins, S. J. Rosner, K. Nauka, D. B. Noble, J. L. Hoyt, C. A. King, C. M. Gronet, and J. F. Gibbons, *Symposium on Thin Films: Stress and Mechanical Properties, MRS Symposium Proceedings*, 179 (1988).
75. C. A. King, J. L. Hoyt, D. B. Noble, C. M. Gronet, J. F. Gibbons, M. P. Scott, S. S. Laderman, T. I. Kamins, and J. Turner, *Symposium on Rapid Thermal Annealing/Chemical Vapor Deposition and Integrated Processing, MRS Symposia Proceedings*, 71 (1989).
76. C. A. King, J. L. Hoyt, D. B. Noble, C. M. Gronet, J. F. Gibbons, M. P. Scott, T. I. Kamins, and S. S. Laderman, *IEEE Electron Dev. Lett.* **10,** 159 (1989).
77. C. A. King, J. L. Hoyt, C. M. Gronet, J. F. Gibbons, M. Scott, and J. Turner, *IEEE Electron Dev. Lett.* **10,** 52 (1989).
78. D. B. Noble, J. L. Hoyt, J. F. Gibbons, M. P. Scott, S. S. Laderman, S. J. Rosner, and T. I. Kamins, *Appl. Phys. Lett.* **55,** 1978 (1989).
79. H. Temkin, M. L. Green, D. Brasen, and J. C. Bean, *Symposium on Rapid Thermal Annealing/Chemical Vapor Deposition and Integrated Processing, MRS Symposia Proceedings*, 65 (1989).
80. J. Gibbons, J. Hoyt, C. King, D. Noble, C. Gronet, M. Scott, S. Laderman, K. Nauka, J. Turner, and T. Kamins, *Extended Abstracts on the 21st Conference on Solid State Devices and Materials*, 5 (1989).
81. P. M. Garone, J. C. Sturm, P. V. Schwartz, S. A. Schwartz, and B. Wilkens, *Rapid Thermal Annealing/Chemical Vapor Deposition and Integrated Processing Symposium, MRS Symposia Proceedings*, 41 (1989).
82. F. Wong, *Solid State Tech.* **32,** 53 (1989).
83. M. L. Green, D. Brasen, H. Temkin, R. D. Yadvish, T. Boone, L. C. Feldman, M. Geva, and B. E. Spear, *Thin Solid Films* **184,** 107 (1989).
84. E. A. Fitzgerald, Y. H. Xie, D. Brasen, M. L. Green, J. Michel, P. E. Freeland, and B. E. Weir, *Thin Solid Films* **184,** 93 (1990).
85. J. L. Hoyt, C. A. King, D. B. Noble, C. M. Gronet, J. F. Gibbons, M. P. Scott, S. S. Laderman, S. J. Rosner, K. Nauka, J. Turner, and T. I. Kamins, *Thin Solid Films* **184,** 93 (1990).

86. S. Reynolds, D. W. Vook, and J. E. Gibbons, *Symposium on Rapid Thermal Processing of Electronic Materials, MRS Symposia Proceedings*, 305 (1987).
87. J. C. Liao, J. L. Crowley, and P. H. Klein, *Proceedings of the 2nd International Conference on Amorphous and Crystalline Silicon Carbide, Recent Developments*, 20 (1989).
88. F. H. Ruddell, D. W. McNeill, B. M. Armstrong, and H. S. Gamble, *ESSDERC 90, 20th European Solid State Device Research Conference*, 357 (1990).
89. D. Levy, P. Delpech, M. Paoli, C. Masurel, M. Vernet, N. Brun, J.-P. Jeanne, J.-P. Gonchond, M. Ada-Hanifi, M. Haond, T. Ternisien d'Ouville, and H. Mingam, *IEEE Trans. Semicon. Manuf.* **3**, 168 (1990).
90. K. Maex, L. P. Hobbs, and W. Eichhammer, *Proceedings of the Third International Symposium on Ultra Large Scale Integration Science and Technology* **91-11**, 254 (1991).
91. H. Norström, K. Maex, and P. Vandenabeele, *Thin Solid Films* **198**, 53 (1991).
92. P. K. Tedrow, V. Ilderem, and R. Reif, *Appl. Phys. Lett.* **46**, 189 (1985).
93. V. Ilderem and R. Reif, *J. Electrochem. Soc.* **135**, 2590 (1988).
94. V. Ilderem and R. Reif, *J. Electrochem. Soc.* **136**, 2989 (1989).
95. J. L. Regolini, D. Bensahel, G. Bomchil, and J. Mercier, *Appl. Sur. Sci.* **38**, 408 (1989).
96. D. Bensahel, J. L. Regolini, and J. Mercier, *Appl. Phys. Lett.* **55**, 1549 (1989).
97. B. W. Shen, G. C. Smith, J. M. Anthony, and R. J. Matyi, *J. Vac. Sci. Tech. B* **4**, 1369 (1986).
98. M. M. Moslehi and K. C. Saraswat, *Symposium on Rapid Thermal Processing of Electronic Materials, MRS Symposia Proceedings*, 295 (1987).
99. M. M. Moslehi, M. Wong, K. C. Saraswat, and S. C. Shatas, *1987 Symposium on VLSI Technology*, Technical Digest, 21 (1987).
100. M. M. Moslehi and K. C. Saraswat, *IEEE Electron Dev. Lett.* **EDL-8**, 421 (1987).
101. J. C. Sturm, C. M. Gronet, C. A. King, S. D. Wilson, and J. F. Gibbons, *IEEE Electron Dev. Lett.* **EDL-7**, 577 (1986).
102. A. F. Tasch, H. Shin, and C. M. Maziar, *Electronics Lett.* **26**, 39 (1990).
103. F. Y. Sorrell, C. P. Eakes, M. C. Öztürk, and J. J. Wortman, *Symposium on Rapid Isothermal Processing, SPIE Proceedings*, **1189**, 55 (1989).
104. F. Y. Sorrell, M. J. Fordham, M. C. Öztürk, and J. J. Wortman, *IEEE Trans. Electron Dev.* **39**, 75 (1992).
105. P. Vandenabeele, K. Maex, and R. D. Keersmaecker, *Proceedings of the MRS Symposium on Rapid Thermal Annealing/Chemical Vapor Deposition and Integrated Processing* **146**, 149 (1989).
106. M. C. Öztürk, F. Y. Sorrell, J. J. Wortman, F. S. Johnson, and D. T. Grider, *IEEE Trans. Semicond. Manu.* **4**, 155 (1991).
107. M. C. Öztürk, M. K. Sanganeria, and F. Y. Sorrell, *Appl. Phys. Lett.* **61**, 2697 (1992).
108. P. Vandenabeele and K. Maex, *Rapid Thermal and Integrated Processing, MRS Symposium Proceedings* **224**, 185 (1991).

5 Extended Defects from Ion Implantation and Annealing

Kevin S. Jones
Department of Materials Science and Engineering
University of Florida
Gainesville, Florida

George A. Rozgonyi
Department of Materials Science and Engineering
North Carolina State University
Raleigh, North Carolina

I. Introduction	123
A. Ion Implantation and Rapid Thermal Annealing	124
B. Ion Implantation Damage	125
C. Amorphous Layer Regrowth	132
II. Defect Formation Kinetics	133
A. Point Defect Sources	133
B. Types of Extended Defects	134
III. Defect Annealing Kinetics	155
A. Type I Defect Annealing Kinetics	156
B. Type II Defect Annealing Kinetics	157
IV. Summary	162
References	163

I. Introduction

During the past 30 years ion implantation has become increasingly important as a method of controllably doping the near-surface region of semiconductors. The major disadvantage of ion implantation is the damage that arises during implantation and the need to activate the dopant. In addition to each ion creating large numbers of Frenkel pairs and recoils via elastic collisions, the implantation process is inherently nonconservative in that it introduces a large concentration of extra atoms. The defects that develop upon annealing, from the displaced host Si crystal atoms and the implanted dopant atoms, can adversely affect both dopant diffusion and device performance [1–8]. As lateral device dimension design rules decrease to 0.3 μm and below, vertical dimensions such as junction depths must decrease to <0.1 μm. In this depth regime interactions between point defects introduced during implantation and extended defects that form upon annealing dominate dopant diffusion processes. The complex nature of these interactions is schematically illustrated in Fig. 1. Current efforts to accurately model dopant diffusion (see Chapter 6) make the study of implantation damage of increasing importance. This chapter will focus on those factors that influence extended defect formation and dissolution kinetics ion order to extend our understanding of how these defects influence dopant diffusion.

A. Ion Implantation and Rapid Thermal Annealing

The initial application of ion implantation in the doping of semiconductors was realized around 1956 [9], and by 1962 the first papers describing its use in the fabrication of thin contacts on nuclear radiation detectors were published [10]. Since these early beginnings a large volume of work has been done and many books and review articles written on ion implantation [9, 11–15]. Today, ion implantation has emerged as the preferred doping technique in semiconductor processing because of the many advantages it has over doping by surface diffusion, including improved control of dose and depth, and the ability to self-align junctions [13, 16]. The principal problem with ion implantation is that during the implantation process the crystalline lattice of the semiconductor is damaged and an annealing (either furnace annealing or rapid themal annealing, RTA) step is necessary to repair the damage as well as to electrically activate the dopant. The kinetics of the damage relaxation process results in a variety of structural morphologies, which will be discussed in depth in the ensuing sections. It suffices to say that the most common defect morphologies consist of dislocations that can form and subsequently dissolve upon annealing.

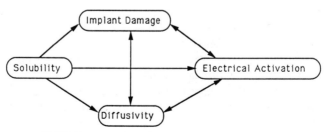

FIGURE 1. Implantation–diffusion interaction matrix.

In recent years much of the interest in ion implantation research has focused on minimizing the amount of dopant redistribution while activating the implanted dopant and removing the damage. There is very good evidence that this is best done by rapid thermal annealing. The use of RTA in the seconds to minutes time regime is being actively investigated because of evidence that indicates that the activation energy for dissolving the secondary defects is greater than the activation energy for impurity diffusion [17-19]. Because the current goal of ultralarge-scale integration technology is to produce defect-free *pn* junctions with the minimum amount of diffusion, the above evidence suggests using shorter time, higher temperature anneals woud be preferred. A serious complication is the fact that much of the inital diffusion is anomalously fast because of the high point defect concentrations in the as-implanted layers. Thus, RTA is also being studied as a method of limiting anomalous diffusion by concentrating on the "seconds" time regime. Laser annealing (in the nanosecond or microsecond time regime) was shown to introduce quenched-in electrically active defect centers as well as being unable to provide a practical large-area heating source [20]. As will be shown, the optimal means of removing the damage upon RTA cannot be realized unless those effects that dominate the defect formation and annealing kinetics are thoroughly understood. Only recently has such an understanding begun to emerge.

B. Ion Implantation Damage

1. Basic Theory

Interaction between an incident ion and the electrons and nuclei of the target substrate account for its loss of kinetic energy upon penetration of the sample. At typical ion implant energies of 25 to 200 keV, light ions such as boron will lose most of their kinetic energy to inelastic energy loss processes, whereas heavier ions such as antimony will lose most of their kinetic energy to elastic collisions between the incoming ion and target atom nuclei. Elastic collisions are primarily responsible for the production of lattice displacements (Frenkel pairs). In order to estimate the number of displaced atoms per incoming ion, Kinchen–Pease [21] formulas were developed. Further developments in displacement calculations led to various methods of estimating the damage density distribution. Brice [22-23] developed a statistical calculation based on extending the transport equations derived by Lindhard *et al.* [24] and Sigmund and Sanders [25], from final ion distributions to energy deposition (damage density)

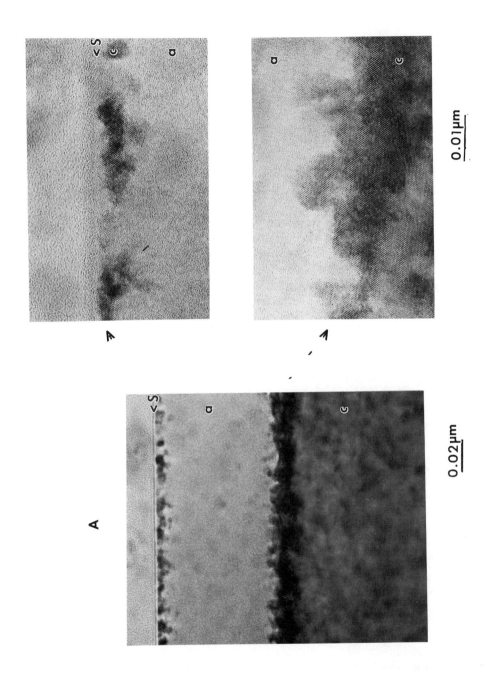

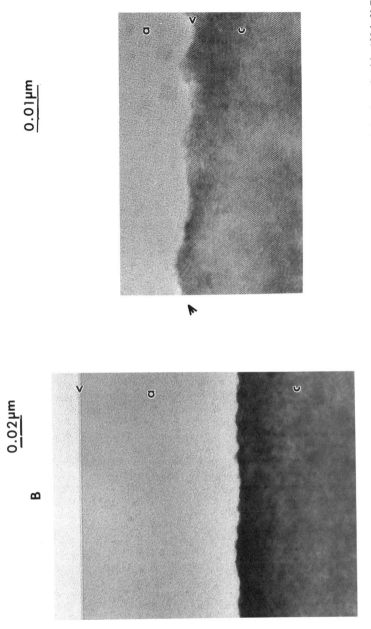

FIGURE 2. Amorphous layer formation: high-resolution cross-sectional TEM micrographs of ⟨100⟩ Si implanted with 100-keV Ge$^+$. (a) Dose = 2×10^{14}/cm^2, resulting in a buried amorphous layer. (b) Dose = 2×10^{15}/cm^2, resulting in a surface amorphous layer.

distributions. The statistical nature of the Brice method makes it much faster than other methods; however, it lacks the flexibility necessary for many studies.

Additional methods for estimating the deposited energy distributions include Monte Carlo type simulations [26, 27] including the TRIM and MARLOWE programs that allow one to accommodate more experimental variables (such as implant angle, etc.). The third method of modeling the implant damage is solving the Boltzmann transport equation (BTE code) [28]. This code, like the TRIM program, is computer-time intensive, but offers the additional flexibility of being able to estimate the net interstitial and vacancy distributions, and has been used to explain damage-related processes in ion-implanted compound semiconductors.

2. Amorphization

For a sufficiently high concentration of damage, it is possible to transform the crystalline silicon into the amorphous phase. The formation of an amorphous phase upon implantation of silicon has been reviewed by Shih [29], and extensive studies of the structure and thermodynamics of implantation-induced amorphous layers in silicon have been conducted in an effort to better understand this material [29–40]. Experimental evidence indicates the crystalline-to-amorphous phase change is a first-order phase transformation and not just a gradual accumulation of point defects [40–46]. The transformation is accompanied by a 1.8% decrease in density [47]. Figure 2A shows cross-sectional transmission electron (TEM) micrographs of a buried amorphous layer produced by 100-keV Ge^+ implantation at a dose of $2 \times 10^{14}/cm^2$, while a dose of $2 \times 10^{15}/cm^2$ (Fig. 2b) results in complete surface amorphization. Also shown in Fig. 2 are high-resolution cross-sectional TEM micrographs of the various amorphous/crystalline (α/c) interfaces. The abruptness of the interface is one indication of the first-order nature of this phase change. Amorphization upon irradiation of semiconductors is believed to be possible because of the highly directional nature of the covalent bonding. However, it has recently been shown that the degree of ionicity cannot solely explain the susceptibility of semiconductors to amorphization [48].

The amount of damage from implantation necessary to convert the crystalline surface to an amorphous state is dependent upon the mass of the implanted species, the water temperature during implantation, as well as the dose rate and implant energy. In an attempt to explain these dependencies, several models have been proposed. These models are usually divided into the heterogeneous and homogeneous nucleation models, which in their purest forms apply to heavy ions at low temperatures and light ions at high implant temperatures, respectively. The heterogeneous model was first proposed by Morehead and Crowder [49] and has been referred to and discussed extensively [12, 14, 29, 41-42, 50]. The model assumes that an amorphous cylinder of radius R is created by each incoming ion and the temperature dependence of amorphization is a result of out-diffusion of vacancies from this cylinder. Experimental evidence for heavier ions confirms the model's premise that amorphous layer formation involves the overlap of amorphous zones [12, 41-42].

The homogeneous nucleation model is useful in explaining light-ion amorphization. This model is based on the premise that when the point defect concentration reaches a critical value the system relaxes to an amorphous state [51]. This model has been further modified by the idea that the creation of small amorphous zones at the end of the light-ion track act as nuclei for the subsequent crystalline to amorphous transformation [29, 41-42]. These modifications may apply to low-temperature (77 K) or high dose rate implementation [29] of boron into silicon; however, amorphization from high-dose boron implantations at room temperature has been reported to be unusual and may be best explained by high-energy recoils [52]. The models above theoretically indicate when the silicon should become amorphous and they correlate qualitatively with electron paramagnetic resonance [53] and Rietherford backscattering RBS [29] data in estimating the temperature dependence of this process. However, the models fail to determine the type (buried or surface) or continuity (isolated pockets, continuous) of the amorphous layer. Understanding the type and continuity of the amorphous layer is crucial in understanding the defects that form upon annealing.

Stein et al. [54] proposed the concept of a critical energy deposition or a threshold damage density. This is the amount of energy deposited into nuclear collisions that is necessary to change the crystalline lattice to an amorphous phase. The threshold damage density can be calculated by comparing the depth of the α/c interface(s) as measured, for example, by cross-sectional TEM [55-57], tapered groove profilometry [26], differential reflectometry [58], etc., with the damage density distribution curve calculated from Brice, Monte Carlo, or BTE simulations. The deposited damage density necessary for amorphous layer formation (the threshold damage density) can be directly read off a graph, as shown in Fig. 3. The calculation method yields the deposited damage in units of electron volts per angstrom and thus must be converted into units of kiloelectron volts per cubic centimeter by multiplying by the dose. The threshold damage density (TDD) is known to be a strong function of ion mass, dose rate, and wafer temperature during the implantation [56, 59-61]. In order to accurately determine

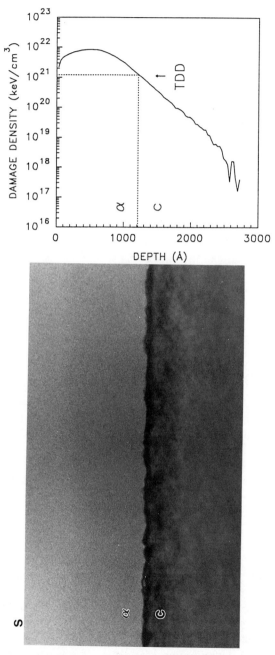

FIGURE 3. Threshold damage density (TDD) determination for 100-keV Ge$^+$, $2 \times 10^{15}/cm^2$ into a water-cooled ⟨100⟩ Si wafer.

Extended Defects from Ion Implantation and Annealing

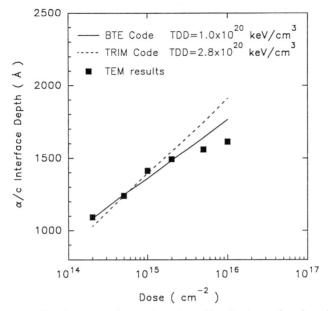

FIGURE 4. Modeling the measured amorphous/crystalline depth as a function of 100-keV Ge$^+$ dose (⟨100⟩ silicon) using a constant TDD from TRIM'88 and BTE simulations [59]. Reprinted with permission of the American Physical Society.

the TDD for a given ion, wafer heating must be suppressed and the TDD fit over a range of energies or doses. Figure 4 shows the measured depth of the α/c interface as a function of dose for a Freon-cooled endstation. Also shown are the results of TRIM'88 Monte Carlo simulations and BTE simulations. The simulations predict reasonably well the α/c depth from a single TDD value, although there is some deviation at the highest doses. This may result from some ion beam induced epitaxial crystallization (IBIEC) occurring during the implant. The absolute value of TDD varies depending on the simulation method, emphasizing the need to be consistent in the calculation method used when attempting to predict amorphous layer depths.

The effective threshold damage density, determined by cross-sectional TEM, as a function of ion mass is shown in Fig. 5 [60]. Brice calculations were used to determine TDD values. The values are for room temperature (Waycool end station [62]) medium-current (265 μA) implant conditions. Use of these data and corresponding damage density distribution simulations allow one to estimate the type (if any) of amorphous layer that will result from implantation of silicon with most species between the energies of 25 and 200 keV. These data only apply to {100} wafers, stabilized at room temperature during the implant and for a single dose rate (265 μA). Even

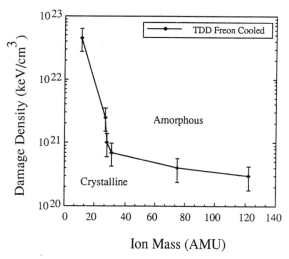

FIGURE 5. Effective threshold damage density (TDD) vs. ion mass for 50–190 keV implants at ~3–5 $\mu A/cm^2$ using a Waycool endstation. Reprinted with permission of Springer-Verlag from Ref. 60.

with these restrictions, the use of the effective threshold damage density still represents the most accurate method of predicting the type and depth of the amorphous layer that is produced and, as will be shown, the types of defects that may result upon annealing.

C. Amorphous Layer Regrowth

Recent work by Roorda *et al.* [63] has shown that the as-implanted amorphous phase exists in a highly disordered state that, upon low-temperature annealing (< 500 °C), transforms to a less defective state with a corresponding heat release measurable by the differential scanning calorimetry. During higher temperature (≥ 500 °C) annealing, solid phase epitaxial (SPE) regrowth of the amorphous layer is initiated via the motion of the amorphous/crystalline interface. There is no evidence of homogeneous nucleation of new crystals within the amorphous layer [64]. The thermal regrowth of the α/c interface has been studied by RBS [65–66], TEM [64, 67–69], TRR [47], and differential reflectometry [70]. The velocity was shown to follow an Arrhenius curve with an activation energy between 2.35 and 2.9 eV depending on a number of factors including wafer orientation ((100) being the fastest) and implanted species (substitutional dopants increase the velocity) [64–76]. Most group III-A and V-A dopants exhibit substitutional incorporation and electrical activation with little redistribution upon solid phase regrowth [77-79]. However, species with lower solubility, such as Pb or F, show marked redistribution upon regrowth [80-81]. There is also evidence that some group V dopants (As and Sb) trap interstitials during SPE, which become a possible point defect source during subsequent annealing [19].

It is also possible to regrow the amorphous layer by IBIEC during the implantation process [82-84]. Elliman *et al.* have shown that regrowth or continued amorphization is controlled by the formation and migration of divacancies at the α/c interface and that these point defects, which are provided by the implantation process, thereby greatly reduce the activation energy for SPE (0.24 eV versus 2.7 eV) [85-86]. As will be discussed below, IBIEC can also play a major role in subsequent extended defect formation.

II. Defect Formation Kinetics

A. Point Defect Sources

The dislocation loops that form and evolve upon annealing ion-implanted silicon are always extrinsic (i.e., interstitial). There are four obvious implantation related sources of interstitials. The first is the incoming ions themselves, which are introduced in numbers greater than the equilibrium vacancy concentration and, therefore, result in an increase in the density of the lattice (see Fig. 6a). A second interstitial source is the recoiling of atoms from the surface into the crystal, or if an amorphous layer is formed, interstitials could be recoiled from the amorphous layer (as shown in Fig. 6b and c). Third, recoils from the amorphous layer have recently been shown to give a better quantitative explanation of the trends in the concentration of interstitials in the dislocation loops than an increased density in the lattice from the incoming ions, as will be discussed [59]. The fourth implantation source is the Frenkel pairs illustrated in Fig. 6d. If the interstitial and vacancy profiles are sufficiently displaced, it is possible for the deep interstitials to feed the extrinsic end-of-range dislocation loops, while the near-surface vacancies could form either stacking fault tetrahedra [87] or else diffuse to the surface. Interstitial sources associated with the dissolution of small crystalline islands near the α/c interface have been proposed [88]. However, this source is not consistent with various experimental observations, which show that as the interface becomes more planar (as shown in Fig. 2 for increasing doses) the concentration of atoms in the loops continues to increase [59-60]. Finally, Pennycook *et al.* [19] have shown that for certain high dose implants, interstitials can be trapped during

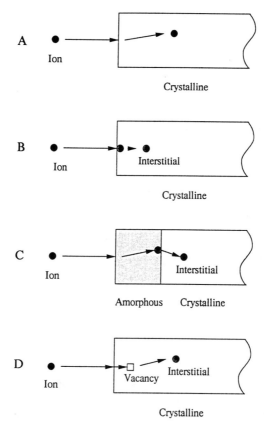

FIGURE 6. Possible sources of excess interstitials upon implantation: (a) the implanted ion; (c) recoils from the surface; (c) recoils from the amorphous layer; (d) separated Frenkel pairs.

solid phase epitaxy by dopant atoms. Upon annealing these interstitials are released, resulting in enhanced diffusion and subsequent formation of extrinsic dislocation loops.

B. TYPES OF EXTENDED DEFECTS

Comparison of the damage density distribution curve with the concept of an effective threshold damage density leads to three possible conditions arising upon implantation. First, no amorphous layer will form if the dose is too low or the effective threshold damage density is too high. With increasing dose or decreasing threshold damage density, a *buried* amorphous layer forms, which can, upon further implantation, grow to the surface, forming a *surface* amorphous layer. These transitions are important in determining the type of secondary defects that arise from annealing. A classification scheme has been developed for the secondary defects that arise or develop during annealing of implanted silicon [55, 60, 89]. This classification scheme successfully groups all secondary defects into five types based upon the origin of the damage as detailed schematically in Fig. 7. The categories are closely related to the morphology and position of the amorphous layer (if any) resulting from implantation. In brief, type I defects form above a critical dose but before amorphization has occurred. Type II defects are the so-called end-of-range defects and form just beyond the α/c interface whenever an amorphous layer is formed. Type III defects are a type of "threading" defect resulting from irregularities in the initial stages of solid phase epitaxial regrowth. Type IV defects arise if a buried amorphous layer is formed and result when the advancing upper and lower α/c interfaces meet upon solid phase epitaxy. Finally, type V defects result when the solid solubility of the implanted dopant is exceeded. It should be stated that this classification scheme appears to apply quite well to implantation energies ≤ 200 keV; however, it is unclear how well it applies to million electronvolt implant energies.

1. Type I Damage

This damage forms when the dose exceeds a critical value and simultaneously no amorphous layer is formed. Another term for this damage is "sub-threshold" damage. Figure 7A shows the schematic relationship between the damage density distribution, the effective threshold damage density, and the location of the type I extrinsic dislocation loops that form upon annealing. Type I defects are typically located at a depth corresponding to the projected range of the implanted species; see R_p in Fig. 8a [60, 90]. The evolution, upon annealing, from a supersaturation of point defects through the formation of intermediate defect configurations to a layer of perfect dislocation loops has been extensively studied [87, 91–103]. Generally, the resulting perfect dislocation loops [104–106] lie on {111} habit planes, are elongated in ⟨110⟩ directions (Fig. 8b) and have a/2[110] type Burgers vectors perpendicular to the long axis of the loops. The aforementioned studies [107] of these elongated perfect dislocation loops indicate that the Burgers vector evolves through a sequence from a⟨100⟩ to a/6⟨411⟩ to a/3⟨111⟩ to a/2⟨110⟩, with the evolutionary steps involving the nucleation of specific Shockley partial dislocations. In addition to the dislocation loops, upon annealing B- and Ne-implanted Si at <700 °C,

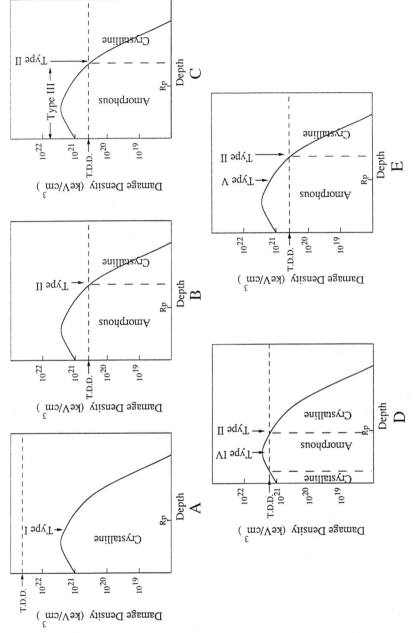

FIGURE 7. Defect classification scheme for the five types of defects. Reprinted with permission of Springer-Verlag, from Ref. 60.

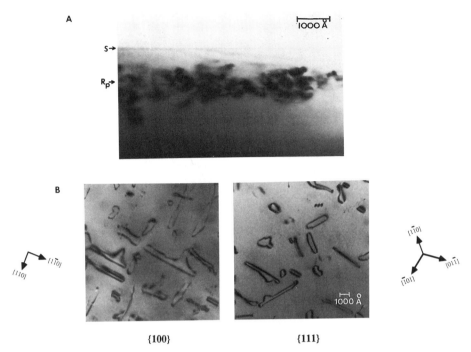

FIGURE 8. Type I defects. (a) Cross-sectional TEM micrograph of 30-keV B$^+$, 2×10^{15}/cm^2 implant after 900 °C 30-min annealing. (b) Plan-view TEM micrographs of 50-keV B$^+$, 5×10^{14}/cm^2 implant after 900 °C 30-min annealing. Reprinted with permission of Springer Verlag, from Ref. 60.

rod-like defects have been observed to grow [101, 108–109]. High-resolution TEM studies indicate these defects lie on (113) or (001) planes and have displacement vectors of $ã/4[116]$ or $a/4[001]$ respectively [96]. They were also shown to be extrinsic. Their dissolution kinetics implies they may contain boron [101] and may also supply point defects for type I defect growth [101, 110]. Other studies indicate that rod-like defects transform to perfect type I dislocation loops upon nucleation of a shear dislocation, $b = 3a/22\langle 332 \rangle$ [103].

The dependence of the concentration of type I defects on the implant dose and energy can be seen in the series of TEM micrographs in Fig. 9. The micrographs show how the concentration of atoms bound by the dislocations varies with dose. There is a strong correlation between the concentration of trapped atoms and the dose, which would imply that the point defect source for the type I defects is associated with the increased lattice density from the implant. This explanation is consistent with both the concentration of atoms bound by the loops and the location of the loops

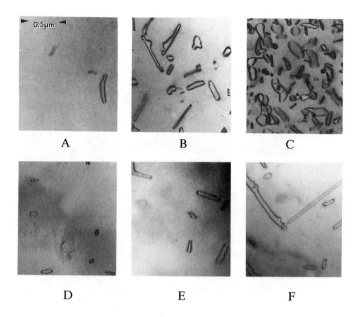

FIGURE 9. Effect of dose and energy on type I defect densities. Plan-view TEM micrographs of room temperature B^+ implants into $\langle 100 \rangle$ Si annealed at 900 °C for 30 min. For 50-keV implants the doses were (a) $2 \times 10^{14}/cm^2$, (b) $5 \times 10^{14}/cm^2$, (c) $1 \times 10^{15}/cm^2$. For $2 \times 10^{14}/cm^2$ implants the energies were (d) 36 keV, (e) 72 keV and (f) 107 keV.

with respect to the projected range. However, the energy dependence implies a second source of interstitials is also significant. The more likely source is Frenkel pair separation, because surface recoils are confined to depths close to the surface and would not tend to form defects centered around the projected range. If Frenkel pair separation is an important source of interstitials, the excess vacancies must be closer to the surface and could either annihilate or form vacancy defects such as stacking fault tetrahedra [87]. It appears probable that both mechanisms are operative. To better understand the source, it is necessary to investigate when type I dislocation loops first form.

For room-temperature implantation, type I loops are typically associated with light ions such as ^{11}B, because the effective threshold damage density is high (i.e., the threshold dose for amorphization is greater than $2 \times 10^{16}/cm^2$) and the critical dose for type I defect formation ($\sim 1 \times 10^{14}/cm^2$) can be exceeded prior to formation of an amorphous layer. This concept of a "critical dose" is an empirical observation based on experiments and reports in the literature [60]. The critical dose apparently represents the point where the concentration of point defects (interstitials) exceeds the homogeneous nucleation barrier for stable extended defect formation. Figure 10 is a graph of dose vs. ion mass, showing the conditions under which type I and type II defects are usually observed. This graph is based on experimental evidence for "room temperature" (Waycool endstation), ≤200-keV implants. The critical dose for type I defects shows a slight energy dependence. For example, B^+ implants at 37 and 50 keV show a critical dose of $2 \times 10^{14}/cm^2$, whereas implants at 72 and 107 keV exhibit a critical dose of $1 \times 10^{14}/cm^2$. Previous reports have shown the critical dose at 190-keV $^{11}B^+$ to be $2 \times 10^{14}/cm^2$ [60].

It has also been found [111] that implanting in a channeling orientation (on axis) reduces the concentration of type I defects and increases the critical dose for defect formation by reducing the displaced atom concentration. Recent results by Raueri et al. [112] indicate that the total displaced atom concentration, as measured by integrating RBS spectra, necessary for type I dislocation loop formation varies from $2.5 \times 10^{17}/cm^2$ for heavy ions (e.g., As) to $3-4 \times 10^{16}/cm^2$ for lighter ions (e.g. Si^+). This displaced atom concentration includes both interstitial dopant atoms and Si atoms. Because the dose ($<5 \times 10^{14}/cm^2$) is much less than these values, the dominant source of interstitials for type I loop formation would appear to be the displaced atoms. However, the total concentration of interstitials bound by the loops is $<5 \times 10^{14}/cm^2$; therefore most of the displaced atoms must

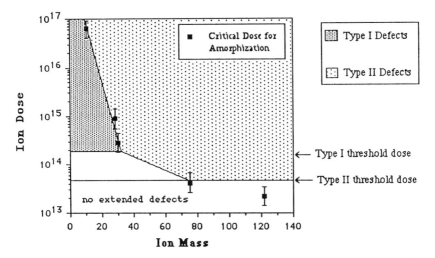

FIGURE 10. Type I and type II defect regimes as a function of ion dose and mass for 50-190 keV room temperature implants. Dose rate was $\sim 5 \mu A/cm^2$ into ⟨100⟩ wafers, subsequently annealed at 900 °C for 30 min.

recombine. As this concentration is comparable to the dose, it is still unclear which source of interstitials—those from Frenkel pairs or those from an increased lattice density—is the most important. It appears likely that both are important. Another method of reducing the displaced atom concentration is by multiple implant and anneal treatments [113]. If the number of displaced atoms never exceeds the critical concentration necessary for type I loop formation, then the total dose of B^+ implanted before loop formation is observed can be increased by a factor of four or possibly more. As is apparent from Fig. 10, for heavier species amorphization begins below the critical dose for type I defect formation; thus these defects are not observed for these species unless very high (million electronvolt) energies are used [113].

2. Type II Damage

Type II defects, also known as "end-of-range" defects, form below the amorphous/crystalline interface in the damaged crystalline material. Figure 7b shows the schematic relationship between the damage density distribution and the effective damage density that results in a surface amorphous layer. The location of the type II defects is also indicated. It is important to stress that type II defects arise only when an amorphous layer is formed by implantation. It is relatively easy to produce an amorphous layer and avoid formation of type III, IV, and V defects by operating at the right implant energy, dose, and temperature. However, type II defects cannot be avoided by paying careful attention to the details of the implantation [111]. Figure 11b and c shows a typical layer of type II defects that formed following a 900 °C anneal. Their location is just below the amorphous/crystalline interface illustrated in Fig. 11a.

The type II defect region, after implantation, is supersaturated with interstitials. This has been confirmed by both TEM observations of the resulting extrinsic defects and X-ray rocking curve studies [114–115]. The type II defects evolve into extrinsic dislocation loops (150–1000 Å in diameter) after regrowth of the amorphous layer at 550 °C and additional higher temperature annealing (i.e., 900 °C, 30 min). The evolution from clusters of excess interstitials to prismatic and Frank (faulted) dislocation loops has been previously reviewed and is believed to be similar to type I defect evolution—via intermediate defect configurations such as {113} stacking faults [91, 116]. By 900 °C most of the type II dislocation loops are either faulted ($a/3[111]$) or perfect ($a/2[110]$) on {111} planes. Figure 11c shows some perfect $a/2[110]$ type II loops that formed after 100-keV, $1 \times 10^{15}/$cm^2 Ge$^+$ implantation and subsequent annealing at 900 °C for 30 min.

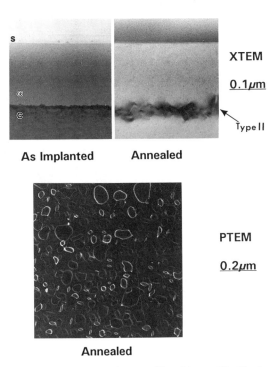

FIGURE 11. Type II defects. 150-keV Ge$^+$, $2 \times 10^{15}/\text{cm}^2$ into $\langle 100 \rangle$ Si using a water-cooled endstation. Annealed 900 °C 30 min. (a) Cross-sectional TEM micrographs. (b) Plan-view TEM micrographs. Reprinted with permission of the American Physical Society, from Ref. 59.

In order to better understand the source of interstitials, the effect of implant dose, mass, and wafer temperature on type II defect formation in {100} wafers has been studied and several interesting trends reported [59–60, 117–118]. Increasing the implant energy or dose results in an increase in the concentration of atoms bound by the type II dislocation loops. It is possible to estimate the number of atoms bound by the dislocation loops using plan-view TEM [59]. It was shown that the concentration of atoms bound by the type II loops is not well modeled by the amount of damage deposited below the α/c interface, thereby excluding Frenkel pair formation below the α/c interface as the principal source. Two other models for the source of the interstitials include the implanted ion concentration below the α/c interface and the recoil of interstitials from the amorphous layer. It was previously reported that the integrated ion concentration explains the trends in atom concentration bound by type II defects with increasing dose for lower mass ions (e.g., Si^+); however, it fails to explain the heavier mass (e.g., Ge^+) results [60]. Gannin and Marwick [117] proposed that the recoil concentration was the dominant source. A subsequent thorough investigation using plan-view and cross-sectional TEM along with TRIM simulations concluded that for Ge^+ implants the integrated recoil model best explains the trends in type II defect concentration with increasing dose and energy [59]. Figure 12 shows the effect of increasing the dose using a Freon-cooled endstation, and Fig. 13 shows the results graphically. The integrated recoil model appears to be better than the transmitted ion model in describing the

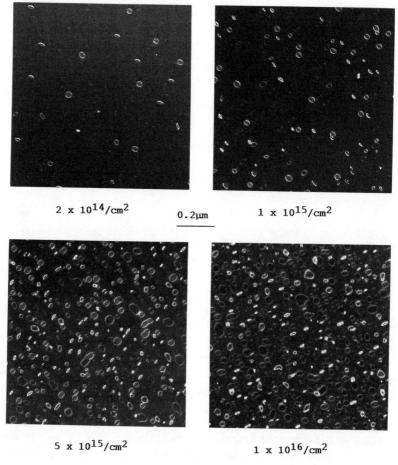

FIGURE 12. Effect of dose on type II defect concentration. 100-kEV Ge into ⟨100⟩ Si annealed 900°C 30 min, plan-view TEM micrographs. Reprinted with permission of the American Physical Society, from Ref. 59.

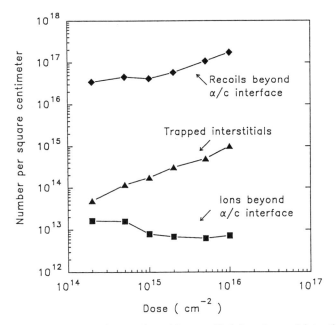

FIGURE 13. Concentration of atoms bound by type II defects (trapped interstitials) as a function of 100-keV Ge dose. Also shown are attempts to model the dose dependence. Reprinted with permission of the American Physical Society, from Ref. 59.

observed trends. In addition, the recoil model can quantitatively account for the atom concentrations in the type II defects, whereas the integrated ion model does not supply a sufficient concentration of interstitials. The slope of the integrated recoil model is not as steep as the measured curve, but despite this failure to accurately model quantitatively the dose dependence of type II defects, the integrated recoil model still appears to be the best model for the source of interstitials for type II defect formation in heavier ion implants.

The wafer temperature is also known to influence the concentration of atoms within the type II loops. It has been shown that reducing the implant temperature lowers the effective threshold damage density (increases the amorphous layer thickness), and also reduces the interstitial concentration for all models, thereby leading to less type II damage [119–121]. It has been shown that reducing the implant temperature to 77 K, where dynamic annealing and point defect migration processes are inhibited, dramatically reduces the type II defect concentration [60]. However, because the threshold damage density appears to be independent of temperature below 77 K, no further decrease in concentration would be expected at lower implant temperatures. For the integrated recoil model, decreasing the effective threshold damage density reduces the concentration of interstitials coming to rest beyond the α/c interface by increasing the thickness of the amorphous layer. The opposite effect has also been observed [59]. If the wafer is allowed to heat up and IBIEC occurs during implantation, then two effects are observed. First, the type II defects formed at a depth slightly deeper than the deepest position that the α/c interface attained, implying that this is the position of the most stable defect nuclei. Second, the concentration of atoms bound by the loops increases, again consistent with an increase in the integrated recoil concentration.

With respect to the behavior of type II defects the strain field associated with a dislocation loop can behave as a gettering site for impurities and implanted dopants [122-128]. The detrimental aspects of such gettering include pipe diffusion and higher leakage currents if the defects are in the depletion region of a *pn* junction. Implantation through an oxide layer or the use of BF_2 has been shown to result in gettering of recoiled oxygen or implanted mobile F atoms by the type II dislocation loops [122, 129-130]. Such gettering can result in severe pinning of the dislocation loops, which restricts the dissolution kinetics upon subsequent higher temperature annealing. Because one of the goals in contact formation is to remove the damage with the least amount of concomitant impurity redistribution, such pinning is not desirable. It may be useful if the damage is to be used as a gettering site due to its increased stability.

As mentioned, formation of an amorphous layer generally results in type II defect formation. Despite this, preamorphization [131-133] is commonly used as a means of preventing the deep channeling tails associated with random channeling of lighter ions such as boron [134]. The surface is usually preamorphized with a high-dose, high-energy Si^+ or Ge^+ implant into which a second, lower energy B^+ implant is done. Regrowth of a preamorphized layer implanted with boron results in electrical activation of the boron, avoids the problem of reverse annealing, and reduces the leakage current [135, 106]. However, enhanced diffusion of boron is observed if the tail of the boron implant overlaps with the type II defect layer [7].

3. Type III Damage

Type III defects are associated with imperfect solid-phase epitaxial regrowth of any amorphous layer produced during implantation. One important aspect of such defects is that they generally can be avoided for {100} wafers. The major forms of type III defects are hairpin dislocations, microtwins, and segregation related defects [81, 135]. Examples of each are shown in

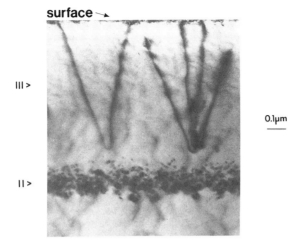

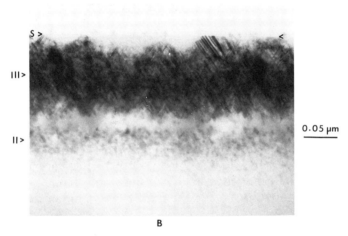

FIGURE 14. Type III defects (a) Hairpin dislocation (III) formation after 50-keV, $1 \times 10^{15}/\text{cm}^2$ BF_2^+ implant annealed 950 °C 10 sec. Reprinted with permission from the Materials Research Society. (b) High density of stacking faults and microtwins (III) after solid phase regrowth (annealed 600 °C 17 min) of 190-keV, $1 \times 10^{13}/\text{cm}^2$ Ar^+ implant followed by a 50 keV, $1 \times 10^{15}/\text{cm}^2$ Ar^+ implant [183].

Fig. 14. Hairpin dislocations were extensively studied by Sands et al. [136] using high-resolution TEM. It was shown that hairpin dislocations nucleate when the regrowing α/c interface encounters small microcrystalline regions that are misoriented slightly with respect to the bulk crystalline material. As the microcrystalline pocket is incorporated into the single-crystal bulk, a hairpin dislocation is nucleated that forms a "V" and propagates with the advancing α/c interface until it intersects the surface, thereby creating two threading dislocations.

To avoid hairpin dislocations, it is necessary either to not form the misoriented microcrystalline regions or to remove them prior to solid phase epitaxial growth of the amorphous layer. Implantation with lighter ions (Si^+ versus Ge^+) results in a broader α/c transition region (2-15 times wider). The broader transition region contains a large number of misoriented microcrystallites and subsequently can result in a greater number of hairpin dislocations [81, 136-138]. This observation has led several groups to propose using Ge^+ instead of Si^+ ions for preamorphization of Si [136, 139]. In addition, if the amount of dynamic annealing that occurs during implantation is reduced either by stabilizing the wafer at room temperature or by cooling to 77 K during implantation, then hairpin dislocation formation can be avoided [60, 136, 140]. Many studies noting hairpin dislocations involved BF_2^+ implants, where the combination of a light amorphizing ion (F) and a significant amount of dynamic annealing led to conditions ideal for type III hairpin defect formation [81, 141].

Using very low temperature annealing (VLTA, 250-450 °C) prior to solid phase epitaxial growth [140, 142], it is possible to reduce the concentration of misoriented microcrystallites by localized rearrangement of the α/c interface. This in turn was shown to result in a reduction in the concentration of hairpin dislocation upon annealing [142]. VLTA is a processing step that allows preamorphization conditions to be expanded while still avoiding type III defects.

There have been a large number of published reports showing a correlation between use of {111} oriented wafer and microtwin formation upon amorphous layer regrowth [42, 67, 109, 119, 143-150]. Various crystallization models have been proposed to explain microtwinning [67, 74, 151], Washburn [67] discusses a model that accounts for both microtwin formation and the slower SPE regrowth rate of the amorphous layer for {111} versus {100} oriented wafers. Their model uses differences in the bonding arrangements of the different {111}, {110}, and {100} surfaces to explain these observations. In the model, formation of two undistorted bonds defines the difference between an atom in the amorphous phase and one that is part of the crystal. On a {001} face an atom can add anywhere and form two undistorted bonds, whereas the {111} face requires simultaneous attachment of three adjacent atoms for the nucleation of a growth step. These three atoms can add in the correct position or with a twin orientation leading to microtwin formation. Microtwin formation does not always occur for {111} oriented wafers [87]; however, it rarely occurs for {100} oriented wafers. Microtwin and staking fault formation during SPE of

{100} wafers generally occurs when the conditions favor type III segregation related defects, for example after Ar^+ or BF_2^+ implants [135, 152].

The third form of type III defect arises from the rejection of an implanted species by the moving α/c interface [81, 135, 141, 152-154]. These defects generally extend from the surface to a depth of 300-500 Å. They have been observed in samples implanted with fluorine (BF_2) and high doses of indium. Because of the low solid solubility of these species and their high diffusivity in amorphous silicon, the ion is rejected into the amorphous layer as long as is kinetically possible. However, at some maximum concentration, the regrowing α/c interface breaks down and defect formation occurs. These defects include misoriented microcrystallites, microtwins [141], fine clusters (15-40 Å in diameter), and small stacking faults [135-153].

4. Type IV Damage

A comparison of the damage density distribution with the effective threshold damage density shows that a buried amorphous layer will be produced prior to a surface amorphous layer, as shown schematically in Fig. 7. The regrowth of this buried amorphous layer results in a layer of defects labeled type IV defects [55], which have a zipper or clamshell appearance [87]. These defects, which form at the junction of the two advancing α/c interfaces, have been extensively studied [60, 87, 126, 155-158, 182]. One of the first comprehensive studies of type IV defects was done by Sadana *et al.* [87], who showed that 950 °C annealing of {111} wafers implanted with $^{31}P^+$ at high energies (120 keV) and low doses ($5 \times 10^{14}/cm^2$) resulted in type IV defect formation. These defects consisted of both perfect and faulted dislocation loops, ~400-600 Å in diameter after 750 °C annealing. Jones *et al.* [60] studied type IV defect formation in {100} silicon and showed that after 550 °C SPE regrowth (see Fig. 15) large (500-2000 Å diameter) shear type (Burgers vector parallel to the habit plane) dislocation loops formed with their habit plane parallel to the surface. The shear nature of these dislocation loops is expected because the anneal temperature was too low for significant dislocation climb to occur.

Further annealing at 900 °C for 30 min results in the movement of some of the large type IV dislocation loops out of the plane parallel to the surface, while other type IV loops remained in the plane parallel to the surface and retained their shear nature. The type II damage evolved into small (100-300 Å diameter) extrinsic perfect and faulted dislocation loops. These results indicate that the type II loops have the same morphology for both {111} [87] and {100} [60] oriented wafers. However, because shear

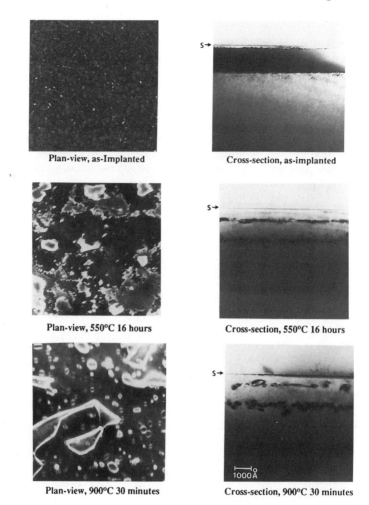

FIGURE 15. Type IV defects for 190-keV As^+, $1 \times 10^{15}/cm^2$ in $\langle 100 \rangle$ silicon. Plan-view TEM micrograph, centered dark field g_{220}, cross-sectional TEM micrographs, bright field g_{220}. Reprinted with permission of Springer-Verlag, from Ref. 60.

dislocation loops parallel to the surface are not possible for (111) wafers, the type IV defects for {111} wafers tend to be smaller than in {100} wafers immediately after regrowth of the amorphous layer ($\leq 600\ °C$). It has also been shown that type IV dislocation loops (as with other types of defects) can getter dopant atoms, thereby reducing the electrical activation of the implanted dopant [87, 160].

Type IV defects are most easily avoided by forming a surface amorphous layer rather than buried layer upon implantation. This is possible by several means, including decreasing the implant energy, increasing the ion dose or mass, and using sequential multiple-energy implants.

5. Type V Damage

Because ion implantation is a nonequilibrium process, it is possible to exceed the solid solubility of a species in a given target material. Upon solid phase epitaxy, dopant concentrations in excess of the retrograde maximum solid solubility can be substitutionally incorporated. Upon further annealing, precipitation occurs and defects arise from either the precipitation/precipitate dissolution process or upon the release of trapped point defects. Extended defects are *not* observed for all implanted species, as will be discussed. The type V defects include precipitates and dislocation loops that are generally centered around the projected range [161]. There have been numerous observations of type V defects by both RBS channeling and TEM experiments on high dose indium [117], antimony [19, 61, 161], arsenic [163], gallium [60], aluminum [160], phosphorus [164], boron [154], BF_2^+ [135, 150, 165], and bismuth [19] implants after annealing. For In, Sb, Ga, Al, and B (preamorphized) the type V defect morphology includes large (50–500 Å) precipitates of the implanted impurity in silicon. These precipitates may be either purely elemental (Sb, In, Ga, Al) silicides (B), or gas bubbles (F from BF_2^+). Some examples of Ga, Sb, and B precipitates are shown in Fig. 16.

In addition to precipitates, several ions (i.e., Sb, As, and Al) have been reported to form type V dislocation loops upon annealing. Pennycook *et al.* [19, 162] have reported that group V-A implants (i.e., Sb, As) result in enhanced diffusion of the dopant as a result of the release of interstitials trapped during solid phase epitaxy. It was proposed that these interstitials are the point defect source for subsequent type V dislocation loop formation. For heavier group V-A implants (e.g., Bi) the lack of type V dislocation loop formation is attributed to insufficient interstitial trapping. On the basis of this model, group III-A implants do not trap interstitials, therefore the diffusion is less enhanced and type V dislocation loops are not observed. Results published by Jones *et al.* [60] include several observations that are not entirely consistent with this model. First, high-dose Al implants (group III-A) resulted in type V dislocation network formation, implying interstitials might be trapped by some group III-A elements. Second, phosphorus implants exhibited no type V dislocation loop formation, which, depending on diffusivity measurements, may also be explained in the same manner as the Bi results. These results imply that the interstitial phenomina is probably more complex than originally proposed.

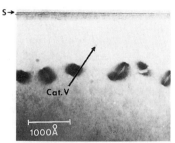

100 keV ^{70}Ga 1 x 10^{15}/cm^2, {100}
Annealed 900°C 2 hours

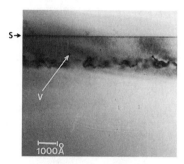

190 keV ^{122}Sb 2 x 10^{15}/cm^2, {100}
Annealed 900°C 30 minutes

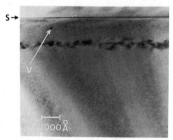

17 keV ^{11}B 1 x 10^{16}/cm^2 into {100} preamorphized Si,
(70 keV + 30 keV ^{28}Si, 5 x 10^{15}/cm^2 each)
Annealed 900°C 1 hour

FIGURE 16. Precipitate forms of type V defects. Reprinted with permission of Springer-Verlag, from Ref. 60.

Extended Defects from Ion Implantation and Annealing

In a separate, detailed study of high-dose (100 keV, $\leq 5 \times 10^{15}/\text{cm}^2$) As^+-implanted silicon, type V defect (half-loop dislocation) formation kinetics were investigated by Jones et al [60], and a summary of their model for type V half-loop dislocation formation and growth follows. Upon recrystallization of the amorphous layer at 550 °C, most of the implanted arsenic is electrically active. At the same time, according to Pennycook's model, silicon interstitials are trapped by substitutional arsenic [19]. Upon isothermal annealing at 900 °C the following sequence occurs. For times less than 1 min, interstitials are released and enhanced diffusion results in clustering or precipitation. This is confirmed by a decrease in electrical activation. TEM and channeling results indicate the atomic position of As is nearly substitutional and the clusters are too small to be observed by even high-resolution TEM unless the concentration is significantly increased [166]. These clusters, which may be as small as dimers and trimers [167–168], may also trap some of the released interstitials. After 2 min at 900 °C declustering begins, and after 4 min type V defects are observed, as shown in Fig. 17.

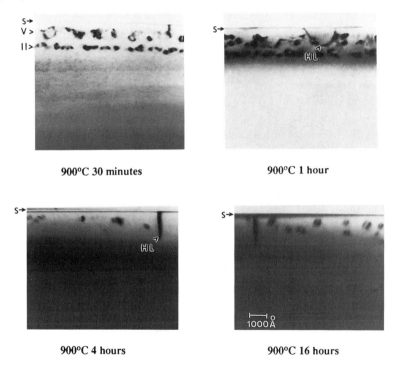

FIGURE 17. Cross-sectional TEM micrographs of type V half-loop dislocation formation. 100-keV As^+, $1 \times 10^{16}/\text{cm}^2$ into $\langle 100 \rangle$ silicon. Reprinted with permission of Springer-Verlag, from Ref. 60.

The type V defects may exist after shorter anneals but this was the shortest time investigated. After 30 min elongated loops that are the type V half-loop nuclei are observed as shown in plan-view TEM in Fig. 18. Upon annealing for 1 hour, half-loop dislocations form when the elongated type V loops intersect the surface. The half-loops are extrinsic in nature with Burgers vectors of $a/2\langle 110 \rangle$ perpendicular to the $\{110\}$ type habit plane and parallel to the surface. For annealing times between 2 and 72 hours, the half-loops were observed to grow.

Quantitive analysis of the concentration of atoms bound by *all* of the dislocation loops that are not half-loops indicates that the dissolution of the other dislocation loops is *not* responsible for the observed half-loop growth. However, it was shown that there is a strong correlation quantitatively between the electrical activation of the As (from Hall effect measurements)

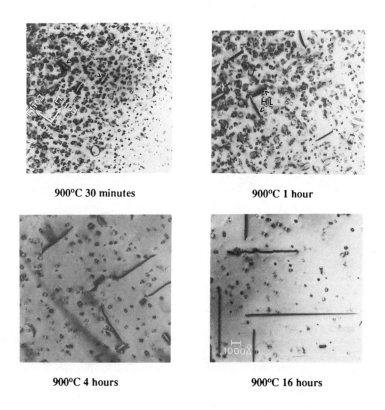

FIGURE 18. Plan-view TEM micrographs of type V half-loop dislocation growth. 100-keV As^+, $1 \times 10^{16}/cm^2$ in $\langle 100 \rangle$ silicon. Reprinted with permission of Springer-Verlag, from Ref. 60.

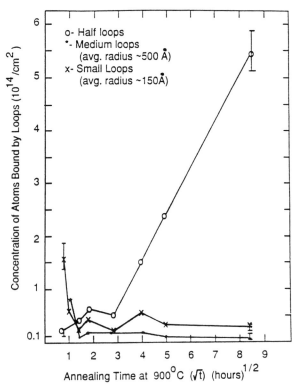

FIGURE 19. Effect of annealing time on the concentration of atoms bound by the various types (II and V) of defects. 100-keV As^+, $1 \times 10^{16}/cm^2$ in ⟨100⟩ silicon. Reprinted with permission of Springer-Verlag, from Ref. 60.

and the concentration of atoms bound by the half loop dislocations (see Fig. 19). This implies that the source of the point defects resulting in half-loop growth is associated with the declustering of arsenic precipitates via either the formation of interstitials or the release of retrapped interstitials upon dopant activation.

Type V defects are therefore shown to be either simply precipitates or precipitates plus dislocation loops, depending on the species and furnace annealing. Finally, the difference between rapid thermal annealing at 1100 °C and furnace annealing at 900 °C is shown in Fig. 20 for high-dose ($5 \times 10^{15}/cm^2$) 100-keV As^+ implanted silicon. Taking into account the concentration-enhanced diffusivity at the high peak concentration ($\sim 1 \times 10^{21}/cm^3$), annealing at 900 °C for 120 min and 1100 °C for 64 sec results in equivalent diffusion lengths ($2\sqrt{Dt}$) at the peak concentration.

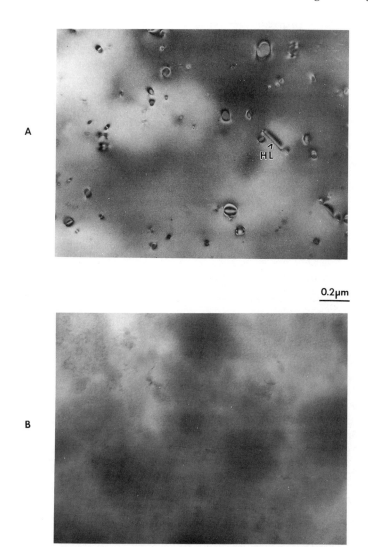

FIGURE 20. Difference between (a) furnace annealing 900 °C 120 min and (b) rapid thermal annealing 1100 °C 64 sec. 100-keV As^+, $5 \times 10^{15}/cm^2$ implants into ⟨100⟩ silicon.

Figure 20 shows that after 120 min at 900 °C half-loop dislocations have formed. Figure 18 showed that these dislocation continued to grow at 900 °C. However, after 64 sec at 1100 °C all type II and type V have dissolved (or never formed). In fact all defects dissolved after only *two* seconds at 1100 °C. These results indicate that for high-dose As^+ implants it is possible to prevent the formation of half-loop dislocations by use of rapid thermal processing. Thus, in addition to *dissolving* defects prior to significant dopant diffusion, RTA can be useful in *preventing* defect formation.

III. Defect Annealing Kinetics

Most ion implantation damage can be avoided or reduced by adjusting the implant conditions. Type I damage can be avoided by either decreasing the dose below the critical level ($\sim 2 \times 10^{14}/\text{cm}^2$) necessary for extended defect formation or, when possible, increasing the damage density above the threshold damage density, thereby creating an amorphous layer. For the case of light ions such as B^+, where self-amorphization is not feasible, implantation into a preamorphized substrate can be used to avoid type I defects. The option to amorphize then shifts the defect control burden to type II defects, because they generally form whenever an amorphous layer is produced. Because amorphization is desirable for many reasons, including reduction of random channeling tails, type II defect annealing kinetics has been the subject of much study, which will be further elaborated on below.

Type III microtwin defects can be avoided by using (100) wafers instead of (111) wafers and by avoiding implantation of insoluble species such as Ar^+ as previously discussed. It may be possible to remove microtwins by a two-step annealing cycle [143], although other studies [60, 147, 169] show that upon annealing at higher temperatures (900 °C) the microtwin defects evolve into a tangle of dislocations [147]. These dislocations are difficult to dissolve even with furnace annealing of 1100 °C [169]. Thus microtwins should be avoided if possible. The type III hairpin dislocations can be avoided by using heavier ions (Ge) or low implant temperatures (77 K) for preamorphization or VLTA [142] to remove the misoriented microcrystallites. When hairpin dislocations have formed they were observed to be less stable than type II defects [170] and are removed by glide toward the surface within the glide cylinder defined by the inclined Burgers vector and the dislocation line direction [171].

Type IV defects can be avoided by producing surface amorphous layers. When they do form they have been shown to be less stable than type II defects [170]. Type V defects can be avoided by not exceeding the solid solubility (when feasible) or by using RTA, as previously discussed. Thus type III, IV, and V defects can, in principle, be readily avoided or dissolved. However, for doses above $1 \times 10^{14}/\text{cm}^2$ in general, either type I or type II defects will form upon annealing unless one thoroughly understands the defect formation process and the options for defect suppression or annihilation.

A. Type I Defect Annealing Kinetics

For type I defects formed by B^+ implants at doses below $1 \times 10^{15}/cm^2$ the dissolution kinetics has been studied by Mader [172]. Figure 21 shows an Arrhenius plot of the amount of time necessary to eliminate type I defects as a function of annealing temperature for both 35- and 350-keV $^{11}B^+$ implants at a dose of $4 \times 10^{14}/cm^2$. The activation energy for the process is shown to be approximately 5 to 5.4 eV, which is on the order of the activation energy for self-diffusion in Si. This is consistent with the dissolution occuring by positive climb as will be discussed further in the section on type II defect dissolution. Recall from Fig. 9 that the defect stability will increase with either the energy or the dose, because they both increase the concentration of atoms bound by the loops. If dissolution occurs by climb, then an increase in the atoms bound by the dislocations (either by increasing the energy or the dose) would be expected to increase the annealing time for total defect elimination. From the graph at a dose of $4 \times 10^{14}/cm^2$ (35 keV) the defects dissolve after 47 sec at 1100 °C. When the energy is kept constant and the dose is doubled defects are still observed after 64 sec, consistent with the model. In addition, Ganin and Marwick [117] show that decreasing the net concentration of atoms bound by the

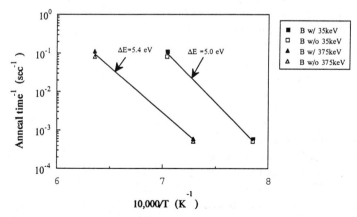

FIGURE 21. Dissolution of type I defects upon annealing 35- and 375-keV B^+ implants at $5 \times 10^{14}/cm^2$. Courtesy of S. Mader (unpublished).

type II loops decreases the annealing time necessary for complete defect elimination, again consistent with the model.

The total concentration of atoms bound by the type I extrinsic dislocation loops was shown in Fig. 9 to be approximately equal to the dose [60, 67, 104]. At high doses, above $2 \times 10^{15}/cm^2$, the concentration of atoms bound by the extrinsic dislocation loops exceeds the concentration of atoms in a monolayer of the {111} plane of silicon ($\sim 1.4 \times 10^{15}/cm^2$), whereupon dislocation network formation becomes possible upon annealing, via dislocation–dislocation interactions [104–105, 109]. The network structure was very stable, and no dissolution curves on this defect morphology have been reported to the authors' knowledge.

B. Type II Defect Annealing Kinetics

For technological reasons, especially shallow p-type junctions, light ions such as B^+ are generally implanted into preamorphized surfaces. Most other species of technological interest (P, As, etc.) implanted at room temperature are self-amorphizing at higher doses ($\geq 2 \times 10^{14}/cm^2$); thus type II defects are the most common and in many cases the most stable extended defects. For this reason the effects of a number of factors on the defect stability have been investigated and are listed as follows:

1. Annealing ambient/surface chemistry (i.e., oxidation or silicidation).
2. Initial defect concentration.
3. Surface proximity.
4. Stress at the mask edge.
5. Implant species, concentration and solubility.

Many of these effect overlap, which has made careful experimental design imperative. Without the influence of one or more of these factors the type II dislocation loops are quite stable. For example, type II defects 1000 Å deep formed by either Si^+ and Ge^+ implants have been observed to still exist after 24 hours at 1075 °C [173]. This would imply that the thermodynamic instability of a dislocation in the crystal is an insufficient driving force to overcome the kinetic limitations to defect dissolution. Although the dissolution process can be accelerated by optimizing one of the aforementioned factors, it has been shown [170] that any initial increase in the concentration of atoms bound by the type II dislocation loops will retard the defect dissolution process.

1. Process-Induced Point Defect Control

Because the source of interstitials for type II defect nucleation and growth are process related, Ajmera and Rozgonyi [112] have systematically reduced the interstitial concentration by lowering the implant energy and dose, as well as by suppressing the oxidation of the surface during activation annealing. They were then able to achieve complete type II defect elimination. The instability of the type II defects was explained via a model that included both the minimization of the local point defect concentration and optimization of the role of the free surface. The results can be qualitatively explained by Fig. 22, where C_I^B is defined as the bulk equilibrium concentration of interstitials. Upon implantation that results in amorphization, excess interstitials are introduced via mechanisms discussed previously in Fig. 6. After annealing in a reducing ambient, the excess interstitial concentration above C_I^B reaches a maximum at a depth C_I^L corresponding to the location of the type II dislocation loops. If an oxidizing ambient is used, then the injection of interstitials from the Si/SiO$_2$ interface results in a profile represented by C_I^O, which serves to stabilize or even expand the type II dislocation loops. The model suggested that reducing the implant energy and dose would achieve two results. First, the excess interstitial concentration C_I^L would be reduced (possibly below the equilibrium value, C_I^B). Second, the loops would be closer to the surface where enhanced defect elimination might occur if the surface provided a sink for interstitials, or if image forces could promote dislocation loop elimination via glide [174]. It was demonstrated

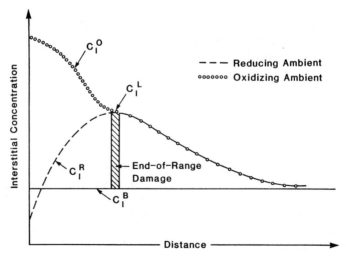

FIGURE 22. Schematic of self-interstitial concentration profiles near oxidizing and reducing silicon surfaces. Reprinted with permission of the American Institute of Physics, from Ref. 173.

that reduction of the Ge$^+$ implant energy from 150 to 40 keV and the dose from 9×10 to $2 \times 10^{14}/cm^2$ resulted in the depth of the α/c boundary being reduced from 1100 to 700 Å, while the concentration of interstitials was significantly lowered. Thus, upon rapid thermal annealing at 1050 °C for 10 sec in a nonoxidizing ambient, the type II defects for the lower energy, lower dose sample were completely removed, whereas the higher energy higher dose sample still had a well-formed layer of type II dislocation loops.

Although this experiment showed that reducing the implant energy and dose was important in defect elimination, it did not indicate which phenomenon—the reduction in point defect concentration or the effect of the proximity of the surface—was more important. This question was answered by an experiment conducted by Ganin and Marwick [113]. Silicon wafers were amorphized upon implantation with In. The implant energies were 40 and 200 keV. In both cases the dose was adjusted to produce a surface amorphous layer. For the 40 and 200 keV implants the amorphous layers were 560 and 1600 Å thick, respectively. Anodic oxidation was used to reduce the amorphous layer thickness for the 200-keV implanted sample to ~560 Å. Upon annealing at 950 °C for 10 sec in an argon ambient, the type II defects in the 40-keV implanted sample dissolved whereas the type II defects for the thinned 200-keV implanted sample were quite stable. Thus, with respect to the type II dislocation loop elimination process, at these depths it appears that the reduction in the interstitial point defect concentration, affected by changing the implant conditions, is more important than the surface proximity effects. For additional discussion see Chapter 6.

Finally, it has been found [175] that type II interstitial defects will dissolve at relatively low temperatures, less than 1000 °C, if vacancies are injected during silicidation processing. Subsequent work [176] extended the silicidation procedure to selectively enhance the diffusion of a buried Sb layer in the direction of the silicided surface. The silicidation processing is discussed in more detail in Chapter 7 by Osburn.

2. Species, Concentration, and Solubility Effects

It has been shown that the effect of implant species on defect stability is quite dramatic. If the implant dose is such that the solid solubility is exceeded at the annealing temperature, and the concentration of type II defects is minimized (e.g., by using a Freon-cooled endstation), then for certain species (As, P, and Ga) repid dissolution of the type II defects is observed [173], Fig. 23 illustrates this phenomenon. Cross-sectional TEM confirmed the type II defects form at depths of ~1100 Å (deep enough to avoid any surface proximity effects). After annealing for 60 min at 900 °C,

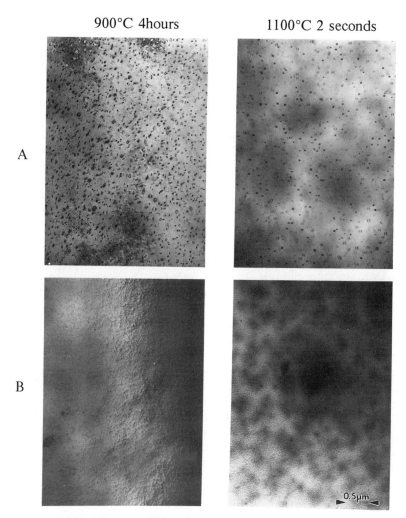

FIGURE 23. Effect of solubility of type II defect stability. 50-keV P^+ implants into $\langle 100 \rangle$ silicon. (a) $1 \times 10^{15}/cm^2$, below solid solubility, (b) $1 \times 10^{16}/cm^2$, above solid solubility. Reprinted with permission of the American Institute of Physics, from Ref. 173.

type II defects are observed for both the 50-keV, $1 \times 10^{16}/cm^2$ P implant and the 100-keV, $1 \times 10^{15}/cm^2$ Ga implant. After 8 hours at 900 °C all of the type II loops have completely dissolved. It was also shown that when the peak concentration was decreased below the solid solubility limit (by decreasing the dose) then the type II defects did not dissolve, even after 72 hours at 900 °C.

It is possible to determine the activation energy for removal of the type II defects by producing an Arrhenius plot ($t_D = t_D^0 \exp(E_a/kT)$) of the inverse of the time necessary to remove all category II defects versus the annealing temperature. Seidel et al. [17] first reported an activation energy of ~5 eV for removal of type II defects in high-dose As$^+$ implants. Similar Arrhenius plots were made for P and Ga implants above solid solubility and are shown in Fig. 24 [173]. An activation energy of ~5 ± 0.5 eV was observed for all of the species. If category II dislocation loops are composed primarily of Si atoms, then the activation energy of ~5 eV can be explained in the context of self-diffusion away from the loops or diffusion of vacancies to the loops [177]. The y-intercept or t_D^0 reflects the concentration of vacancies or the magnitude of the interstitial sink that is enhancing the type II dislocation positive climb rate. The shift in t_D^0 appears to imply that P accelerates the defect dissolution slightly more than As and Ga. A similar shift in t_D^0 has also been noted upon increasing the dose for As [170] and Ga [178] investigations. It should be emphasized that this enhanced type II defect dissolution was only reported for As, P, and Ga implants. Other species such as Sb and Al did not show this effect when the implanted concentrations exceeded the solubility limit. For species in which the solubility

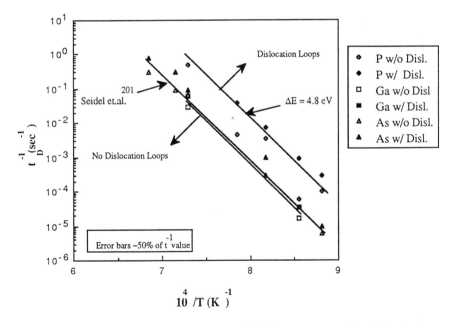

FIGURE 24. Arrhenius curve of type II defect dissolution upon annealing. All implants are doses resulting in peak concentrations above solid solubility. Reprinted with permission of the American Institute of Physics, from Ref. 173.

limit could not be exceeded (Si^+ and Ge^+) no enhanced dissolution was observed [170]. Also, when the implants were done into a preamorphized substrate complete dissolution was not realized for any of the species including B [170, 179].

Because this phenomenon appears to be confined to the time-temperature regime when the precipitates are dissolving [173], it is postulated that the precipitates are acting as vacancy sources or interstitial sinks during their dissolution. Precipitation of SiP has been studied extensively by Bourret and Schroter [180] for high concentration (exceeding the solid solubility) P predeposition into Si. Their results indicate that precipitation is an excellent source of interstitials. This supports the model that the dissolution of precipitates is a source of vacancies or is an interstitial sink. The observation that preamorphized samples do not exhibit the same dissolution kinetics implies that the position of the type II defects to the precipitates or their position within the diffusion profile may also be an important factor.

An important point made by Sedgwick [181] and Seidel et al. [17] is that the activation energy for defect dissolution (~ 5 eV) is greater than the activation energy for diffusion of the major dopants (i.e., ~ 4 eV for As). Thus, effective removal of the defects with the minimum amount of dopant redistribution would occur at high temperatures for short times, that is, rapid thermal annealing (RTA). In addition it was previously noted that RTA at 1100 °C is necessary to avoid half-loop formation in high-dose As implants. Thus in order to completely dissolve both the type V and type II defects for As implants and to minimize dopant diffusion it is best to use rapid thermal processing conditions (1100 °C, 2 sec).

IV. Summary

A systematic classification of defects that arise from ion implantation and annealing based on the as-implanted morphology and the solubility has been presented. Five types of defects were discussed as well as point defect sources for these defects. It was shown that there is a strong dependence of type I defects on the implant energy and dose and that their stability also depends on the implant conditions. Type II defects were shown to form whenever amorphization occurs and their concentration also depends on the implant conditions. Type II defect stability is shown to be influenced by the surface proximity, implant species and solubility, and surface chemistry/annealing ambient. Type III defects are shown to have several different morphologies including hairpin dislocations, microtwins, and stacking faults. Type IV defects are shown to form as either large or small dislocation loops depending upon the wafer orientation. Both type III and IV defects are easily avoided by adjusting the implant conditions. When they cannot be avoided they are generally less stable than type II defects. Finally type V defects form when the solubility has been exceeded and are generally precipitates. For certain species, dislocation loop formation is also possible when interstitials trapped by the high-concentration species are released. Half-loop dislocations observed to form upon annealing high-dose ($\geq 5 \times 10^{15}/cm^2$) arsenic implants nucleate from type V defects and are avoided via rapid thermal annealing at 1100 °C.

References

1. J. Fletcher and G. R. Booker, *Electrochemical Society Proceedings*, 1155 (1980).
2. K. H. Nicholas, *J. Phys. D.* **9**, 393 (1977).
3. C. Bull, P. Ashburn, G. R. Booker, and K. H. Nicholas, *Solid State Electronics* **22**, 95 (1979).
4. M. Finetti, R. Galloni, and M. Mazzone, *J. Appl. Phys.* **50**, 1381 (1979).
5. D. K. Sadana, J. Fletcher, and G. R. Booker, *Elect. Let.* **13**, 632 (1977).
6. P. Ashburn, C. Bull, K. H. Nicholas, and G. R. Booker, *Solid State Electronics* **20**, 731 (1977).
7. Y. Kim, H. Z. Massoud, U. M. Gosele, and R. B. Fair, MCNC Technical Report TR90-30 (1990).
8. Y. M. Kim, G. Q. Lo, H. Kinoshita, D. L. Kwong, H. H. Tseng, and R. Hance, *J. Electrochem. Soc.* **138**, 1122 (1991).
9. G. Carter and W. A. Grant, *Ion Implantation of Semiconductors*, Halstead Press, New York (1976).
10. G. Alvager and N. J. Hansen, *Rev. Sci. Instrum.* **33**, 567 (1962).
11. J. F. Ziegler, *Ion Implantation: Science and Technology*, Academic Press, New York (1984).
12. J. F. Gibbons, *Prof. IEEE,* 1062 (1972).
13. K. H. Nicholas, *J. Phys. D.* **9**, 939 (1976).
14. *Ion Implantation and Beam Processing* (J. S. Williams and J. M. Poate, eds.), Academic Press, New York (1984).
15. S. Wolf and R. N. Tauber, *Silicon Processing for the VLSI Era*, Lattice Press, Sunset Beach, CA (1986).
16. S. Prussin, private communication.
17. T. E. Seidel, D. J. Lischner, C. S. Pai, R. V. Knoell, D. M. Maher, and D. C. Jacobson, *Nucl. Inst. Meth. Phys. Res. B* **7/8**, 251 (1985).
18. D. Baither, R. Koegler, D. Panknin, and E. Wieser, *Phys. Stat. Sol. (a)* **94**, 767 (1986).
19. S. J. Pennycook, R. J. Culbertson, and J. Narayan, *J. Mater. Res.* **1**, 476 (1986).
20. L. C. Kimerling and J. L. Benton, in *Laser and Electron Beam Proc. of Materials* (C. W. White and P. S. Pearcy, eds.), Academic Press, New York, p. 385 (1980).
21. G. H. Kinchin and R. B. Pease, *Rep. Prog. Phys.* **18**, 2 (1955).
22. D. K. Brice, *Ion Implantation Range and Energy Deposition Distribution*, Plenum Press, New York, p. 1 (1975).
23. D. K. Brice, *J. Appl. Phys.* **46**, 3385 (1975).

24. J. Lindhard, M. Scharff, and H. E. Schiott, *Mat. Fys. Medd. Dan. Vid. Selsk.* **33**, 1 (1963).
25. P. Sigmund and J. B. Sanders, *Proc. Int. Conf. Appl. of Ion Beams to Semiconductor Technology,* (P. Glotin, ed.), Centre d'Etudes Nucleaires, Genoble, p. 215 (1967).
26. J. B. Biersack and L. G. Haggmark, *Nucl. Inst. Meth.* **174**, 257 (1980).
27. J. F. Ziegler, J. P. Biersack, and U. Littmark, *The Stopping and Range of Ions in Solids*, Pergamon Press, New York, p. 1 (1984).
28. L. A. Christel, J. F. Gibbons, and S. Mylroie, *J. Appl. Phys.* **51**, 6176 (1980).
29. Y. Shih, *The Formation of Amorphous Si by Light Ion Damage* Ph. D. Diss., University of California, Berkeley (1985).
30. D. E. Polk, *J. Non-Cryst. Solids* **5**, 365 (1971).
31. W. H. Zachariasen, *J. Am. Chem. Soc.* **54**, 3841 (1932).
32. D. E. Polk and D. S. Bondreaux, *Phys. Rev. Lett.* **31**, 92 (1973).
33. S. Veprek, Z. Iqbal, H. R. Oswald, F. A. Sarott, J. J. Wagner, and A. P. Webb, *Solid St. Comm* **39**, 509 (1981).
34. C. Kittel, *Introduction to Solid State Physics* 4th ed., Wiley, New York (1971).
35. K. P. Jain, A. K. Shukla, R. Ashokan, S. C. Abbi, and M. Balkanshi, *Phys. Rev. B.* **32**, 6688 (1985).
36. D. Kirillov, R. A. Powell, and D. T. Hodul, *Proceedings of Fall Mat. Res. Soc. Meeting*, Symposium B (1985).
37. S. Veprek, Z. Iqbal, and F. A. Sarrott, *Phil. Mag. B* **45**, 137 (1982).
38. M. H. Brodsky and R. S. Title, *Phys. Rev. Lett.* **23**, 581 (1969).
39. M. H. Brodsky, R. S. Title, K. Weiser, and G. D. Pettit, *Phys. Rev. B* **1**, 2632 (1970).
40. J. Narayan, D. Fathy, O. S. Oen, and O. W. Holland, *Mat. Let.* **2**, 211 (1984).
41. J. Washburn, C. S. Murty, D. K. Sadana, P. Byrne, R. Gronsky, N. Cheung, and R. Kilaas, *Nucl. Inst. Meth.* **209**, 345 (1983).
42. J. Washburn, C. S. Murty, D. K. Sadana, P. Byrne, R. Gronsky, N. Cheung, and R. Kilaas, *Int. Conf. on Ion Beam Modification of Materials,* Grenoble (1982).
43. J. Narayan, D. Fathy, O. S. Oen, and O. W. Holland, *J. Vac. Sci. Tech. A* **2** (3), 1303 (1984).
44. Z. Iqbal and S. Verprek, *J. Phys. C.* **15**, 377 (1982).
45. T. Kamiya, M. Kishi, A. Ushirokawa, and T. Katoda, *Appl. Phys. Lett.* **38**, 377 (1981).
46. W. E. Spear, G. Willeke, P. G. LeComber, and A. G. Fitzgerald, *Physical B, C* **117-118**, 908 (1983).
47. J. M. Poate, S. Coffa, D. C. Jacobson, A. Polman, J. A. Roth, G. L. Olson, S. Roorda, W. Sinke, J. S. Custer, M. O. Thompson, F. Spaepen, and E. Donovan, *Nucl. Inst. Meth, B* **55**, 533 (1991).
48. K. S. Jones and C. J. Santana, *J. Mater. Res.* **6**, 1048 (1991).
49. F. F. Morehead, Jr. and B. L. Crowder, *Rad. Eff.* **6**, 27 (1970).
50. G. Carter and W. A. Grant, *Ion Implantation of Semiconductors*, Halstead Press, New York, p. 109 (1976).
51. J. F. Gibbons, Lectures on Ion Implantation and Proton Enhanced Diffusion, University of Tokyo, Tokyo, Japan (Jap. Soc. Appl. Phys.), Aug (1977).
52. K. S. Jones, D. Sadana, S. Prussin, J. Washburn, E. R. Weber, and W. Hamilton, *J. Appl. Phys.* **63**, 1414 (1988).
53. F. F. Morehead, B. L. Crowder, and R. S. Title, *J. Appl. Phys.* **43**, 1112 (1972).
54. H. J. Stein, F. L. Vook, D. K. Brice, J. A. Borders, and S. T. Picraux, *Rad. Eff.* **6**, 19 (1970).
55. S. Prussin and K. S. Jones, *Materials Issues in Silicon Integrated Circuit Processing* (M. Strathman, J. Stimmel, and M. Wittmer, eds.), Proc. Mat. Res. Soc., 71 (1986).

56. W. P. Maszara and G. A. Rozgonyi, *J. Appl. Phys.* **60**, 2310 (1986).
57. A. Claverie, C. Vieu, J. Faure, and J. Beauvillain, *J. Appl. Phys.* **64**, 4415 (1988).
58. R. E. Hummel, Wei Xi, P. H. Holloway, and K. A. Jones, *J. Appl. Phys.* **63**, 2591 (1988).
59. K. S. Jones and D. Venables, *J. Appl. Phys.* **69**, 2931 (1991).
60. K. S. Jones, S. Prussin, and E. R. Weber, *Appl. Phys. A.* **45**, 1 (1988).
61. J. Narayan and O. W. Holland, *J. Electrochem. Soc.* **131**, 2651 (1984).
62. Varian/Extrion Div., Gloucester, Mass. 01930.
63. S. Roorda, J. M. Poate, D. C. Jacobson, D. J. Eaglesham, B. D. Dennis, S. Dierker, W. C. Sinke, and F. Spaepen, *Solid State Commun.* **75**, 197 (1990).
64. B. Drosd and J. Washburn, *J. Appl. Phys.* **51**, 4106 (1980).
65. L. Cspregi, E. F. Kennedy, T. J. Gallagher, J. W. Mayer, and T. W. Sigmon, *J. Appl. Phys.* **48**, 4234 (1977).
66. L. Cspregi, E. F. Kennedy, T. J. Gallagher, J. W. Mayer, and T. W. Sigmon, *J. Appl. Phys.* **49**, 3906 (1978).
67. J. Washburn, *Defects in Semiconductors* (J. Narayan and T. Y. Tan, eds.), North Holland Publishing, Amsterdam, p. 209 (1981).
68. C. W. Neih and L. J. Chen, *J. Appl. Phys.* **63**, 575 (1988).
69. C. W. Neih and L. J. Chen, *J. Appl. Phys.* **60**, 3546 (1986).
70. S. Weixi Feng, Ph.D. Diss., University of Florida (1991).
71. S. S. Lau, *J. Vac. Sci. Technol.* **15**, 1656 (1978).
72. L. Csepregi, E. F. Kennedy, J. W. Mayer, and T. W. Sigmon, *J. Appl. Phys.* **49**, 3906 (1978).
73. M. Tamura, K. Yagi, N. Sakudo, K. Tokiguti, and T. Tokuyama, *Rad. Eff.* **48**, 109 (1980).
74. R. Drosd and J. Washburn, *J. Appl. Phys.* **53**, 397 (1982).
75. L. Csepregi, E. F. Kennedy, T. J. Gallagher, J. W. Mayer, and T. W. Sigmon, *J. Appl. Phys.* **48**, 4234 (1977).
76. L. J. Chen, C. W. Nieh, and C. H. Chu, *Solid State Phen.* **1, 2**, 45 (1988).
77. B. L. Crowder and F. F. Morehead, *Appl. Phys. Lett.* **14**, 313 (1969).
78. W. M. Lomer, *Phil. Mag.* **42**, 1327 (1951).
79. H. P. Fredrichs and S. Kalbitzer, *Rad. Eff.* **83**, 135 (1984).
80. C. E. Christodoulides, G. Carter, and J. S. Williams, *Rad. Eff.* **48**, 87 (1980).
81. C. Carter, W. Maszara, D. K. Sadana, G. A. Rozgonyi, J. Liu, and J. Wortman, *Appl. Phys. Lett.* **44**, 459 (1984).
82. R. G. Elliman, J. S. Williams, D. M. Maher, and W. L. Brown, *Proceedings of Fall Mat. Res. Soc. Meeting*, Symposium A (1985).
83. F. F. Komarov, A. P. Novikov, T. T. Samoilyuk, V. S. Solov'yev, and S. Yu. Shiryaev, *Rad. Eff* **90**, 307 (1985).
84. J. Linnros and G. Holmen, *Proceedings of Fall Mat. Res. Soc. Meeting,* Symposium A (1985).
85. R. G. Elliman, J. S. Williams, W. L. Brown, A. Lieberich, D. M. Maher and R. V. Knoell, *Nucl. Inst. Meth. B* 19/20, 435 (1987).
86. J. Linnros, R. G. Elliman, and W. L. Brown, *MRS Proceed.* **74**, 477 (1987).
87. D. K. Sadana, J. Washburn, and G. R. Booker, *Phil. Mag. B* **46**, 611 (1982).
88. N. R. Wu, P. Ling, D. K. Sadana, J. Washburn, and M. I. Current, *Proc. of the Electrochemical Society,* (W. M. Bullis and L. C. Kimerling, eds.), ECS Press, Pennington, NJ, **83-9**, 366 (1983).
89. K. S. Jones, S. Prussin, and E. R. Weber, *Proceedings of the 14th International Conference on Defects in Semiconductors* (H. J. von Bardelben, ed.), Trans Tech Publications, Aedermannsdorf 1986 (Materials Science Forum Vol. 10), p. 751.

90. W. Vandervorst, D. C. Houghton, F. R. Shepherd, M. L. Swanson, H. H. Plattner, and G. J. C. Carpenter, *Can. J. Phys.* **63**, 863 (1985).
91. T. Y. Tan, *Phil. Mag.* **44**, 101 (1981).
92. T. Y. Tan, H. Föll and W. Krakow, *Inst. Phys. Conf. Ser. No. 60*, section 1, 1 (1981).
93. W. Krakow, T. Y. Tan, and H. Föll, in *Defects in Semiconductors* (J. Narayan and T. Y. Tan, eds.), North Holland Publishing, Amsterdam, p. 183 (1981).
94. J. Narayan and J. Fletcher, in *Defects in Semiconductors* (J. Narayan and T. Y. Tan, eds.), North Holland Publishing, Amsterdam, p. 191 (1981).
95. H. Föll, T. Y. Tan, and W. Krakow, in *Defects in Semiconductors* (J. Narayan and T. Y. Tan, eds.), North Holland Publishing, Amsterdam, p. 173 (1981).
96. M. Pastermann, D. Hoehl, A. L. Aseev, and O. P. Pchelyakov, *Phys. Stat. Sol.* (a) **80**, 135 (1983).
97. H. Bartsch, D. Hoehl, and G. Kastner, *Phys. Stat. Sol. (a)* **83**, 543 (1984).
98. I. G. Salisbury and M. H. Loretto, *Rad. Eff.* **59**, 59 (1981).
99. I. G. Salisbury and M. H. Loretto, *Phil. Mag. A* **39**, 317 (1979).
100. C. A. Ferreira Lima and A. Howie, *Phil. Mag.* **34**, 1057 (1976).
101. W. K. Wu and J. Washburn, *J. Appl. Phys.* **48**, 3742 (1977).
102. F. F. Komarov, V. S. Solovev, and S. Yu. Shiryaev, *Rad. Eff. Lett.* **58**, 177 (1981).
103. D. Hoehl and H. Bartsch, *Phys. Stat. Sol. (a)* **112**, 419 (1989).
104. J. J. Comer, *Rad. Eff.* **36**, 57 (1978).
105. V. V. Kalinin and N. N. Gerasimenko, *Phys. Stat. Sol. (a)* **76**, 65 (1983).
106. L. J. Chen and I. W. Wu, *J. Appl. Phys.* **52**, 3310 (1981).
107. T. Y. Tan, H. Foll, and S. M. Hu, *Phil. Mag. A.* **44**, 127 (1981).
108. V. V. Kalinin and N. N. Gerasimenko, *Phys. Stat. Sol. (a)* **86**, 185 (1984).
109. S. M. Davidson and G. R. Booker, *Proceedings of the 1st International Conference on Ion Implantation* (L. T. Chadderton and F. H. Eisen, eds.), Gordon and Breach, p. 51 (1971).
110. S. M. Davidson, *Proc. of European Conf. on Ion Implantation* (Reading, England), 238 (1970).
111. A. C. Ajmera, G. A. Rozgonyi, and R. B. Fair, *Appl. Phys. Lett.* **52**, 813 (1988).
112. A. C. Ajmera and G. A. Rozgonyi, *Appl. Phys. Lett.* **49**, 1269 (1986).
113. E. Ganin and A. Marwick, *Mat. Res. Soc. Symp. Proc.* Vol. 147, MRS, Pittsburgh, PA, p. 13 (1989).
114. M. Nemiroff and V. S. Speriosu, *J. Appl. Phys.* **58**, 3735 (1985).
115. P. Zaumseil, U. Winter, F. Cembali, M. Servidori, and Z. Sourek, *Phys. Stat. Sol. (a)* **100**, 95 (1987).
116. C. F. Cerofolini, L. Meda, M. L. Polignano, G. Ottaviani, H. Bender, C. Claeys, A. Armigliato, and S. Solmi, *Semiconductor Silicon*, Proc. of Electrochemical Society, ECS Press, Pennington NJ (1986).
117. E. Ganin and A. Marwick, in *Ion Beam Processing of Advanced Electronic Materials* (N. W. Cheung, A. D. Marwick, and J. B. Roberto, eds.), MRS, Pittsburgh, PA, Mat. Res. Soc. Symposium Proceedings, **147**, 13 (1989).
118. F. Cembali, M. Servidori, and A. Zani, *Solid State Electronics* **28**, 933 (1985).
119. J. Narayan and O. W. Holland, *Phys. Stat. Sol. (a)* **73**, 225 (1982).
120. W. K. Hofker, W. J. M. Josquin, D. P. Oosthoek, and J. R. M. Gijsbers, *Rad. Eff.* **47**, 183 (1980).
121. M. Servidori and I. Vecchi, *Solid State Electronics* **24**, 329 (1981).
122. D. K. Sadana, J. Washburn, M. Strathman, M. Current, and M. Maenpaa, *Inst. Phys. Conf. Ser. No. 60*, section 9, 453 (1981).
123. W. K. Chu, M. R. Poponiak, E. I. Alessandrini, and R. F. Lever, *Rad. Eff.* **49**, 23 (1980).

124. S. Albin, R. Lambert, S. M. Davidson, and M. I. J. Beale, *Inst. Phys. Conf. Ser.* No. 67, section 4, 241 (1983).
125. D. K. Sadana, J. Washburn, and C. W. Magee, *J. Appl. Phys.* **54**, 3479 (1983).
126. S. Prussin, David I. Margolese, and Richard N. Tauber, *J. Appl. Phys.* **56**, 915 (1984).
127. H. Leibel, *J. Vac. Sci. Tech.* **12**, 385 (1975).
128. Y. Kim, H. Z. Massoud, and R. B. Fair, *Appl. Phys. Lett.* **53**, 2197 (1988).
129. D. K. Sadana, N. R. Wu, J. Washburn, M. Current, A. Morgan, D. Reed, and M. Maenpaa, *Nucl. Inst. Meth.* **209-210**, 743 (1983).
130. K. Sumino and M. Imai, *Phil. Mag. A* **47**, 753 (1983).
131. L. D. Glowinski, K. N. Tu, and P. S. Ho, *Appl. Phys. Lett.* **28**, 312 (1976).
132. K. N. Tu, S.I. Tan, P. Chaudhari, K. Lai, and B. L. Crowder, *J. Appl. Phys.* **43**, 4262 (1972).
133. M. H. Brodsky, R. S. Title, K. Weisser, and G. D. Pettit, *Phys. Rev. B* **1**, 2632 (1969).
134. S. B. Felch and R. A. Powell, *Proc. 6th Int. Conf. on Ion Implantation Technology*, Berkeley, *Nucl. Inst. Meth. Phys. Res., B*, 1986, in print.
135. T. Sands, J. Washburn, R. Gronsky, W. Maszara, D. K. Sadana, and G. A. Rozgonyi, *Appl. Phys. Lett.* **45**, 982 (1984).
136. T. Sands, J. Washburn, R. Gronsky, W. Maszara, D. K. Sadana, and G. A. Rozgonyi, *13th Int. Conf. on Defects in Semiconductors*, Coronado (1984).
137. D. K. Sadana, W. Maszara, J. J. Wortman, G. A. Rozgonyi and W. K. Chu, *J. Electrochem. Soc.* **131**, 943 (1984).
138. T. E. Seidel, R. Knoell, F. A. Stevie, G. Poli, and B. Schwartz, *Proc. of the Electrochemical Society, VLSI Science and Technology*, (K. E. Beale and G. A. Rozgonyi, eds.), ECS Press, Pennington, NJ (1984).
139. D. K. Sadana, E. Myers, J. Liu, T. Finstad, and G. A. Rozgonyi, *Mat. Res. Soc. Symp. Proc.* Vol. 23, 303 (1984).
140. G. F. Cerofolini, L. Meda, G. Queirolo, A. Armigliato, S. Solmi, F. Nava, and G. Ottaviani, *J. Appl. Phys.* **56**, 2981 (1984).
141. W. Maszara, C. Carter, D. K. Sadana, J. Liu, V. Ozguz, J. Wortman, and G. A. Rozgonyi, *Mát. Res. Soc. Symp. Proc.* Vol. 23, 285 (1984).
142. G. A. Rozgonyi, E. Myers, and D. K. Sadana, *Semiconductor Silicon, Proc. of Electrochemical Society* (ECS Press, Pennington, NJ) (1986).
143. Y. Shih, J. Washburn, and S. C. Shatas, *Advanced Semiconductor Processing and Characterization of Electronic and Optical Materials*, SPIE **463**, 93 (1984).
144. H. Muller, W. K. Chu, J. Gyulai, J. W. Mayer, T. W. Sigmon, and T. R. Cass, *Appl. Phys. Lett.* **26**, 293 (1975).
145. M. D. Rechtin, P. P. Pronko, G. Foti, L. Csepregi, E. F. Kennedy, and J. W. Mayer, *Phil. Mag. A* **37**, 605 (1978).
146. P. P. Pronko, M. D. Rechtin, G. Foti, L. Csepregi, E. F. Kennedy, and J. W. Myer, *Ion Implantation in Semiconductors, 1976* (F. Chernow, J. A. Borders, and D. K. Brice, eds.), Plenum Press, New York, p. 503 (1977).
147. M. I. J. Beale and G. R. Booker, *Inst. Phys. Conf. Ser.* No. 67, Section 4, 235 (1983).
148. F. F. Komarov, V. S. Solov'yev, and S. Yu. Shiryayev, *Rad. Eff.* **42**, 169 (1979).
149. P. I. Gaiduk, F. F. Komarov, V. A. Pilipenko, V. S. Solov'yev, and N. I. Sterzhanov, *Rad. Eff. Lett,* **86**, 213 (1984).
150. C. W. Nieh and L. J. Chen, *J. Appl. Phys.* **62**, 4421 (1987).
151. J. Narayan, *J. Appl. Phys.* **53**, 8607 (1982).
152. Z. Lu, C. Zhang, S. Li, Y. Luo, and H. Zhang, *Nucl. Instr. Meth. B* **43**, 46 (1989).
153. T. Sands, J. Washburn, E. Myers, and D. K. Sadana, *Nucl. Inst. Meth. Phys. Res. B* **7/8**, 337 (1985).

154. G. Queirolo, C. Bresolin, D. Robba, M. Anderle, R. Canteri, A. Armigliato, G. Ottaviani, and S. Frabboni, *J. Electronic Mater.* **20**, 373 (1991).
155. D. K. Sadana, M. Strathman, J. Washburn, and G. R. Booker, *J. Appl. Phys.* **52**, 744 (1981).
156. B. J. Masters, J. M. Fairfield, and B. L. Crowder, *Rad. Eff.* **6**, 57 (1970).
157. D. K. Sadana, J. Fletcher, and G. R. Booker, *Elect. Lett.* **13**, 631 (1977).
158. D. K. Sadana, M. Strathman, J. Washburn, and G. R. Booker, *J. Appl. Phys.* **51**, 5718 (1980).
159. N. C. Tung, *J. Electrochem. Soc.* **132**, 914 (1985).
160. D. K. Sadana, M. H. Norcott, R. G. Wilson, and U. Dahmen, *Appl. Phys. Lett.* **49**, 1169 (1986).
161. N. R. Wu, D. K. Sadana, and J. Washburn, *Appl. Phys. Lett* **44**, 782 (1984).
162. S. J. Pennycook, J. Narayan, and O. W. Holland, *J. Electrochem. Soc.* **132**, 1962 (1985).
163. R. Angelucci, G. Celotti, D. Nobili, and S. Solmi, *J. Electrochem. Soc.* **132**, 2728 (1985).
164. D. Nobili, A. Armigliato, M. Finetti, and S. Solmi, *J. Appl. Phys.* **53**, 1484 (1982).
165. C. W. Nieh and L. J. Chen, *J. Appl. Phys.* **60**, 3114 (1986).
166. A. Armigliato, D. Nobili, S. Solmi, A. Bourret, and P. Werner, *J. Electrochem. Soc.* **133**, 2560 (1986).
167. R. O. Schwenker, E. S. Pan, and R. F. Lever, *J. Appl. Phys.* **42**, 3195 (1971).
168. R. B. Fair and G. R. Weber, *J. Appl. Phys.* **44**, 73 (1973).
169. M. Tamura, *Phil. Mag.* **35**, 663 (1977).
170. K. S. Jones, Ph.D. Thesis, Lawrence Berkeley Laboratory Publication, LBL-23180, University of California, Berkeley, CA (1987).
171. O. Scherzer, *J. Appl. Phys.* **20**, 20 (1949).
172. S. Mader, private communications.
173. K. S. Jones, S. Prussin, and E. R. Weber, *J. Appl. Phys.* **62**, 4114 (1987).
174. K. Jagannadham and J. Narayan, *J. Appl. Phys.* **62**, 1698 (1987).
175. D. S. Wen, P. L. Smith, C. M. Osburn, and G. A. Rozgonyi, *Appl. Phys. Lett* **51**, 1182 (1987).
176. J. Honeycutt and G. A. Rozgonyi, *Appl. Phys. Lett.* **58**, 1302 (1991).
177. U. Gosele and W. Frank, in Defects in Semiconductors (J. Narayan and T. Y. Tan, eds.), North Holland, New York, Oxford, p. 55 (1981).
178. K. S. Jones, S. Prussin, and D. Venables, *Mat. Res. Symp. Proc.* **100**, 277 (1988).
179. D. K. Sadana, W. Maszara, J. J. Wortmann and G. A. Rozgonyi, *J. Electrochem. Soc.* **131**, 943 (1984).
180. A. Bourret and W. Schroter, *Ultramicroscopy* **14**, 97 (1984).
181. T. O. Sedgwick, *J. Electrochem. Soc.* **130**, 484 (1983).
182. D. K. Brice, *J. Appl. Phys.* **62**, 4421 (1987).
183. T. Sands, J. Washburn, R. Gronsky, W. Maszara, D. K. Sadana, and G. A. Rozgonyi, *13th International Conference on Defect and Semiconductors*, 531, TMS, Warrendale, PA (1984).

6 Junction Formation in Silicon by Rapid Thermal Annealing

Richard B. Fair

MCNC
Center for Microelectronic Systems Technologies
Research Triangle Park
North Carolina
and
Department of Electrical Engineering
Duke University, Durham
North Carolina

I. Rapid Thermal Annealing of Ion-Implanted Junctions	174
A. Implantation-Induced Damage	174
B. Transient Diffusion of Doping Impurities	180
II. Dopant Activation	213
A. Theory	214
B. Activation of Boron	215
C. Activation of Phosphorus	218
D. Activation of Arsenic	219
III. Summary and Conclusions	220
References	221

Silicon processing technology has depended heavily on impurity diffusion since the 1950s. The idea of using diffusion techniques to form *pn* junctions was disclosed in a 1952 patent by Pfann [1]. Since that time, the development of integrated circuits (ICs) has been the technology driver that has pushed advances in diffusion equipment and practices. For example, the evolution of ICs is driven by increased circuit performance and reduced cost. Although some of this improvement can be attributed to novel processing and manufacturing techniques, certainly the bulk of the improvement is the direct result of the continual reduction in device dimensions. In order to maintain suitable electrical characteristics of devices, the vertical dimensions must be reduced along with the lateral dimensions. Failure to scale properly in the vertical dimension inevitably leads to performance degradation through such familiar effects as short-channel behavior in metal oxide semiconductor (MOS) field effect transistors. While the reduction in lateral dimensions is primarily an exercise in lithography and etch control, the concomitant reduction in vertical dimensions requires the ability to form and maintain stable, shallow junctions. For future 0.15-μm complementary MOS (CMOS) technology, full scaling will require junction depths as shallow as 300 Å. This is a tremendous technological challenge, particularly for p-channel transistors in which the dopant for the source/drain is boron, a relatively small and fast diffusing atom [2].

In modern very large scale integrated (VLSI) circuit manufacturing, ion implantation and annealing have supplanted predeposition by diffusion as the primary method for junction formation and doping. The ability to maintain precise control of dopant profiles over a large operating energy range (equivalent to a large range in junction depth) coupled with excellent specificity of implant species has brought ion implantation to its current place of prominence. However, in the realm of ultrashallow junctions the issue of defects resulting from implantation has begun to dominate the discussion, and what used to be the obvious choice for junction formation may appear speculative at best. A frequently asked question is, "How can implantation damage be removed (and good mobility and low junction leakage currents be maintained) while still preserving the required junction depths?" Such trade-offs can be difficult. As annealing times and temperatures are increased in an effort to remove implantation damage, junction depths inevitably increase. Junction displacement is further complicated by defect-assisted dopant atom diffusions which ties the distribution, type, and annealing rate of defects to the ultimate junction depth. An examination of recent data obtained at very low implantation energies (i.e., 1 keV and lower) addresses the question of defect removal [3], and in doing so provides strong evidence that ion implantation is, in fact, a viable candidate technology for shallow junction formation [2].

Any modern junction formation process must be capable of creating good quality junctions without excessive process complexity. In a good quality junction the final junction depth must be controllable and reproducible. Additionally, the junctions must have low reverse-bias leakage current and good (near ideal) forward characteristics. This requirement translates into a need for well-controlled junction-region defect densities. Both the number and the location of any process-induced defects (i.e., end-of-range damage from implantation) are important. Diffused or implanted regions must have low sheet resistance, thus high carrier mobility and high dopant activation level are required. Finally, there are compatibility issues. An advanced

Junction Formation in Silicon by Rapid Thermal Annealing

junction process should be as compatible as possible with existing, well-established processing steps. A suitable contact technology must be available, either independent of the function formation process or as a part of it.

Clearly, ion implantation is not the only technology capable of meeting the requirements for good junctions. Several alternative approaches to junction formation in the ultra large-scale integration (ULSI) regime are currently under investigation. The prominent contenders appear to be the following: diffusion from implanted polysilicon [4], diffusion from implanted silicide (or salicide) [5, 6] (see Chapter 7), and a form of laser doping often referred to as GILD (gas immersion laser doping) [7, 8]. Diffusion from implanted polysilicon or silicide is attractive for several reasons. First, the implantation is performed into the layer, not into the underlying silicon. This process reduces or eliminates the production of damage in the silicon substrate and simultaneously permits higher, more convenient implantation energies to be used. Second, each of the techniques results in the formation of the critical contact concurrently with the junction diffusion. However, both approaches induce enhanced dopant diffusion into the underlying Si substrate, which reduces the available time to produce a junction at certain depths at the annealing temperatures. The so-called "thermal budget" will be measured in seconds for 0.25-μm MOS technology, compared to minutes or hours for today's typical VLSI processes.

A comparison of the thermal budget (time at temperature) as a function of process temperature for several junction-formation methods for forming 700-Å p^+ junctions is shown in Fig. 1 [2]. Thermal budget calculations were made using the PREDICT process simulator [9] and measured data from $CoSi_2$ diffusion sources [6]. For temperatures below 1000 °C, diffusion from shallow $CoSi_2$ provides some advantage relative to traditional implantation into silicon. However, a clear advantage is seen for implants in polysilicon followed by diffusion. The larger thermal budget occurs because part of the budget is consumed by dopant diffusion within the polysilicon layer itself and, in addition, no implant damage is produced in the crystalline substrate to cause enhanced diffusion.

Besides controlling junction depth and eliminating implantation damage, the thermal budget is also constrained by the stability of the layers used in the fabrication of devices. Degradation of silicide layers occurs as the material agglomerates at high temperature, causing a substantial increase in silicide resistivity. Silicides formed over polysilicon degrade an order of magnitude faster than silicides on amorphous or crystalline silicon [10, 11] (see Chapter 7). The implication of this degradation on thermal budget for silicide processing is depicted in Fig. 1. It has been suggested that this limitation would be eased if amorphous Si gates were used instead of polysilicon gates [11].

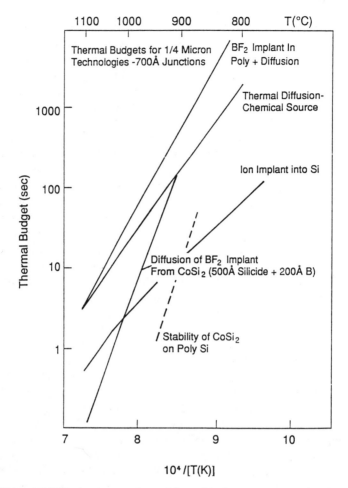

FIGURE 1. Available time at temperature (thermal budget) to produce p^+ scaled junctions for 0.25-μ MOS technology. Curves are shown for 700-Å junction formation using BF_2 implants in polysilicon, thermal source diffusion, ion implantation into silicon, and diffusion from an implanted $CoSi_2$ layer. (Reprinted with permission from R. B. Fair, *Proc. IEEE* **79**(11), 1687–1705. © 1990 IEEE.)

As mentioned above, the device scaling requirement of reduced junction depths ultimately affects the thermal budget available for a given technology. In terms of junction formation the thermal budget is typically limited by dopant diffusion, which can be purely thermal or enhanced by nonequilibrium levels of point defects. The reduction in process thermal budget required when going from a 2 μm to a 0.5 μm MOS technology was recently discussed by Osburn and Reisman [12]. Further analysis shows that if

Junction Formation in Silicon by Rapid Thermal Annealing

one assumes that the surface doping is maintained at a $1 \times 10^{20}/\text{cm}^3$ for a p^+ junction, then the thermal budget can be determined as a function of temperature for a given maximum junction depth. An example of such a calculation is shown in Fig. 2 [2]. The calculations are based on the PREDICT 1.5 process simulator [9] and assume fully scaled technologies for the various minimum feature sizes indicated in the figure. These results translate into p^+n junction depths of 0.5, 0.25, 0.15, and 0.07 μm for the 2, 1, 0.5, and 0.25 μm technologies respectively. From Fig. 2 it can be seen that the total permissible process time at 1000 °C for 700-Å source/drain junctions must be limited to only 24 sec. The calculations assume chemical source diffusion, which is the best-case situation for

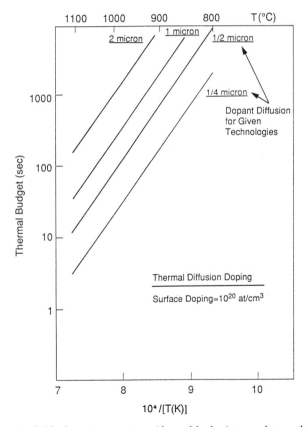

FIGURE 2. Available time at temperature (thermal budget) to produce scaled p^+ junction depths for four MOS technologies. Thermal diffusion sources are assumed with a fixed surface concentration of 1×10^{20} atoms/cm^3. (Reprinted with permission from R. B. Fair, *Proc. of the IEEE* **79**(11), 1687–1705. © 1990 IEEE.)

thermal budget. For the case of low-energy ion implantation, the calculations must take into account the effects of ion implantation damage on enhanced dopant diffusion, pushing the available thermal budget down by a factor of ten! These calculations are bolstered by current experience in the semiconductor memory business, where it is now widely held that 256 MB dynamic memory chips will need to be fabricated with rapid thermal annealing (RTA).

In the following sections the science and technology of forming n^+p and p^+n junctions will be addressed. Section I discusses implantation damage and important variables controlling transient diffusion. Section II is a discussion of dopant activation and the roles played by damage and annealing temperatures. In Section III conclusions are presented.

I. Rapid Thermal Annealing of Ion-Implanted Junctions

As we have seen, submicron ULSI technologies require low thermal budgets for shallow junction formation. However, it has been observed that low thermal budget anneal cycles in conjunction with ion implantation produce junctions that are considerably deeper than expected [13–18]. For example, low-temperature anneals of ion-implanted boron, phosphorus, and arsenic in silicon can exhibit substantial diffusion, depending upon the completeness of activation of the implant [14]. In addition, RTA of similarly implanted layers produces diffusion transients that can exceed the equilibrium diffusivity values of the dopants by more than a factor of 1000. The origin of such diffusion transients is the annealing of ion implantation damage in the silicon subtrate. The magnitude of diffusion enhancement and the time duration of the enhancement are related to the type of damage that is produced [19, 20].

A. Implantation-Induced Damage

Nuclear stopping processes of ions implanted into Si are responsible for producing displacement damage. The types and amount of damage produced depend on implant species, energy, dose, wafer temperature and orientation, dose rate, and materials covering the substrate. The as-implanted defect morphologies then change during wafer annealing depending on temperature, time, furnace ambient, and ramp-up/ramp-down rates. Much of the point-defect generation (vacancies and self-interstitials) occurs during the changes from as-implanted damage to stable or dissolved damage structures as a result of annealing [16, 19, 21, 22].

Junction Formation in Silicon by Rapid Thermal Annealing

Damage-assisted or retarded diffusion of ion-implanted dopants in silicon depends on (1) the type of damage introduced by implantation and point-defect generation characteristics during annealing [18], (2) the location and distribution of the damage [19], and (3) the dominant mechanism by which the dopant diffuses. For instance, B and P are believed to diffuse primarily via a self-interstitialcy mechanism that involves Si self-interstitials, Sb diffusion is dominated by a mechanism involving vacancies as manifested by the oxidation-retarded diffusion phenomena, and As diffuses via a combination of vacancy and self-interstitialcy mechanisms [23–25].

Depending on the dose and species, ion implantation can produce different types of damage as shown in Fig. 3, which summarizes implant damage depending on the implant dose and mass [18]. Each of the damage types will now be described briefly so that transient diffusion during RTA can be understood as a function of implant dose energy and species. A complete discussion of implantation damage appears in Chapter 5.

1. Low-Dose Implants

When the implant dose is below the critical level required for the formation of extended dislocations (on the order of 2×10^{14} cm^{-2}), the dominant damage species are isolated point defects and point-defect clusters.

Multiple-crystal X-ray diffraction has been applied to the detection of these very small lattice defects in otherwise perfect single crystals [26, 27]. These defects are often too small to be observed by conventional transmission electron microscopy (TEM), but they are able to produce appreciable effects on the X-rays intensity profiles so that their distribution can be determined [28].

The presence of interstitial-type isolated defects is manifested by the introduction of strain in the silicon lattice. These strain profiles are obtained from computer simulations of experimentally obtained rocking curves using the dynamic theory of X-ray diffraction [29]. For low-dose implants, these interstitial-type defects are distributed throughout the implanted region as shown schematically in Fig. 3a.

During low-temperature annealing, the strain distributions shrink because of point-defect cluster dissolution and the subsequent diffusion of released self-interstitials [30]. Annealing of the point-defect clusters has been found to have ~5-eV activation energy on the basis of the electrical activation of implanted boron in the isothermal annealing study by Seidel and MacRae [31]. They attributed the activation of the implanted B to the annealing of point-defect clusters. An average annealing rate for clusters is depicted in Fig. 4 as a function of temperature.

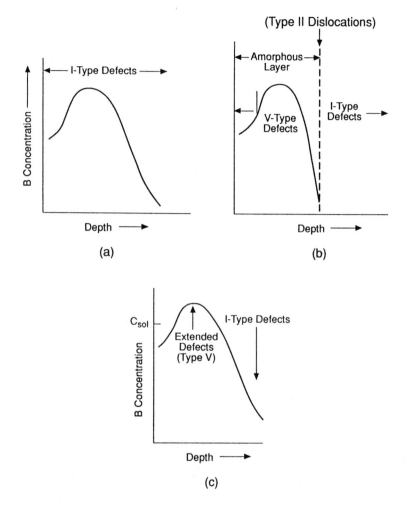

FIGURE 3. Ion implantation defect production for three classes of implants, shown in terms of boron concentration vs. depth. I-type defects are of the self-interstitial type and V-type defects are of the vacancy type.

2. Subthreshold Damage Production

In order to classify general types of implant damage, Jones *et al.* [32] performed a systematic study of characterization (see Chapter 5). For purposes of the current chapter we will focus only on the point-defect generation and recombination aspects of damage essential for our discussion on junction formation using RTA.

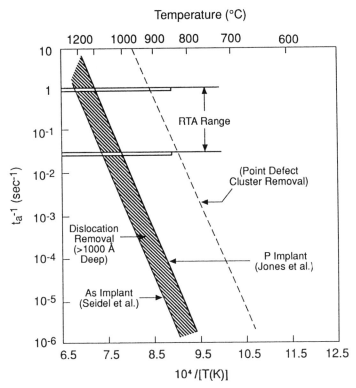

FIGURE 4. Inverse damage annealing times for end-of-range dislocations and point-defect clusters. (After Ref. 22, reprinted with permission from *J. Elec. Mat.* **18** (2), 143 (1989), a publication of The Minerals, Metals & Materials Society, Warrendale, PA 15086.)

Category I defects are typically extrinsic dislocation loops distributed around projected ranges. These defects are expected to form when the implant damage is insufficient to produce an amorphous layer and a critical dose for damage formation is exceeded. This critical value for the dose to form the category I defect is found to be 2×10^{14} cm^{-2}, which corresponds to the critical peak concentration of $\sim 1.6 \times 10^{19}$ cm^{-3} [33]. This critical peak concentration is independent of implanted species and wafer orientation, indicating that the higher density in the silicon lattice due to the nonconservative nature of the implantation process may be the primary cause for the category I defects.

After the recombination of vacancies and self-interstitials during ion implantation, the net point defects remaining are predominantly self-interstitials. During annealing, a supersaturation of self-interstitials will evolve to a layer of perfect extrinsic dislocation loops via the intermediate defect configurations proposed by Tan [33]. The total concentration of atoms bound by these extrinsic dislocation loops is found to be approximately equal to the implanted dose [32]. When the concentration of atoms bound by the extrinsic dislocation loops exceeds the concentration of atoms in a monolayer of the {111} plane of silicon ($\sim 1.4 \times 10^{15}$ cm^{-2}), dislocation network formation becomes possible upon annealing via dislocation-dislocation interactions [32]. The category I extrinsic dislocation loops are quite stable and difficult to remove even at 1100 °C furnace anneals, and the network structure is also quite stable.

3. Amorphous Layer Defects

Category II defects arise whenever an amorphous layer is generated, and they are located just beyond the amorphous/crystalline (α/c) interface in the heavily damaged but still crystalline substrate (see Fig. 3b). They are also called "end-of-range" (EOR) damage [32]. Other types of defects related to the amorphous layer regrowth may be avoided by choosing implantation and annealing conditions carefully, but the category II defect is unavoidable once an amorphous layer is formed. Therefore, when the dose exceeds 2×10^{14} cm^{-2}, implantation at room temperature of any Group III and V elements would result in the formation of at least either the category I or the category II defect (extrinsic dislocation loops) upon annealing [34].

Annealing characteristics of the category II defects may be the most important process to understand, because the preamorphization of the Si surface is performed prior to low-energy dopant implantations in order to prevent channeling. Also, high-dose implantations to reduce sheet resistance are likely to form surface amorphous layers when implantation is done in a crystalline substrate. The location and annealing properties of the category II defects are important factors in determining the implanted dopant diffusion and the quality of implanted junctions. The removal of the category II dislocation loops was studied by Seidel *et al.* [35] by isothermal experiments using TEM. The category II dislocation loops were produced by 100-keV, 5×10^{15} As/cm^2 implants and annealed at temperatures varying from 850 to 1200 °C. The annealing rate and the inverse of time required to completely remove the category II dislocation loops were found to have $\sim$5-eV activation energy, indicating that the self-diffusion of Si atoms limits the annealing process (see annealing curves in Fig. 4).

The annealing characteristics of the EOR dislocation loops are found to be greatly affected by the surface proximity. Ajmera *et al.* [36] have reported the fast removal of the EOR damage located near the surface,

Junction Formation in Silicon by Rapid Thermal Annealing

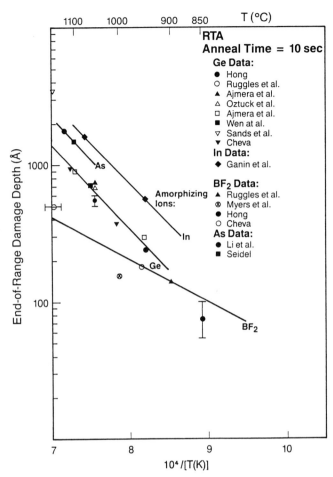

FIGURE 5. Depth of end-of-range damage removed by a 10-sec RTA, as a function of RTA temperature, for four implant species. (Reprinted with permission from R. B. Fair, *IEEE Trans. Electron Dev.* **37**, 2237. © 1990 IEEE.)

whereas the deeper EOR damage still remained. Fair has examined the depth dependence of the anneal characteristics of the EOR dislocation loops for different amorphizing species and showed that the heavier ions exhibited a faster annealing rate [37]. Annealing data for EOR dislocation loops in preamorphized junctions are shown in Fig. 5. The amorphizing ions included BF_2, Ge, As, and In. The data represent 10-sec RTA processing at temperatures 850–1150 °C [3, 36, 38–47]. Each datum represents a processing condition in which the 10-sec RTA was just sufficient to remove EOR loops at the depth indicated.

As seen in Fig. 4, EOR damage anneals with approximately a 5-eV activation energy. Using this information as a basis for damage annealing in B-implanted, Ge-preamorphized Si and using the Ge data in Fig. 5 yields the following relationship between depth and temperature-dependent annealing time constant:

$$t_{Ge}(x_{EOR}) = 2940 \exp(5 \text{ eV}/kT)(x_{EOR})^{4.14} \quad (\text{sec}) \quad (6.1)$$

where x_{EOR} is the EOR damage depth in cm from the Si surface to the leading edge of the dislocated region. Thus the annealing time constant has a fourth-power dependence on depth. Embodied in this strong effect is the energy dependence of damage density as well as some less-important surface effects [37].

4. Projected Range Damage

Projected range damage, known as category V damage, is caused by formation of precipitates of the implanted species distributed near the projected range when the solid solubility of the implanted species at the annealing temperature is exceeded (see Fig. 3c and Chapter 5). These defects may not be found for all implanted species but they are found for high dose In, As, Sb, B, and Ga implants after annealing [32]. The dissolution of these precipitates is proposed to enhance annealing of the category II dislocation loops. The category V defects can be in the form of extrinsic dislocation loops when the implanted species are Group V elements [48]. Also, half-loop dislocations extending from the projected range to the surface have been reported for high-dose arsenic implantations [32].

Projected range dislocations appear to be very stable at high annealing temperatures. During RTA, a crossgrid dislocation network has been observed to form at the peak of a high-dose B implant profile [49]. During this process, enhanced diffusion has been observed outside of the dislocation band, indicating that self-interstitials were generated.

B. Transient Diffusion of Doping Impurities

Observation of implantation-damage annealing and the response of impurity profiles to annealing have been important tools in the understanding RTA of implanted layers. It has been widely observed that the diffusivity of dopants in Si is greatly enhanced during implant-damage annealing because of the generation of excess point defects [22, 50–53]. The extent of the observed transient diffusion enhancement is believed to be related to the dominant dopant diffusion mechanism and the type of point defects generated during the implant-damage annealing. During annealing, a supersaturation of vacancies or self-interstitials is established, which is associated with the net increase in the density of silicon following implantation. It is important to understand the annealing characteristics of the various types of implant damage and the point-defect generation processes during RTA in order to model accurately the dopant diffusion in the transient regime. The annealing time and the excess concentration of point defects generated during residual implant-damage annealing are required for the successful modeling of transient-enhanced diffusion. The relative location and distribution of the implantation damage with respect to the impurity profiles are also important because damage perturbs the point-defect distributions and the electrical properties of the implanted junctions [54].

1. Diffusion from Low-Dose Implants

If the implant dose is below the critical dose for forming extended dislocations (category I damage), then the dominant defects produced are isolated point-defect clusters. RTA produces a large self-interstitial supersaturation from cluster dissolution, which decreases rapidly by diffusion. The number of clusters produced depends on the energy and dose of the implant through the production of displaced atoms per incident ion, $N(E)$:

$$N(E) = E_n/2E_{00}, \tag{6.2}$$

where E_n is the energy going into nuclear collisions and E_{00} is the energy required to displace a silicon lattice atom.

This simplified view can be refined to account for recoil events that create a vacancy at the original collision site and a self-interstitial at some distance away at the end of the recoil path. This separation of vacancies and self-interstitials would increase as the energy of the implanted species increases with the result that vacancy–interstitial recombination would be slowed by the time it takes for the two species to reach each other. The net result should be an increase in transient diffusion of the implanted species.

a. Dose and Energy Effects. The energy effect on the time-averaged transient junction displacement, Δx_J, can be seen for low-dose phosphorus implants in Fig. 6. Data are shown for both RTA and furnace anneals [55–58]. The magnitude of Δx_J defined at a P concentration of 1×10^{17} cm^{-3} varies as $E^{1/2}$. Since the P diffusion coefficient is proportional to Δx_J^2, the direct energy dependence for transient diffusivity is observed. Similar results have been found for B implants.

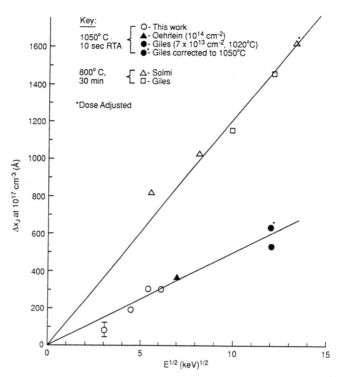

FIGURE 6. Transient junction displacement for low-dose P implants versus implant energy.

The dependence of transient diffusivity on implantation dose is harder to understand. In Fig. 7 Δx_J data for B implants are plotted versus dose for furnace anneals and RTA. Below the damage threshold dose ($\simeq 2 \times 10^{14}$ cm^{-2}), the transient diffusion coefficient shows a (dose)$^{1/4}$ dependence. Results from experiments with low-dose P implants that were annealed in an RTA furnace at 1050 °C for 10 sec are shown in Fig. 8. Most of the data shown were obtained from 20-keV P implants. All other data were normalized to 20 keV by $E^{1/2}$ scaling (see Fig. 6). The resulting dependence of P transient diffusivity on P dose is $D_P \propto$ (dose)$^{0.2}$, in close agreement with the B results.

In other studies, Packan and Plummer [59] studied the effect of low-dose Si implants on B diffusion. They found that when the Si dose was below 10^{13} cm^{-2}, transient B diffusion was independent of Si dose. However, above 10^{13} Si/cm^2 the B displacement increased with Si dose in the proportion (Si dose)$^{1/4}$, i.e., B diffusivity $\propto$ (Si dose)$^{1/2}$. Thus it can be concluded that a sublinear dependence of diffusivity on dose exists in the subthreshold dose regime.

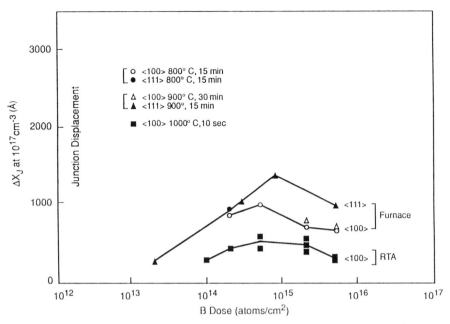

FIGURE 7. Boron dose dependence of transient junction displacement for RTA and furnace annealing.

By contrast, it is found that the crystalline damage produced by sub-threshold implants increases linearly or superlinearly with implant dose. Analysis of Rutherford backscattering data on displaced atom densities produced by low-dose As implants (10^{13}-10^{14} cm^2, 100 keV) showed a factor of 30 increase in displacement damage for a factor of 10 increase in As dose [60]. In another study, integrated damage measurements of low-dose BF_2 implants showed an almost linear dependence on dose [61].

In an attempt to model the dose dependence of low-dose P implants, Giles [55, 56] concluded that it was unnecessary to include the large numbers of self-interstitials and vacancies that make up displacement damage in the Si. Rather, he assumed that one additional self-interstitial atom is created for each implanted dopant atom during annealing as the dopant atom kicks out an Si atom so that it may itself become substitutional. Indeed, it is known that the vast majority of implanted atoms do not end up on substitutional sites until after annealing at high temperature.

The situation after implantation is shown in Fig. 9a. Each recoil event can be considered as the generation of a vacancy at the original lattice site and an interstitial at the stopping point of the recoil path [56]. The nonlattice site location of the implanted atom is shown schematically. The situation

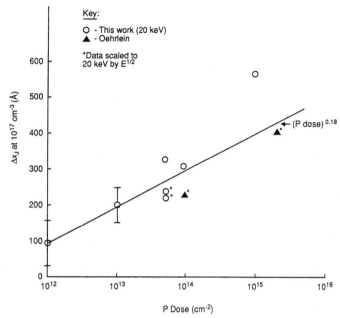

FIGURE 8. The dose dependence of transient displacement for P-implanted layers annealed at 1050 °C, 10 sec.

during annealing is shown in Fig. 9b. While vacancies and interstitials recombine or diffuse into the substrate, each implanted atom produces a single self-interstitial without a corresponding vacancy. These self-interstitials must diffuse away before they find a suitable recombination site, providing excess self-interstitials that can enhance dopant diffusion. By assuming that only a small fraction of the dopant is paired with a self-interstitial at any given time, Giles was able to simulate the proper dose dependencies of enhanced P diffusion. However, Giles's model cannot explain the energy dependence of enhanced diffusion shown in Fig. 6, which requires implantation-induced damage annealing effects. Thus additional fundamental work is required to understand what supposedly is the simplest case of transient diffusion.

When the implant dose exceeds the damage threshold level, extended dislocations form that act to quench the self-interstitial supersaturation responsible for transient diffusion. This effect is visible in Fig. 7 where the peak values of Δx_J occur near a dose of 5×10^{14} cm^{-2}. Thus the dislocations act as sinks for the interstitials, causing dislocation growth. Also shown is a wafer-orientation effect in which more B diffusion is observed in ⟨111⟩ Si wafers than in ⟨100⟩ wafers.

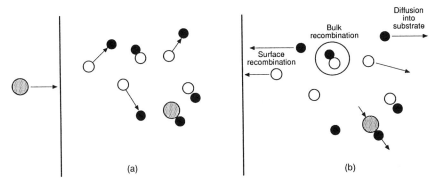

FIGURE 9. (a) Model of point-defect generation during ion implantation. Open circles are vacancies, solid black circles are self-interstitials, and the larger shaded circle is the implanted atom. (b) Point-defect diffusion/recombination during annealing. (After Ref. 55).

b. Transient Time Constant. The time constant for transient diffusion of low-dose B implants (1×10^{14} cm^{-2}, 150 keV) was studied by Miyake et al. [62]. Data from this study were shown to fit an exponential function associated with the decay of excess point defects with annealing time. Thus,

$$D(t) = D_i + D_0 \exp(-t/\tau), \qquad (6.3)$$

where D_i is the intrinsic B diffusion coefficient, D_0 is the enhanced diffusion coefficient at $t = 0$, and τ is the decay time of point defects. The expressions for D_0 and τ were found to be

$$D_0 = 1.4 \times 10^{-7} \exp(-1.1 \text{ eV}/kT) \quad (\text{cm}^2/\text{sec}) \qquad (6.4)$$

$$\tau = 2.9 \times 10^{-6} \exp(1.57 \text{ eV}/kT) \quad (\text{sec}). \qquad (6.5)$$

Miyake's data are plotted in Fig. 10 along with the dissolution rate curves for category II dislocations and point-defect clusters. Other diffusion transient data are also shown for low-dose B implants annealed at temperatures as low as 750 °C and as high as 1100 °C. The latter datum represents a B implant into Si$^+$-preamorphized Si [63]. These results suggest that short-time anneals are dominated by relatively rapid dissolution of point-defect clusters. At higher temperatures enhanced diffusion is assisted by the annealing of extended dislocations.

Solmi et al. [64] studied enhanced B diffusion following ion implantation. Using new data as well as published data from higher dose implants (2×10^{14}–1×10^{15} cm^{-2}, 20–60 keV), they found the time duration of enhanced B diffusion to be

$$t_e = 4.1 \times 10^{-15} \exp(3.7 \text{ eV}/kT) \quad (\text{sec}). \qquad (6.6)$$

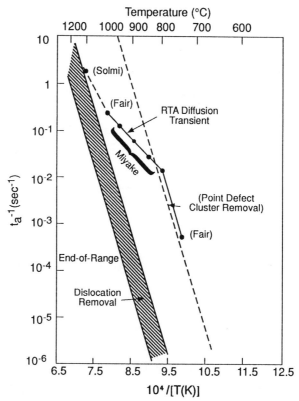

FIGURE 10. RTA boron diffusion transients and damage annealing times in ion-implanted Si. (Reprinted with permission after Ref. 113.)

These data are plotted in Fig. 11. Also plotted are the results of Michel [65], Solmi and Servidori [66], Fair [67], Cowern et al. [68], and Miyake [62]. It can be seen that at temperatures below 950°C the three sets of data diverge. Michel's data were all obtained from single dose and energy implants (2×10^{14} cm^{-2}, 60 keV) to yield a transient activation energy of 4.7 eV. It is unclear why there should be such significant differences in the activation energies. However, it is likely that the transients in B diffusion are caused by multiple sources of point defects such as distributions of point-defect clusters and extended dislocations, all annealing at different rates to produce an effective activation energy to t_e. Above 950 °C defects such as clusters anneal during the RTA ramp-up time. Thus, in this region a simplified transient diffusion time curve results, which is dominated by the annealing of extended dislocations.

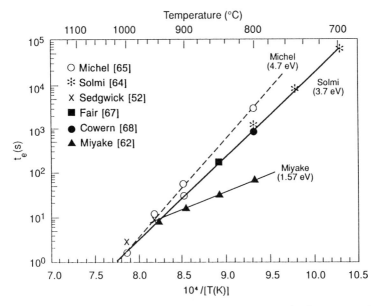

FIGURE 11. Enhanced diffusion time, t_e, for low-dose boron implants versus inverse temperature for postimplantation annealing.

c. Summary of Low-Dose Case. Low-dose implants in Si produce displacement damage that most likely takes the form of point-defect clusters distributed throughout the implanted region. Annealing produces a significant supersaturation of self-interstitials by the dissolution of clusters. This supersaturation decreases rapidly through the diffusion of self-interstitials and the fast dissolution of clusters. The result is a large diffusion transient for B and P over a short time. The transient magnitude depends on implant dose and energy. For the case of B the transient saturates in the dose range of 5×10^{14} cm^{-2} because of the formation of dislocations that act as sinks for the self-interstitial supersaturation. No saturation in transient diffusion is observed for P because of the formation of an amorphous layer (see later section on diffusion from amorphized layers).

As a practical result, enhanced diffusion of B and P during RTA can be reduced for a given dose by decreasing the implantation energy. This effect is demonstrated in Fig. 12 for implants of $1-2 \times 10^{14}$ B atoms/cm^2 performed at energies from 1 to 60 keV. The previously published data of Sedgwick [69] are included, and the solid curves were simulated using PREDICT 1.3 [70]. The dashed curve represents junction movement by thermally assisted diffusion only. Any diffusion above this line is due to transient effects. As the energy of the implant is reduced, the dopant

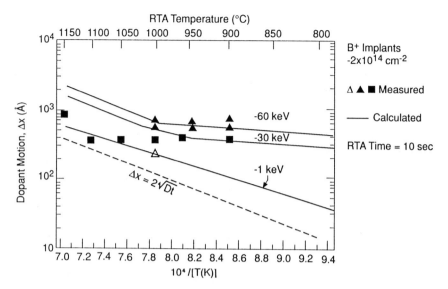

FIGURE 12. Dopant motion during RTA of low-dose B implants in Si showing data and PREDICT 1.3 simulations. (Reprinted with permission from R. B. Fair, in *Microelectronics Processing* (D. W. Hess and K. F. Jensen, eds.), p. 265. © 1989 American Chemical Society.)

motion decreases. The annealing temperature above which thermally assisted diffusion dominates also decreases. Thus at 60 keV, after an initial transient, subsquent diffusion above 1000 °C occurs by thermal diffusion. At 1 keV the transient is small and thermal diffusion dominates above 800 °C.

2. High-Dose B Implants with Amorphization

When high-dose B implants are annealed, two important effects occur: (1) dopant precipitation or clustering and (2) projected range defects that produce a long-term source of excess point defects. Boron diffusion during RTA occurs in two stages. An initial diffusion transient occurs because of the dissolution of point-defect clusters and the resulting supersaturation of self-interstitials. This supersaturation also drives the growth of extended defects, which decreases the concentration of excess point defects. When dislocation growth ceases dislocation dissolution begins, which pumps self-interstitials back into the implantation region, producing a long-term diffusion transient. If the RTA temperature is greater than 1100 °C, there is no extended defect formation, just a short diffusion transient [71].

Junction Formation in Silicon by Rapid Thermal Annealing

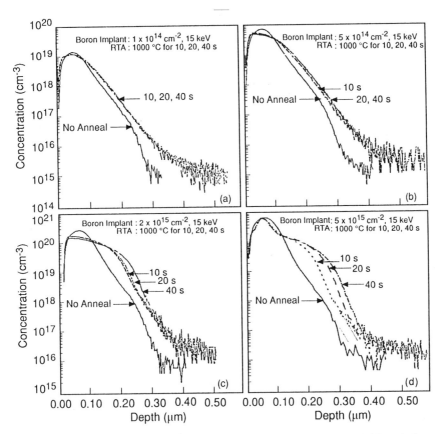

FIGURE 13. Boron SIMS profiles before and after RTA at 1000 °C for 10, 20, and 40 sec (a) 1×10^{14} cm^{-2}, 15 kV; (b) 5×10^{14} cm^{-2}; (c) 2×10^{15} cm^{-2}; (d) 5×10^{15} cm^{-2}. (After Y. M. Kim, G. Q. Lo, H. Kinoshita, and D. L. Kwong, *J. Electrochem. Soc.* **138**, 112, 1991. This figure is reprinted by permission of the publisher, The Electrochemical Society, Inc.)

The onset of the longer diffusion transient time with increasing B dose is shown in Fig. 13 [71]. At 1×10^{14} cm^{-2} the enhanced diffusion time saturates before 10 sec at 1000 °C. As the dose reaches 5×10^{14} cm^{-2} lattice damage increases and $t_e \simeq 20$ sec. At a dose of 2×10^{15} cm^{-2}, $t_e > 40$ sec and dislocation annealing produces a long-term effect. For the highest dose case a region of slow diffusion appears near the peak of the implant profile. In addition, the diffusion front continues to move with approximately a 100-fold enhanced diffusivity over thermally assisted diffusion rates.

The slowly diffusing peak in Fig. 13d is coincident with the region containing category V dislocations. An analysis of similar profiles has shown that retarded diffusion in this damage zone is caused by either B clustering or B segregation into interstitial dislocation layers and low point-defect concentrations relative to the damage-free regions [18, 17]. Thus the high dislocation density creates a region in which the lifetimes of excess point defects are very small, and no point-defect supersaturation can be sustained [16]. Such projected range dislocation networks are very stable and can produce excess self-interstitials for enhanced diffusion at the diffusion front for over 1 h at 950 °C [18].

At annealing temperatures below 900 °C it is observed that the region of retarded B diffusion extends down to the lower concentration tail region of the implanted profile. The effect is also observed for high-dose BF_2 implants which create an amorphous layer. Examples of this effect are shown in Fig. 14. BF_2 was implanted through 140 Å of SiO_2 at normal incidence to the plane of the wafer. Furnace anneals were performed on samples at 650, 750, and 850 °C in N_2 [67]. Little B diffusion occurs in the high-concentration region. The location of EOR damage is 550 Å from the surface, and the enhanced diffusion tails begin at different concentrations, C_{enh}, for each temperature.

Several models have been proposed to explain these results [16, 64, 67, 68, 72, 73]. Cowern *et al.* [68] proposed that the enhanced diffusion at low concentrations occurs because the B solubility is reduced below the

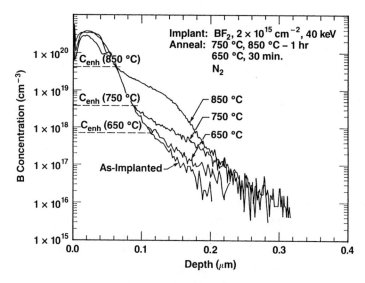

FIGURE 14. Implant damage enhanced B diffusion at low temperatures showing the temperature dependence of C_{enh}. (From R. B. Fair, *J. Electrochem. Soc.* **137**, 667, 1990. This figure is reprinted by permission of the publisher, The Electrochemical Society, Inc.)

equilibrium solubility as a result of nonequilibrium B clustering. The excessive clustering or precipitation of B is caused by the large supersaturation of interstitial B atoms present after ion implantation. Thus, only B atoms below the reduced solubility concentration are mobile and respond to the excess self-interstitials produced by the B clustering process.

Measurements and simulations of the evolution of B dopant and carrier profiles during 800°C annealing have been made by Solmi et al. [64]. An example of this work is shown in Fig. 15. It can be seen that C_{enh} corresponds to the electrically active B concentration, thus supporting the model of Cowern. Solmi also pointed out that the kink concentration in the profile increases from C_{enh} to C_{sol}, the B solubility limit, after the enhanced diffusion transient ends. This can be seen in Fig. 15. However, Cowern's model assumes, without explanation, that the diffusion profile of the mobile B species is Gaussian. It can be seen in Fig. 15 that the initial active B is not Gaussian and its profile shape is not predictable.

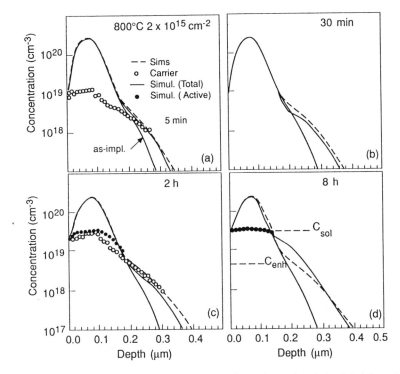

FIGURE 15. Evolution during 800 °C annealing of experimental and simulated dopant and carrier profiles in B-implanted wafers (2×10^{14} cm^{-2}, 20 keV). Values of C_{enh} and C_{sol} are shown. (After Ref. 64.)

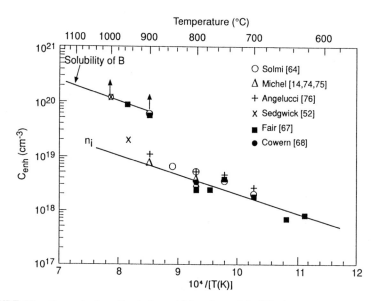

FIGURE 16. Concentration C_{enh} below which enhanced B diffusion occurs versus inverse annealing temperature. Curves of B solubility and n_i are shown for reference. The arrows indicate that all the B underwent enhanced diffusion in experiments with concentrations higher than C_{sol}. (After Ref. 64.)

Fair [67] explained the low concentration transient B diffusion by noting that $C_{enh} \simeq n_i$ for $T < 900\,°C$, where n_i is the intrinsic electron concentration. Data accumulated by Solmi are shown in Fig. 16 to illustrate the relationship [14, 52, 64, 67, 68, 74–76]. Fair also pointed out that C_{enh} is independent of the B implant dose as can be seen in Fig. 17. This result would not be expected from Cowern's nonequilibrium homogeneous clustering model.

The dependence of transient B diffusion on n_i stems from the assumption that B diffuses via I^x and I^+. The neutral and donor self-interstitials respectively. Below 900 °C, B diffusion is dominated by I^x [67]. Now the energy for diffusion of the BI^x pair is [77]

$$Q_B^x = \Delta H_I^x + \Delta H_m^x - E_b^x = 2.85 \text{ eV} \tag{6.7}$$

where ΔH_I^x is the enthalpy of formation of I^x, ΔH_m^x is the migration enthalpy of I^x, and E_b^x is the B–I^x binding energy. When I^x are generated externally during damage annealing, Q_B^x is reduced by ΔH_I^x. Since Q_B^x under conditions of external I generation in the temperature range 525–800 °C is 0.6 eV [78], then

$$\Delta H_I^x = 2.25 \text{ eV}.$$

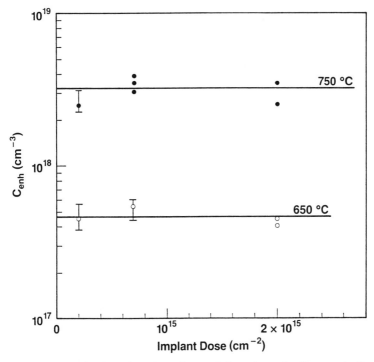

FIGURE 17. Boron implant dose dependence of the enhanced tail concentration C_{enh} after annealing at 650 and 740 °C. (From R. B. Fair, *J. Electrochem. Soc.* **137**, 667, 1990. This figure is reprinted by permission of the publisher, The Electrochemical Society, Inc.)

This result agrees well with previous studies in which the self-interstitial formation enthalpy was assumed to account for the reduction in diffusion activation energy [22, 79, 80].

Under external self-interstitial injection, the B diffusion coefficient is enhanced by the exponential of the Gibbs free energy of self-interstitial formation, ΔG_I. This effect is due to the fact that externally generated self-interstitials cause a supersaturation, and the diffusion free energy of B in response to these excess I's does not have to include the energy required to thermally generate them as would be the case under equilibrium conditions. Thus, assuming that the concentration of I^+ is small compared to n_i, the supersaturations of externally generated I^+ and I^x create the ratio of diffusivities given by

$$\left(\frac{D_i^+}{D_i^x}\right)_{inj} = \frac{D_i^+}{D_i^x} \exp \frac{[-(\Delta G_I^+ - \Delta G_I^x)]}{kT} \qquad (6.8)$$

where D_i^+ and D_i^x are the normal diffusivities of B for thermally assisted diffusion of I^+ and I^x, respectively, that is, the B diffusion coefficients when the external I injection stops. Taking into consideration the enthalpy and entropy associated with the formation of I^+ from I^x [77, 81], we may rewrite Eq. 6.8 as

$$\left(\frac{D_i^+}{D_i^x}\right)_{\text{inj}} = 1510 \exp\left[\frac{-\Delta S_{\text{cv}}}{k}\right] \exp\frac{\Delta Q - (\Delta H_m^+ - \Delta H_m^x)}{kT} \quad (6.9)$$

where $\Delta Q = Q_B^+ - Q_B^x = 0.74$ eV is the activation energy difference between D_i^+ and D_i^x, $\Delta H_m^+ - \Delta H_m^x = 0.27$ eV [77] and ΔS_{cv} is the entropy of the Si band gap [81]. With the use of ΔS_{cv} values from Refs. [81] and [82], the inverse of Eq. (6.9) is plotted in Fig. 18 along with D_i^x/D_i^+ for thermally assisted diffusion. It can be seen that diffusion of B via I^x is enhanced by 1-2 orders of magnitude relative to diffusion via I^+ when self-interstitials are externally generated.

Measurements of $(D_i^x/D_i^+)_{\text{inj}}$ obtained from samples annealed at least 1 h are also shown in Fig. 18. These data were inferred from relative displacement of B at concentrations a factor of four above C_{enh} and a factor of four below. These data show a trend similar to that of the calculations. Fair's model for transient enhanced diffusion, based on the previous discussion, is illustrated in Fig. 19 and summarized as follows:

- Damage annealing produces excess self-interstitials whose charge state depends on the local Fermi level position.
- Enhanced tail diffusion at low temperatures below C_{enh} depends on the dominance of B diffusion via I^x relative to diffusion via I^+.
- $C_{\text{enh}} = n_i$ occurs because E_I^+ is near midgap at low temperatures.
- As the annealing temperature exceeds about 800 °C, the probability of I^+ formation increases in intrinsic Si, and the relation between C_{enh} and n_i disappears. Diffusion via I^+ then becomes dominant, and C_{enh} becomes limited by B solid solubility. Normal concentration-dependent B diffusion applied over the duration of the anneal starts to become significant relative to the short transient of enhanced diffusion. For example, at 850 °C the diffusion transient lasts approximately 3 min. This result corresponds to a B displacement of about 400 Å. However, thermally assisted, concentration-dependent diffusion via I^+ for 1 h produces a displacement of about 500 Å according to simulations [37]. Thus, as soon as the transient ends and B diffusion above and below C_{enh} equalizes, thermal diffusion will eventually cause C_{enh} to increase to C_{sol}.
- A similar model would be expected for donor impurity diffusion via I^- (V^-) and I^x (V^x) if the point-defect acceptor lever were deeper below the

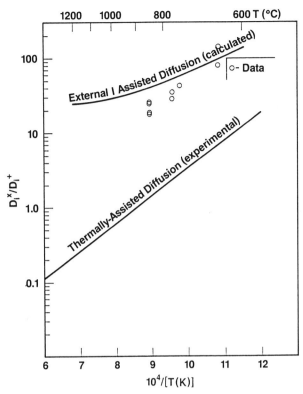

FIGURE 18. Ratio of D_i^x/D_i^+ for thermally assisted diffusion of B for enhanced diffusion in the presence of externally generated self-interstitials. (From R. B. Fair, *J. Electrochem. Soc.* **137**, 667, 1990. This figure is reprinted by permission of the publisher, The Electrochemical Society, Inc.)

conduction band than E_{Fi}. However, $E_I^- = E_c - 0.4$ eV according to Frank [83], and E_{Fi} lies below this level for $T > 900\,°C$. Thus for P, C_{enh} corresponds to the concentration of the kink rather than n_i. In order to reduce the amount of enhanced B diffusion to create very shallow p^+n junctions, it is necessary to screen out the flux of supersaturated I^x formed during damage annealing. Pre- or postamorphization by Si or Ge implantations can serve this purpose. In recent work, Brotherton et al. [84] have shown that if the depth of the amorphous layer exceeds the depth of the B implant, then no enhanced B tail is formed during 800 °C annealing. This result follows because end-of-range dislocations are formed that screen externally generated self-interstitials, and the process of epitaxial regrowth removes sources of excess self-interstitials from within the region that was amorphized.

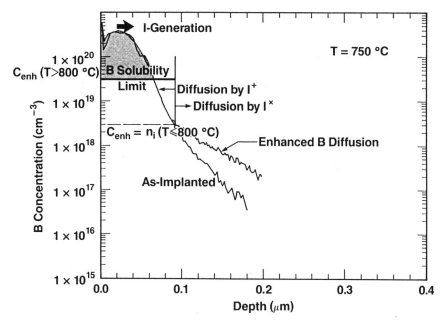

FIGURE 19. Model of enhanced B diffusion at $T < 900\,°C$ by externally generated self-interstitials. (From R. B. Fair, *J. Electrochem Soc.* **137**, 667, 1990. This figure is reprinted by permission of the publisher, The Electrochemical Society, Inc.)

3. Amorphization

Ion implantation of heavy ions into Si at medium-to-large doses produces amorphous surface layers. Amorphization of Si prior to B implantation (preamorphization) is also performed in order to disrupt the penetration of B along crystalline channels, thereby decreasing the depths of the as-implanted profile [85]. In the past, channeling has typically been reduced by offsetting the crystallographic axis at an angle to the incident ion beam. This technique has worked satisfactorily for higher energy implants. However, as the energy of the ions is decreased, the critical angle for scattering into an axial channel increases [86]. Use of large tilt and rotation angles can improve the situation somewhat, but at the price of increased shadowing of implanted impurities by the mask edges (or polysilicon gate). Recent work [21, 87, 88] has clearly shown that low-energy channeling of B in Si cannot be completely eliminated by sample orientation, and 20 to 30% of the implanted B can still be found in the profile tail.

a. Point-Defect Screening. The use of preamorphization has also been shown to suppress transient diffusion during RTA [46, 52, 89], provided that the B implant is contained within the amorphous layer. Any part of the profile that extends beyond the amorphous/crystalline (α/c) boundary into the underlying crystalline region will be subjected to the usual enhanced, transient displacement during RTA [90, 91]. This diffusion is due to excess self-interstitials that exist in the crystalline region. On the other hand, the B inside the regrown amorphous zone is screened from these excess point defects. It is known that the regrown amorphous layer regrows to a single crystal very rapidly upon annealing. The Si interstitials in this layer can easily diffuse to the Si surface during regrowth. In addition, the process that causes EOR damage to form and grow at the α/c interface can be a powerful sink for self-interstitials produced below the α/c region, creating a barrier to interstitial flow to the surface [29, 52].

To illustrate this screening process, a B implant into a preamorphized layer is shown in Fig. 20. The defects associated with the implantation are indicated just after amorphous layer regrowth. The sequence of events

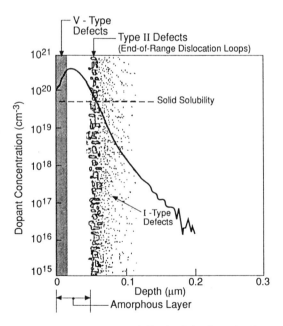

FIGURE 20. Residual implantation damage following B implantation into a preamorphized Si wafer. (From Y. Kim, H. Massoud, U. Gosele, and R. B. Fair, *2nd Int'l. Symp. on Process Physics and Modeling in Semiconductor Technology*, Vol. 91-94, p. 254, 1991. This paper was originally presented at the Spring 1990 Meeting of The Electrochemical Society held in Montreal, Canada.)

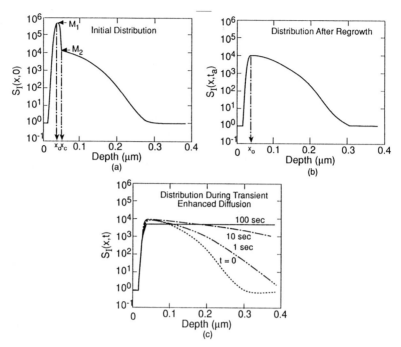

FIGURE 21. The distribution of self-interstitial supersaturation ratio $S_I(x, t)$ (a) prior to the amorphous layer regrowth, (b) at the end of the regrowth, and (c) during enhanced transient diffusion at 750 °C. (After Ref. 92.) (From Y. Kim, T. Tan, H. Massoud, and R. B. Fair, *2nd Int'l. Symp. on Process Physics and Modeling in Semiconductor Technology*, Vol. 91-94, p. 304, 1991. This paper was originally presented at the Spring 1990 meeting of The Electro-Chemical Society held in Montreal, Canada.)

leading to the buildup of Si self-interstitials has been proposed as follows [92]:

- Self-interstitials outdiffusing during the regrowth of the amorphous layer build-up at the α/c interface with some distribution of magnitude M_1, as shown in Fig. 21a. This plot shows the supersaturation ratio $S_I(x, t) = C_I(x, t)/C_I^{eq}$ at time $t = 0$, where $C_I(x, t)$ is the instantaneous concentration of self-interstitials at any point and C_I^{eq} is the equilibrium concentration.
- A second distribution of $S_I(x, 0)$ of magnitude M_2 represents the initial magnitude of the supersaturation ratio distribution due to Frenkel defect pairs and the dissolution of point-defect clusters [19, 93].
- The buildup of M_1 increases and then decreases rapidly during regrowth, driving the initial growth of EOR dislocation loops. The $S_I(x, t)$ distribution after regrowth is shown in Fig. 21b. Note the steep gradient in

$S_I(x, t)$ in Fig. 21a at the original α/c interface. This self-interstitial gradient produces a strong flux term during regrowth, causing uphill B diffusion to occur. This flux is proportional to $C_B \partial C_I/\partial x$ [92].
- After the transient due to regrowth is complete, enhanced B diffusion occurs in the region below the original α/c interface with a time constant dependent on cluster dissolution rate and self-interstitial diffusivity (see Fig. 21c for an example at 750 °C). To the left of the α/c interface, $S_I(x, t) = 1$ because of the screening of EOR dislocations.

The screening effect of EOR damage has also been shown to suppress transient P diffusion within a regrown layer as well as oxidation-enhanced P diffusion [94]. In the latter case the dislocations form a barrier to self-interstitials generated at the Si surface during oxide growth.

High-concentration P diffusion is known to generate point defects leading to enhanced diffusion and the formation of a profile tail [77]. However, implanting Si or Ge to create an amorphous layer in Si, preceded or followed by P implantation, acts to reduce the subsequent diffusion of P [95-98]. An example is shown in Fig. 22 where a 2×10^{15} P/cm^2, 30-keV

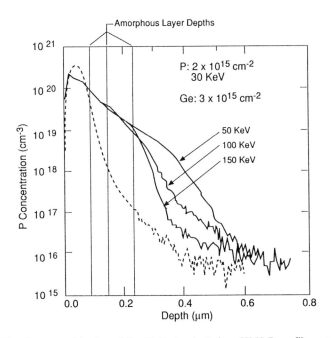

FIGURE 22. Postamorphization of P with Ge implantation: SIMS P profile measurements showing the effect of amorphous layer depth (Ge implant energy) on P diffusion at 900 °C. (After Ref. 96.)

implant was postamorphized with Ge implants at 50, 100, and 150 keV [96]. The depths of the resulting amorphorous layers are shown for reference. The deeper the amorphous layers, the smaller the subsquent P diffusion during the 900 °C, 20-min anneals. This result has been modeled assuming the generation of self-interstitials by P diffusion occurs in the vicinity of the P profile kink [99]. The screening effect of the EOR damage will continue until the P profile kink passes through the damage layer. This model is shown in Fig. 23 where point-defect generation sources on either side of the EOR damage layer are portrayed.

b. Shallow Amorphous Layers. The location of extended dislocations produced by ion implantation relative to the depth of the junctions is important in producing devices with low leakage currents. In older technologies, implanted impurities were diffused deeper than the damaged surface region. However, in VLSI technologies that require RTA, diffusion of dopants is minimized. Thus when an amorphous layer is formed by ion implantation, it should be made as shallow as possible and still preserve

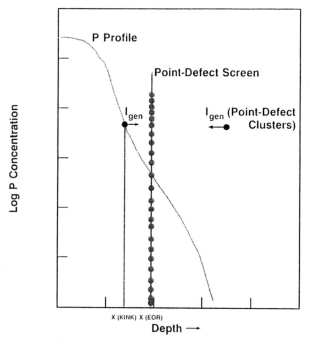

FIGURE 23. Schematic demonstration of the point-defect screening effect of an end-of-range damage layer.

Junction Formation in Silicon by Rapid Thermal Annealing

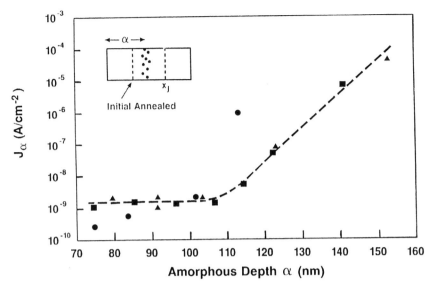

FIGURE 24. Diode area leakage current density J_α as a function of amorphous layer thickness α. (Reprinted with permission after Ref. 100.)

the benefits that an amorphous layer may provide the junction process. Sedgwick [100] has calculated the diode area leakage current as a function of amorphous layer thickness based on Öztürk's data [40], as shown in Fig. 24. The break in the leakage current curve at 1100 Å is believed to occur when the EOR loops reach the junction space-change region.

As the amorphization energy is reduced, the EOR damage density is also diminished (see Fig. 5). As a result, the smaller density of defects does not provide as effective a screen against point defects diffusing toward the Si surface. The result is more enhanced diffusion of dopants in the regrown region during RTA. For example, transient displacement of B completely contained within a shallow amorphous layer is shown in Fig. 25. These results were obtained with an Si preamorphization step that produced an 800-Å deep α/c interface [10]. The dashed line in the figure is the result of a simple B diffusion calculation. It can be seen that enhanced diffusion has occurred on the order of 300 Å in contrast to no excess displacement for B diffusion in thicker amorphous layers [52, 101].

For ultrashallow junctions it is better to place the EOR damage as close as possible to the Si surface [37]. The result is more rapid damage annealing and point-defect transients shorter in time and smaller in magnitude. To illustrate this result, shallow B implants were performed in Ge-preamorphized Si. Ten-second RTA anneals were performed at 1000,

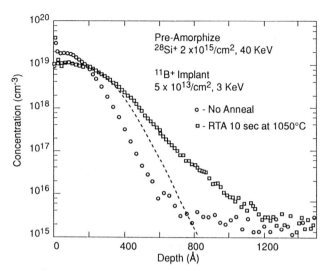

FIGURE 25. Transient displacement of B contained within a shallow amorphous layer. The dashed line is a calculation of dopant movement based solely on thermally assisted diffusion. (From A. E. Michel, *2nd Int'l. Symp. on Process Physics and Modeling in Semiconductor Technology*, Vol. 91-94, p. 254, 1991. This paper was originally presented at the Spring 1990 Meeting of The Electrochemical Society held in Montreal, Canada.)

1050, and 1100 °C. For the samples, the energy of the Ge implant was varied, creating EOR damage from 200 to 1400 Å deep. By observing the amount of transient B diffusion, the average point-defect supersaturation ratios for each anneal were calculated. The results are shown in Fig. 26 [37]. It can be seen that no detectable enhanced B diffusion occurs if the B implant and the EOR damage are less than 550 Å deep.

Design curves that account for implantation damage annealing and B diffusion length are shown in Fig. 27 for B [37]. The Ge data from Fig. 5 are mapped onto curves of total B diffusion length versus $1/T$ for various damage depths to form a damage removal curve. Total diffusion length include all contributions to B diffusion that produce the measured profile. The curves of total B diffusion length versus $1/T$ were obtained from a combination of measurements and simulations. The diffusion data included some results with 550-600 °C preanneals, which cause the final junctions after RTA to be up to 100 Å deeper. To the left of the damage removal curve (higher RTA temperature), EOR damage removal is complete. In Fig. 27 it can be seen that if the EOR damage is 1000 Å deep, then a 10-sec, 1100 °C RTA is required, but the B diffusion length is 800 Å. At lower Ge and B implant energies such that the EOR damage is 200 Å deep, a 900 °C, 10-sec RTA is sufficient with only 130 Å diffusion. For these

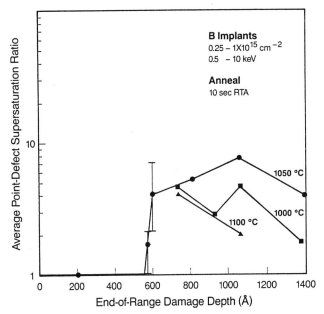

FIGURE 26. Point-defect supersaturation ratios for B implants in Ge-preamorphized Si during 10-sec RTA. (Reprinted with permission from R. B. Fair, *IEEE Trans. Electron Dev.* **37**, 2237. © 1990 IEEE.)

conditions, the surface proximity reduces the point-defect supersaturation almost to the point where the thermal diffusion limit is achieved. The breaks in the diffusion length curves at 1050 °C result from the dominance of thermally assisted B diffusion over damage-assisted B diffusion above this temperature.

The conditions for creating minimum depth, defect-free p^+n junctions have been determined for RTA processing as a function of B or BF_2 implantation energy with and without Ge preamorphization [37]. Using damage-dependent RTA models in PREDICT 1.4 [70] and the curves in Fig. 5, simulated curves of junction depth versus B or BF_2 implant energy for a dose of 2×10^{15} cm^{-2} are shown in Fig. 28. The Ge preamorphization implant energies were adjusted to set the damage depth, x_{EOR}, equal to the as-implanted B depth at 1×10^{17} cm^{-3}. The annealing temperatures were set so that EOR damage would be removed in 10 sec. For B implants in crystalline Si, a 950 °C, 10-sec RTA was used for all B energies. This anneal is sufficient to remove substrate damage in the junction region so that junction leakage currents are minimized [3]. The anneal is not sufficient to remove category V, projected range defects that do not affect leakage currents.

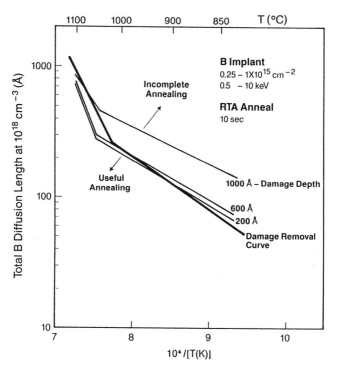

FIGURE 27. Damage removal/diffusion tradeoff for B-implanted, Ge-preamorphized Si. The damage removal curve is the minimum temperature required to anneal damage in 10 sec at the damage depths shown. The corresponding B diffusion lengths are shown for damage 1000, 600, 200 Å deep. (Reprinted with permission from R. B. Fair, *IEEE Trans. Electron Dev.* 37, 2237. © 1990 IEEE.)

In general, B and B + Ge implants produce deeper junctions than BF_2 + Ge for a given implant energy, because for BF_2 implants the effective B implant energy is about 18% of the total BF_2 molecular implant energy. Even taking the energy difference into consideration there still remains a slight difference in the junction depth. A comparison between a 6-keV BF_2 implant and a 0.5-keV B implant, both with the same Ge implant, is shown in Fig. 29. Both implants received a 1050 °C, 10-sec RTA [37]. Even though the effective energy of the BF_2 implant is twice the energy of the B implant, the latter produces a 15% deeper junction. It can be speculated that the EOR damage produced in the B and Ge implants anneals more rapidly, thus exposing the B to the unscreened flux of self-interstitials produced by self-interstitial cluster annealing below the α/c interface. It is apparent that more diffusion occurs in the high-concentration portion for the B + Ge sample, which would be explained by this effect.

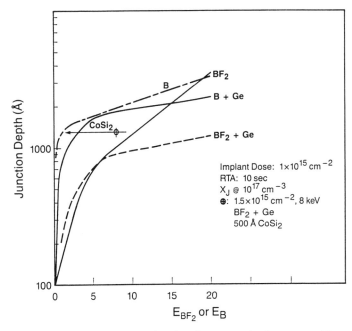

FIGURE 28. Minimum-depth, damage-free junctions versus implant energy. The curves are simulated with annealing temperatures necessary to remove damage in 10 sec from the junction region. (Reprinted with permission from R. B. Fair, *IEEE Trans. Electron Dev.* **37**, 2237. © 1990 IEEE.)

Using Ge preamorphization with BF_2 produces a shallower, defect-free junction than BF_2 alone. This result is due to the fact shown in Fig. 5 that EOR damage produced by Ge implantation is more easily removed than that produced by BF_2 implantation. For example, for $x_{EOR} = 200$ Å, damage annealing requires a 900 °C RTA for Ge implant damage, whereas a 1050 °C RTA is needed for BF_2. Thus, these higher temperatures drive the BF_2-only junctions deeper. This example illustrates the value of defect engineering in which one type of damage can be replaced by another to produce a desired effect.

In Fig. 28 a datum is shown that represents implantation in an existing 500 Å $CoSi_2$ layer with subsequent RTA [102]. Simulations show that because B diffusion in $CoSi_2$ is so rapid, the junction depth after RTA is almost independent of implant energy. Also B diffusion from a silicide is enhanced above intrinsic diffusion at 900 °C or more. The result is that junctions less than 1000 Å formed by silicide outdiffusion may not be practical with a realistic thermal budget requirement [2]. (See Chapter 7 for more recent results.)

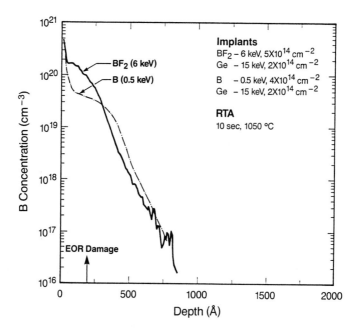

FIGURE 29. Comparison of low-energy B and BF$_2$ implants into Ge-preamorphized Si after 10 sec RTA. The EOR damage depth is 200 Å. (Reprinted with permission from R. B. Fair, *IEEE Trans. Electron Dev.* **37**, 2237. © 1990 IEEE.)

Finally, the real opportunity for creating ultrashallow junctions occurs when the B or BF$_2$ energies are so low that the EOR damage can be rapidly removed, i.e., the fourth-power dependence of τ on x_{EOR} in Eq. 6.1. From Fig. 28 the rapid decrease in defect-free junction depth occurs at implant energies less than 6 keV for BF$_2$ and 2 keV for B implants. In these energy ranges the following additional advantages occur:

- Preamorphization is not needed to secure a shallow junction but may be needed for complete dopant activation [3].
- Implantation directly into Si gives shallower junctions than implantation and diffusion from a silicide.

4. Self-Amorphizing Implants

Heavy implants of As, P, BF$_2$, or Sb will produce amorphous layers that coincide with an impurity concentration in the profile of about 1.6×10^{19} cm^{-3} [32]. During a subsequent RTA step, transient diffusion usually occurs at concentrations below the solid solubility level at the

RTA temperature for As, at the profile kink for P, or at the solubility level above 900 °C and at C_{enh} below 900 °C for B [67]. No transient diffusion is observed for Sb implants because Sb is a vacancy diffuser and the self-interstitial is the dominant point defect in the region where enhanced diffusion is normally observed.

a. Short and Long-Time Diffusion Transients. Reports of anomalous diffusion of As during RTA have been made by numerous workers [16, 103–109]. On the other hand, others have reported no enhanced As diffusion during RTA [46, 110, 111]. Since these early reports, the As transient has definitely been observed at lower temperatures where thermally assisted As diffusion is small [22]. An example from Ref. 22 is shown in Fig. 30a in which a 2×10^{15}-cm^{-2}, 40-keV As implant is annealed at 900 °C for 15, 30, and 60 sec. It can be seen that the transient has saturated before 15 sec of annealing has occurred. If one of these samples (30-sec RTA sample) is then placed in a furnace at 850 °C for 60 min, additional transient diffusion is observed as shown in Fig. 30b. Similar results have also been reported for P and BF$_2$ implants [22], and these results are shown in Figs. 31 and 32 respectively. In all three cases a short transient is observed during RTA at 900 °C that is less than 15 sec in duration. A longer time transient then is seen at 850 °C, which produces additional diffusion with a time constant of around 5 min [22].

The existence of two time constants is believed to be caused by two different sources of point defects. As shown in Fig. 4, a 900 °C anneal would dissolve clusters within 3 sec. The resulting self-interstitial supersaturation in the Si is believed to account for the transient during RTA.

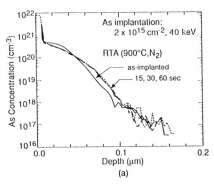
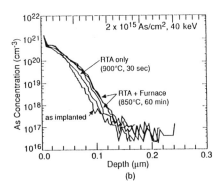

FIGURE 30. SIMS arsenic profiles for 2×10^{15} cm^{-2} implants in ⟨100⟩ Si at 40 keV annealed in N$_2$ (a) RTA at 900 °C for 15, 30, and 60 sec; (b) additional furnace anneal at 850 °C for 60 min. (After Ref. 22, reprinted with permission from *J. Elec. Mat.* **18** (2), 143, a publication of The Minerals, Metals, & Materials Society, Warrendale, PA 15086.)

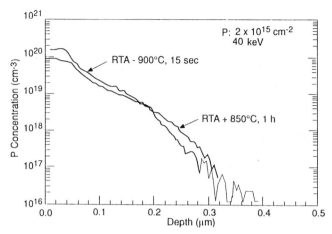

FIGURE 31. SIMS phosphorus profiles for 2×10^{15} cm^{-2} implants in $\langle 100 \rangle$ Si at 40 keV and annealed by RTA and RTA plus furnace annealing.

However, a second and longer transient would be anticipated from the annealing of EOR damage, except for very shallow implants. Also the results in Fig. 4 predict that the second transient would last approximately 2 h at 850 °C. Nevertheless, the results in Figs. 30-32 may be explained on the basis of rapid dissolution of shallow EOR dislocations that exist 300–500 Å from the Si surface. The results of TEM studies of these samples

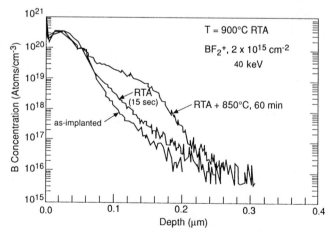

FIGURE 32. SIMS boron profiles for 2×10^{15} cm^{-2} BF$_2$ implants in $\langle 100 \rangle$ at 40 keV and annealed by RTA and RTA plus furnace annealing. (After Ref. 22, reprinted with permission from *J. Elec. Mat.* **18**(2), 143, a publication of The Minerals, Metals & Materials Society, Warrendale, PA 15086.)

Junction Formation in Silicon by Rapid Thermal Annealing

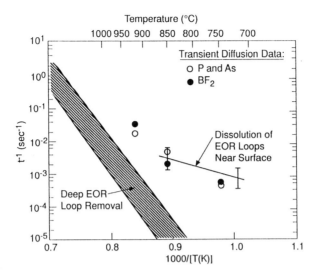

FIGURE 33. Temperature dependence on the time constant of implant damage annealing (deep and shallow EOR loops) and transient diffusion of implanted P, As, and BF_2.

are shown in Fig. 33 where the inverse annealing times of both deep (>1000 Å) and shallow (300–500 Å) EOR damage are plotted versus temperature [22]. The measured temperature dependence of the transient-enhanced diffusion time constant is also shown in Fig. 33. It can be seen that t_e for the shallow P, As, and BF_2 implants coincides approximately with the annealing time constant of the corresponding shallow end-of-range damage. A second important observation is that the activation energy for removal of shallow EOR damage is approximately 2.1 eV, rather than 5 eV for deep EOR damage. This reduced energy is much less than the activation energies for intrinsic dopant diffusion (3.4–4.1 eV), which is, in turn, fundamental for producing damage-free, very shallow junctions at low temperatures. This unexpected benefit from Mother Nature was not anticipated by those who predicted that high-temperature RTA (1100 °C) would be required for defect-free junctions.

b. Activation Energy for Transient Diffusion. The activation energy for enhanced As diffusion during RTA has been reported to be 1.8 eV [104]. For P, the corresponding activation is 2.2 eV [112]. Rapid thermal annealing of self-amorphizing BF_2 implants at temperatures above 900 °C produces transient diffusion with an activation energy of 1.8 eV [113]. On the other hand, thermally assisted diffusion of As, P, and B requires energies of 3.4 to 4.1 eV [77].

The reduction in diffusion activation energy during RTA is a consequence of externally generated self-interstitials that make up the transient supersaturation. Equation 6.7 indicates that under such supersaturation conditions the dopant diffusion energy is $Q - \Delta H_I \simeq \Delta H_m$, where ΔH_m is the dopant self-interstitial pair migration energy. For example, using $Q_{As} = 4.1$ eV and $\Delta H_I = 2.25$ eV gives a pair migration energy of 1.85 eV, in good agreement with the results reported for As above.

The observation that the enhanced diffusion activation energy for As is reduced by the formation enthalpy of self-interstitials has been observed elsewhere. Recently, Tsai et al. [114] saw a 2.5-eV reduction in buried layer As diffusion in the presence of a high-concentration P surface layer. In addition, PREDICT 1.4 simulations show the net effect of As implant damage annealing is a reduction in diffusion energy by 2.1 to 2.5 eV [113].

The results for BF_2 or B implants are complicated by the temperature dependence of the diffusion energy for B. According to Fair [67], below 900 °C thermally assisted B diffusion is dominated by I^x, and the activation energy is 2.85 eV [77]. Thus, the energy for enhanced diffusion from external point-defect injection is 0.6 eV [78]. However, above 900 °C it is believed that BI^+ pair diffusion dominates [67]. Thus, $Q_B^+ - \Delta H_I^+ \simeq$ 1.3–1.6 eV [77]. This range of enhanced diffusion activation energies is near to the reported value of 1.8 eV [113].

c. Two-Time-Constant Model. The combination of RTA and furnace annealing demonstrates that two different point-defect sources exist when an amorphous layer is present. Kim formulated a model for the time constant of transient diffusion based on the annealing characteristics of point-defect clusters, the coarsening of EOR dislocation loops, and the sink action of the Si surface [115, 116].

The point-defect clusters were modeled as small spheres approximately 10 Å in radius, r_c. The concentration of self-interstitials at the surface of a cluster sphere is given by [117].

$$C_I(r_c) = C_I^{eq} \exp(2\sigma\Omega/kTr_c) \qquad (6.10)$$

where σ is the surface energy of the cluster, Ω is the atomic volume of Si, and C_I^{eq} is the equilibrium concentration of self-interstitials in the Si crystal. Cluster dissolution occurs by emission of self-interstitials as depicted in Fig. 34. The rate of generation of self-interstitials is given by

$$R_I = 4\pi r_c^2 J_I(r_c) \qquad (6.11)$$

where $J_I(r_c)$ is the flux from the surface of the spherical cluster. Thus the time required to completely dissolve a cluster can be calculated by summing up the time segments (Dt_i) required to remove a volume element of the

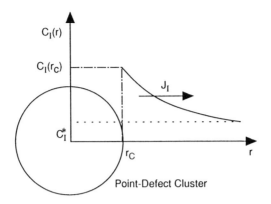

FIGURE 34. Schematic diagram for the dissolution of point-defect clusters by the diffusion of Si self-interstitials to the bulk. (After Ref. 115.)

sphere until the sphere decreases to the radius of a single Si atom [115]. By using reasonable values for σ, $D_I C_I^{eq}$ and r_c, Kim found that the inverse of the dissolution time constant of point-defect clusters, τ_{PD}^{-1}, had a 4.3 eV activation energy. This value is consistent with the Si self-diffusion activation energy below 1000 °C [77].

Results from Kim's work are shown plotted in Fig. 35 and are compared with the measured curve [31] previously shown in Fig. 4.

The annealing of EOR dislocation loops was assumed to commence after the dissolution of clusters was completed [115]. Loop annealing was assumed to occur by a coarsening process similar to the Ostwald ripening process. The concentration of self-interstitials at the loop perimeter of radius r_L is given by

$$C_I(r_L) = C_I^{eq} \exp(\Gamma A/kTr_L), \qquad (6.12)$$

where A is the atomic area of Si and Γ is the line energy of the dislocation loop. It should be noted that the concentration of self-interstitials on a small radius loop is larger than $C_I(r_L)$ on a larger loop. Thus, during the loop dissolution a gradient of self-interstitials will occur between loops of different radii. The effect is illustrated in Fig. 36. The result is a flux from smaller loops to larger loops, causing large loops to grow and coarsen at the expense of small loops.

After cluster dissolution there will exist a distribution of loop sizes, $f(r_L, t = 0)$ and number $N(r_L, t = 0)$, where [115]

$$f(r_L, t) = \lim_{\Delta r_L \to 0} \frac{N(r_L + \Delta r_L, t) - N(r_L, t)}{\Delta r_L} \qquad (6.13)$$

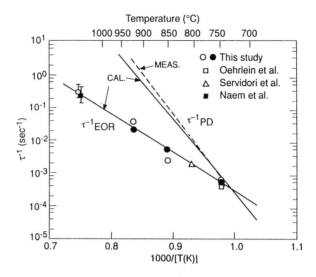

FIGURE 35. Comparison of calculated and measured time constants for EOR damage and cluster dissolution. (After Ref. 115.)

This distribution function is unknown at $t = 0$, but Kim assumed an exponential function for his calculations. Then a continuity equation for $f(r_L, t)$ can be solved [118], given by

$$\frac{\partial f(r_L, t)}{\partial t} + \frac{\partial}{\partial r_L}\left(f(r_L, t)\frac{dr_L}{dt}\right) = 0, \quad (6.14)$$

with the rate of change of the dislocation loop radius expressed as

$$\frac{dr_L}{dt} = -\frac{4D_I A}{\pi}[C_I(r_L) - C_I^{eq}] \quad (6.15)$$

Finally, the supersaturation ration, $S_I = C_I/C_I^{eq}$ can be calculated:

$$\frac{\partial S_I}{\partial t} = \int_0^\infty 8r_L D_I\left[\exp\left(\frac{\Gamma A}{kTr_L}\right) - S_I\right]f(r_L, t)\,dr_L - \frac{S_I - 1}{\tau_s} \quad (6.16)$$

where $\tau_s = L^2/D_I$ and L is the average distance between the dislocation loops and the surface. The time constant for dissolution was taken as the time required for S_I to reach 1.1. The result of solving for S_I and extracting the EOR loop dissolution time constant, τ_{EOR}, is shown in Fig. 35. The calculated curve is compared with data [115]. Values of self-interstitial diffusivity that give the best fit to these data were based on the following expression

$$D_I = 1.8 \times 10^{-3} \exp(-2.1\,\text{eV}/kT) \quad (\text{cm}^2/\text{sec}). \quad (6.17)$$

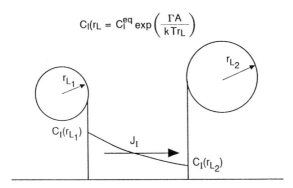

FIGURE 36. Schematic diagram to illustrate the self-interstitial concentration gradient established due to the size difference of dislocation loops, as part of the end-of-the-range loop coarsening process. (After Ref. 115.)

The 2.1-eV activation energy is much less than the 5 eV associated with D_I in the Si bulk. However, 2.1 eV is consistent with the activation energy associated with the shrinkage of surface oxidation–induced stacking faults [119]. The result is also consistent with Si diffusion via vacancies [120] that may become important at low temperatures or near the Si surface.

Further progress in modeling will be made when it becomes possible to calculate the damage state of the Si after ion implantation as well as the distributions of point defects created during stable damage evolution or dissolution. To date, no work has been done to calculate distributions of displacement damage or distributions of excess point defects produced by implantation. The MARLOWE [121] program uses the binary collision approximation to calculate dynamic damage processes during implantation [122]. Alternative methods have also been proposed with the goal of calculating the point-defect profiles at the end of the implant process for postimplant diffusion modeling [123]. However, the difficulty with such calculations is that they may be qualitative at best. It is known that radiation-induced migration of self-interstitials occurs at low temperature [124, 125]. Thus distributions of point defects probably don't remain as implanted, but diffuse away or coalesce into stable clusters.

II. Dopant Activation

The electrical activation of implanted impurities by RTA is an important issue, especially for shallow junctions created by low thermal budget processing. Thus, a key question is: how can a high activation be achieved without loss of junction depth control? It is desirable to achieve 100% electrical activation up to the solubility limit of the dopant. High electrical activity is needed to assure low contact resistances to junctions because contact resistance is sensitive to active surface concentrations. In addition, high activation allows for the lowest possible sheet resistances of the implanted/diffused layers. The attainment of high carrier mobility is also important and requires that residual defects be minimized in the doped layer. In the following sections electrical activation of B, P, and As implants will be described. First, however, a basic theoretical framework will be established to understand the role of damage on activation.

A. Theory

The electrical activation of an implanted impurity will be limited by the solubility of the impurity and the presence of Si defects. Annealing will reduce the areal lattice damage density, $N_d(t)$, from some initial volume density N_d^0 according to the equation [126]

$$\frac{dN_d(t)}{dt} = -v_d N_d^0 \exp(-E_d/kT), \tag{6.18}$$

where v_d is the velocity of damage decay and E_d is the damage decay energy. If it is assumed that the restoration of crystalline order during annealing is responsible for inducing free carrier production, then it can be shown that the change in areal free carrier density, ΔN_s, tracks the decrease in damage through the expression [127]

$$\Delta N_s = N_d(0) - N_d(t) \tag{6.19}$$

$$= N_d^0 v_d t \exp(-E_d/kT). \tag{6.20}$$

Thus, we would expect ΔN_s to increase with annealing time up to the maximum doping density available or to the solubility limit, whichever is less. Solving for E_d, it can be seen that

$$E_d \propto -d(\ln N_s)/d(1/T). \tag{6.21}$$

Since the doped layer sheet resistance, R_s, is inversely proportional to N_s, then

$$E_d \propto -d[\ln(1/R_s)]/d(1/T). \tag{6.22}$$

Thus plotting R_s versus the inverse annealing temperature produces a curve with the slope proportional to E_d. Examples for B implants (2×10^{14} cm^{-2} at 150 keV) and As implants (5×10^{15} cm^{-2} and 1×10^{16} cm^{-2} at 150 keV) are shown in Fig. 37. All samples were annealed for 12 sec in an RTA system [128].

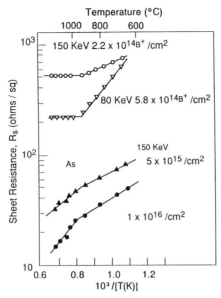

FIGURE 37. The relations between sheet resistance and annealing temperature after B and As implantation and RTA. (After Ref. 128.)

B. ACTIVATION OF BORON

The activation of B in high-dose implants into Si proceeds slowly at low annealing temperatures. An example of an 800 °C anneal is shown in Fig. 15 where it takes up to 8 h to achieve B solid solubility [64]. In addition it can be seen in Fig. 15 that the activation of B proceeds slowly in the region past the projected range of the B implant where a high density of damage exists. Similar results have been obtained from anneals at 850 °C [64]. The cause of this phenomenon is unknown at this time, but some models that speculate on the form of the inactive B are presented at the end of this section.

Improved activation of B is possible by performing RTA at or above 900 °C. An example is shown in Fig. 38 [64]. After a 30-sec RTA at 900 °C a maximum electrical activity is achieved for a 2×10^{15} cm^{-2} implant that is 2-3 times greater than the equilibrium solubility. If the nonelectrically active B is caused by B precipitation or clustering, then during RTA the activation may occur faster than the process of precipitation. However, as the annealing time is increased the free hole concentration decreases to the solubility level. The dissolution of B precipitates then maintains the active dopant at C_{sol} [64]. The observation of very high levels of activation after

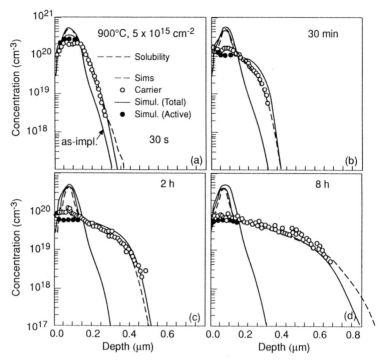

FIGURE 38. Evolution during annealing at 900 °C of experimental and simulated dopant and carrier profiles in B-implanted wafers at 20 keV, 5×10^{15} cm^{-2}. (After Ref. 64.)

RTA of high-dose B implants was also noted by McMahan et al. [129]. The high carrier concentrations were correlated with lower defects densities in the implanted profile peak compared to furnace anneals.

Another approach to increasing the active B concentration is through a preamorphization implant [130]. During solid-phase epitaxial regrowth, implanted B can reach substitutional lattice sites and become active at relatively low temperatures. This result follows because the dopant atoms are incorporated on lattice sites and so not have to displace a Si lattice atom in the process. In addition, the regrown Si is usually defect free.

An example comparing activation of B implanted into crystalline Si and preamorphized Si is shown in Fig. 39. The B implantation was 1×10^{15} cm^{-2} at 50 keV, and all annealings (500–950 °C) were done in a nitrogen ambient for 30 min [131]. The preamorphization was accomplished using a 1×10^{15} Ge/cm^2, 300-keV implant. The sheet resistance versus annealing temperature is shown for the crystalline and preamorphized samples in Fig. 39a, Hall mobilities are shown in Fig. 39b, and percentage

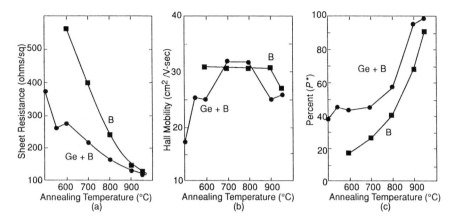

FIGURE 39. Activation versus annealing temperature of B and Ge + B implants: (a) sheet resistance, (b) Hall mobility, and (c) percent activation. (After Ref. 131.)

activation in Fig. 39c. Similar results have been attained during RTA and show improved activation in preamorphized samples up to 1050 °C [41, 54].

A special case exists for low-energy B or BF_2 implants into preamorphized Si. The fraction of activated B as a function of annealing temperature depends on the depth of the amorphous layer relative to the implant depth. Two effects have been observed if the amorphous layer is shallower than the B implant [3]:

- Reduced B activation occurs if EOR loops exist in the implanted region. A decrease in electrically active B occurs as the dislocation loops coalesce from defect clusters at moderate annealing temperatures (600–800 °C).
- Decreasing the preamorphization depth decreases the fraction of activated dopant because fewer B atoms reside within the regrown layer.

The effect of 550–600 °C preanneals on B activation has been shown to be relatively minor [3]. The process of regrowing the amorphous layer at low temperatures yields only marginal improvements in sheet resistance if followed by additional annealing above 950 °C.

For medium-to-high dose B implants into crystalline Si it was previously pointed out in Fig. 15 that for $T < 900$ °C transient diffusion was only observed at B concentrations below $C_{enh} = n_i$. The static peak of the profile above n_i is known to be electrically inactive [64, 65, 132]. It has been proposed that the B interstitials produced during ion implantation become trapped on implantation damage centers [65, 75]. However, the damage profile is different from the impurity profile, whereas the inactive B follows the as-implanted B distribution above n_i. Other models include the formation of B pairs or clusters in the presence of a large self-interstitial supersaturation [133]. Thus, self-interstitials are created that displace substitutional B atoms to form migrating B interstitials. The concentration of B interstitials is so high that homogeneous nucleation of B-containing defects may occur, which can effectively trap additional B interstitials. At the end of the point-defect transient the B defects are able to dissolve forming other, more stable defects [134].

It should be pointed out that none of these models can explain why the active B concentration tracks with n_i as a function of temperature (below 900 °C). It is hard to imagine a defect-driven, B-trapping process that is related to an intrinsic electrical property of the Si crystal, except by coincidence.

C. Activation of Phosphorus

Activation limits of P-doped Si have been defined in terms of a solubility of electrically active P [135]. Subsequent studies have shown that the electrically inactive P is in the form of phosphide precipitates [136, 137] rather than electrically compensating P-point defect pairs [138, 139]. However, it has also been shown that 100% activation of P implants is possible using laser annealing [137]. Free-carrier concentrations of up to 5×10^{21} cm^{-3} were obtained in laser-annealed layers free of structural imperfections. All of the P was on substitutional lattice sites [137, 140]. Subsequent thermal anneals reduced the carrier densities to equilibrium values dependent only on temperature.

For the annealing time associated with high-temperature RTA, the equilibrium, active P concentrations are rapidly achieved for diffusions from spin-on phosphosilicates glasses [141]. However, a higher activation is generally observed for RTA samples when compared with furnace-annealed samples [142–144]. This result is believed to be caused by the rapid wafer cooling in an RTA furnace, which helps "freeze in" a higher carrier concentration [141].

Activation of P-implanted layers at lower temperatures is complicated by the presence of lattice defects. In general, the following observations have been made:

- Implanted P in very low–damage regions can be activated by annealing at temperatures as low as 400 °C [145].
- For damage densities on the order of 20%, carrier trapping centers may be formed to compensate the activation. Temperatures above 500 °C are needed to achieve electrical activity [146].

- At damage densities up to 75%, amorphous clusters formed by ion implantation overlap, producing larger damage regions. Decomposition of these regions during annealing may produce vacancies that enhance activation of P atoms onto substitutional sites [146].
- For fully amorphized regions (100% damage density), movement of the α/c interface during epitaxial regrowth incorporates P atoms on lattice sites. However, the activation of P atoms deeper than the α/c interface depends on the annealing temperature used [147].

D. Activation of Arsenic

The activation of As is similar to that of P. However, controversy still remains regarding the form of electrically inactive As [148]. It has been proposed that complex point defects exist in thermal equilibrium that result in eletrically inactive As species. Two such equilibrium reactions are

$$3As^+ + e^- \rightleftharpoons As_3^{++} \xrightarrow{25\,°C} As_3 \qquad (6.23)$$

proposed by Tsai *et al.* [149], and the formation of an As–vacancy complex by Fair [150, 151]:

$$2As^+ + V + 2e^- \rightleftharpoons As_2V \qquad (6.24)$$

Indeed, two As pairs have been observed directly by Wickert and Swanson using perturbed angular correlation measurements [152]. However, there is convincing evidence that As precipitation is also important [148]. Arsenide precipitates in the form of thin platelets of radius 40 Å have been observed.

The effect of low thermal budget anneals and furnace ramps on the electrical activation of As was recently studied by Orlowski *et al.* [153]. The equilibrium between active and inactive As can generally be described by a rate equation such as

$$\frac{\partial C_{\text{active}}}{\partial t} = K_D F(C_{\text{inactive}}, C_{\text{chemical}}) - K_C R(C_{\text{active}}, C_{\text{chemical}}), \qquad (6.25)$$

where K_D is a declustering or deprecipitation coefficient, K_C is a clustering or precipitation coefficient, and F and R are functions of the concentrations indicated and depend on the model chosen. Computer simulation of activation or deactivation requires that Eq. 6.25 be solved simultaneously with the diffusion equation for As [153] given as follows:

$$\frac{\partial C_{\text{chemical}}}{\partial t} = \frac{\partial}{\partial x} D \frac{\partial C_{\text{active}}}{\partial x} + Z u C_{\text{active}} \frac{\partial \psi}{\partial x}, \qquad (6.26)$$

where D is the mobile As diffusivity, Z is the charge state, and Ψ is the crystal potential. The second term on the right-hand side of Eq. 6.26 is the electric field dependence.

Table I Effect of ramp-down rate on sheet resistance (the thermal cycle at 1000 and 950 °C was different from the thermal cycle at 900 °C) (from Orlowski et al., Ref. 153).

Temperature	Ramp	R_s (calculated)[a]	R_s (measured)	% change from rapid
900 °C	Slow	42.7	41.5	18.2%
	Intermediate	39.4	40.8	16.2%
	Rapid	35.1	35.1	—
950 °C	Intermediate		45.9	31%
	Rapid		35.0	—
1000 °C	Intermediate		29.6	29.2%
	Rapid		22.9	—

[a] Calculated values of sheet resistance were obtained using the model of Tsai and co-workers (Ref. 149).

Depending on the initial conditions chosen, the solution to Eq. 6.25 will yield different solutions for active As concentration versus anneal time. Using the activation model of Tsai et al. [149], it was shown that for high As doping less than 5 min is required to achieve equilibrium at 1000 °C. However, at 900 °C the time approaches 60 min. Thus, during most common RTA cycles, the activated As concentration for high-dose As implants is in transition from the initial value after epitaxial regrowth to a final equilibrium value. This also means that the activation depends on the ramp-up and ramp-down rates of the RTA cycle [153].

The dependence of ramp-down rate on the sheet resistance of 2×10^{15}-cm^{-2}, 80-keV As implants annealed at 900, 950, and 1000 °C is shown in Table I [153]. The rapid ramp-down rates yielded samples with lower sheet resistance by "freezing in" the higher electrical activity. This result is due to the clustering/precipitation process having a slow time constant relative to the cooling rate.

III. Summary and Conclusions

The feasibility of using isothermal RTA in annealing ion implanted layers for forming junctions has been investigated for the past 10 years. While many of the scientific details surrounding defect formation, transient diffusion, and dopant activation remain to be clarified, RTA intrinsically is a viable annealing process that is essential for fabricating advanced silicon devices. The key remaining issues for insertion of RTA into manufacturing are addressed in Chapters 3, 8, 9, and 10.

Two of the keys to fabrication of shallow, scaled junctions for ULSI devices are low-energy ion implantation and short-time (< 10 sec) annealing at moderate temperatures ($\simeq 900\,°C$). Both approaches address the physical phenomena responsible for creating defect-free junctions. In addition, the issue of dopant activation can be addressed by utilizing the beneficial effects of amorphization on electrical activity following epitaxial regrowth.

This chapter has focused on mechinisms of transient enhanced diffusion of ion-implanted dopants in Si during low thermal budget processing. In general, the magnitude and duration of diffusion transients are determined by the types of implant damage produced, the damage location, and the supersaturation of point defects generated during annealing. Two types of implant damage seem to dominate junction formation for the majority of cases: end-of-range dislocation loops at an α/c interface and point-defect clusters. Dissolution of point-defect clusters is known to occur at temperatures as low as 550 °C, with the resulting supersaturation of self-interstitials driving enhanced diffusion. Extended dislocations can perturb the supersaturation distribution by acting as point-defect sinks. When cluster dissolution ends, extended dislocations may then become point-defect generators, providing a second transient for enhanced diffusion of dopants. If these dislocations are in the vicinity of the junction space-charge region, then annealing must continue until the damage is either removed or the junction depth passes the location of the damage. Junction reverse-leakage currents are directly affected by the concentration and position of residual defects relative to the junction itself.

Finally, relatively few computer-aided design tools are available to assist the device or process engineer in the brave new world of RTA processing. Simulation software based on first-principle physics is years away. Phenomenologically based simulators such as PREDICT 1.5 contain RTA diffusion and defect models based on experimental data. While the debate goes on regarding the usefulness of such tools, nevertheless they are pragmatic and capture the observations of hundreds of experiments in the form of physically based models with macroscopically observed variables. Thus, such descriptive models embedded in software can be very useful in design while all of the details and explanations of RTA are being worked out.

References

1. W. G. Pfann, "Semiconductor Signal Translating Device," U.S. Patent No. 2,597,028 (1952).
2. R. B. Fair, *Proc. IEEE* **79**(11), 1687–1705 (1990).
3. S. N. Hong, G. A. Ruggles, J. J. Wortman, and M. C. Ozturk, *IEEE Trans. Electron Dev.* **38**, 476 (1991).

4. H. J. Bohm, H. Wendt, H. Oppolzer, K. Masseli, and R. Kassing, *J. Appl. Phys.* **62**, 2784 (1987).
5. V. Probst, P. Lippens, L. Van den Hove, K. Maex, H., Schaber, and R. De Keersmaecker, *Proc. of the European Solid State Dev. Res. Conf.* Bologna, Italy (September, 1987), p. 437.
6. H. Jiang, C. M. Osburn, P. Smith, D. Griffis, G. McGuire, and G. A. Rozgonyi, Meeting of Electrochem. Soc., Los Angeles, Recent News Paper #737 (May 1989).
7. P. G. Carey, T. W. Sigmon, R. L. Press, and T. S. Fahlen, *IEEE Electron Dev. Lett.* **EDL-6**, 291 (1985).
8. T. Sameshima, S. Usui, and M Sekiya, *J. Appl. Physics* **62**, 711 (1987).
9. PREDICT—*PR*ocess *E*stimator for the *D*esign of *I*ntegrated *C*ircuit *T*echnologies, Microelectronics Center of North Carolina, Research Triangle Park, NC.
10. S. Nygen, Ph.D. Thesis, The Royal Institute of Technology (KTH), Stockholm, Sweden (1989).
11. R. D. J. Verhaar, A. A. Bos, H. Kraaji, R. A. M. Wolters, K. Maex, and L. Van den Hove, *19th European Solid State Device Res. Conf.*, Berlin (September 1989).
12. C. M. Osburn and A. Reisman, *J. Supercomputing* **1**(2), 149 (1987).
13. J. R. Marchiando, P. Roitman, and J. Albers, *IEEE Trans. Electron Dev.* **ED-32**, 2322 (1985).
14. A. E. Michel, in *Rapid Thermal Processing* (T. O. Sedgwick, T. E. Seidel, and B. Y. Tsaur, eds.), p. 3. Mat. Res. Soc., Pittsburgh (1986).
15. R. T. Hodges, J. E. E. Baglin, A. E. Michel, S. M. Mader, and J. C. Gelpey, in *Energy Beam-Solid Interactions and Transient Thermal Processing* (J. C. C. Fan and N. M. Johnson, eds.), p. 253. Mat. Res. Soc., Pittsburgh (1984).
16. R. B. Fair, J. J. Wortman, and J. Liu, *J. Electrochem. Soc.* **131**, 2387 (1984).
17. W. K. Hofker, *Philips Res. Rep. Suppl.* **8** (1975).
18. R. B. Fair, *IEEE Trans. Electron Dev.* **35**, 285 (1988).
19. M. Servidori, R. Angelucci, F. Cembali, P. Negrini, S. Solmi, P. Zaumsel, and U. Winter, *J. Appl. Phys.* **61**, 1834 (1987).
20. S. Solmi, R. Amngelucci, F. Cembali, M. Servidori, and M. Anderle, *Appl. Phys. Lett.* **51**, 331 (1987).
21. K. Cho, M. Numan, R. G. Finstad, W. K. Chu, J. Liu, and J. J. Wortman, *Appl. Phys. Lett.* **47**, 1321 (1985).
22. Y. Kim, H. Z. Massoud, and R. B. Fair, *J. Electron Mater* **18**, 143 (1989).
23. D. A. Antoniadis, A. M. Lin, and R. W. Dutton, *Appl. Phys. Lett.* **33**, 1030 (1978).
24. S. Mizuo and H. Higuchi, *Jpn. J. Appl. Phys.* **20**, 739 (1981).
25. R. M. Harris and D. A. Antoniadis, *Appl. Phys. Lett.* **43**, 937 (1983).
26. F. Cembali, M. Servidori, and A. Zani, *Solid State Electronics* **28**, 933 (1985).
27. V. S. Speriosu, *J. Appl. Phys.* **52**, 6094 (1981).
28. P. Zaumseil, U. Winter, F. Cembali, M. Servidori, and Z. Sourek, *Phys. Stat. Sol. A* **100**, 95 (1987).
29. M. Servidori, Z. Sourek, and S. Solmi, *J. Appl. Phys.* **62**, 1723 (1987).
30. M. Servidori, P. Zaumseil, U. Winter, F. Cembali, and A. M. Mazzone, *Nucl. Inst. Meth. Phys. Res. B* **22**, 497 (1987).
31. T. E. Seidel and A. U. MacRae, *Rad. Effects* **7**, 1 (1971).
32. K. S. Jones, S. Prussin, and E. R. Weber, *J. Appl. Phys.* **62**, 4114 (1987).
33. T. Y. Tan, *Phil. Mag.* **44**, 101 (1981).
34. J. Narayan, D. Fathy, D. S. Wen, and O. W. Holland, *J. Vac. Sci. Tech. A* **2**, 1303 (1984).
35. T. E. Seidel, D. J. Lischner, C. S. Pai, R. V. Knoell, D. M. Maher, and D. C. Jacobson, *Nucl. Inst. Meth. Phys. Res. B* **7/8**, 251 (1985).

36. A. C. Ajmera and G. A. Rozgonyi, *Appl. Phys. Lett.* **49**, 1269 (1986).
37. R. B. Fair, *IEEE Trans. Electron Dev.* **37**, 2237 (1990).
38. E. Ganin and A. Marwick, in *Ion Beam Processing of Advanced Electronic Materials* (N. W. Cheung, A. D. Marwick, and J. B. Roberto, eds.), Vol. 147, Materials Research Society, Pittsburgh, PA (1989), pp. 13-18.
39. A. C. Ajmera, G. A. Rozgonyi, and R. B. Fair, *Appl. Phys. Lett.* **52**, 813 (1988).
40. M. C. Öztürk, J. J. Wortman, C. M. Osburn, A. Ajmera, G. A. Rozgonyi, E. Frey, W. K. Chu, and C. Lee, *IEEE Trans. Electron Dev.* **35**, 659 (1988).
41. G. A. Ruggles, S-N. Hong, J. J. Wortman, M. Öztürk, E. R. Myers, J. J. Hren, and R. B. Fair, in *Processing and Characterization of Materials Using Ion Beams*, Vol. 128, Materials Research Society, Pittsburgh, PA (1989), pp. 611-616.
42. D. S. Wen, P. L. Smith, C. M. Osburn, and G. A. Rozgonyi, *J. Electrochem. Soc.* **136**(2), 446 (1989).
43. T. Sands, J. Washburn, E. Meyers, and D. K. Sadana, *Nucl. Instr. Math. Phys. Res. B* **7/8**, 337 (1985).
44. E. Meyers, J. J. Hren, S-N. Hong, and G. A. Ruggles, in *Ion Beam Processing of Advanced Electronic Materials* (N. W. Cheung, A. D. Marwick, and J. B. Roberts, eds.), Vol. 147, Materials Research Society, Pittsburgh, PA (1989), pp. 27-32.
45. X. Li, T. Psihsin, and L. Zhijian, in *Fundamentals of Beam-Solid Interactions and Transient Thermal Processing* (M. J. Aziz, L. E. Rehn and B. Stritzker, eds.), Vol. 100, Materials Research Society, Pittsburgh, PA (1987), p. 683-688.
46. T. E. Seidel, C. S. Pai, D. J. Lischner, D. M. Maher, R. V. Knoell, J. S. Williams, B. R. Penumalli, and D. C. Jacobson, in *Proceedings of the Materials Research Society* (D. K. Biegelsen, G. A. Rozgonyi, and C. V. Shank, eds.), Mat. Res. Soc., Pittsburgh, PA (1985), p. 329.
47. S. Prussin and K. S. Jones, "Role of Ion Mass and Implant Dose on End-of-Range Defects," paper #176 presented at Electrochemical Society meeting, Atlanta, GA (1988).
48. R. J. Culbertson and S. J. Pennycook, *Mat. Res. Soc. Symp. Proc.* **79**, 391 (1987).
49. D. K. Sadana, S. C. Shatas, and A. Gat. in *Microscopy of Semiconducting Materials*, Institute of Physics Conference Series, Instit. of Phys., London (1983).
50. W. K. Hofker, H. W. Werner, D. P. Oosthoek, and H. A. M. de Grefts, *Appl. Phys.* **2**, 265 (1973).
51. G. S. Oehrlein, S. A. Cohen, and T. O. Sedgwick, *Appl. Phys. Lett.* **45**, 417 (1984).
52. T. O. Sedgwick, A. E. Michel, V. R. Deline, S. A. Cohen, and J. B. Lasky, *J. Appl. Phys.* **63**, 1452 (1988).
53. I. D. Calder, H. M. Nagiub, D. Houghton, and F. R. Shepherd, *Mat. Res. Soc. Symp. Proc.* 35, 353 (1985).
54. R. B. Fair, in *Semiconductor Silicon 1990* (H. R. Huff, K. G. Barraclough, and J. Chikawa, eds.), Electrochemical Society, Pennington, NJ (1990), p. 429.
55. M. D. Giles, *Electronic Materials Conference*, Santa Barbara, CA (1990).
56. M. D. Giles, *J. Electrochem. Soc.* **138**, 1160 (1991).
57. G. S. Oehrlein, S. A. Cohen, and T. O. Sedgwick, *Appl. Phys. Lett.* **45**, 417 (1984).
58. S. Solmi, F. Cembali, R. Fabbri, M. Servidori, and R. Canteri, *Appl. Phys. A* **48**, 255 (1989).
59. P. A. Packan and J. D. Plummer, *Appl. Phys. Lett.* **56**, 1787 (1990).
60. T. Hara, H. Hagiwara, R. I. Chikawa, S. Nakashima, K. Mizoguchi, W. Smith, C. Wells, S. K. Hahn, and L. Larson, *IEEE Electron Dev. Lett.* **11**, 485 (1990).
61. M. C. Paek, O. J. Kwon, J. Y. Lee, and H. B. Im, *J. Appl. Phys.* **70**, 4176 (1991).
62. M. Miyake, S. Aoyama, and K. Kurchi, *Meeting of the Electrochemical Society*, Abs. #691, Honolulu (1987).

63. S. Guimares, E. Landi, and S. Solmi, *J. Appl. Phys.* **62**, 587 (1986).
64. S. Solmi, F. Baruffaldi, and R. Canteri, *J. Appl. Phys.* **69**, 2135 (1991).
65. A. E. Michel, W. Rausch, P. A. Ronsheim, and R. H. Kastl. *Appl. Phys. Lett.* **50**, 416 (1987).
66. S. Solmi and M. Servidori, in *Ion Implantation in Semiconductors* (D. Stievenard and J. C. Bourgoin, eds.), Trans. Tech., Aedermannsdorf, Switzerland (1989), p. 65.
67. R. B. Fair, *J. Electrochem. Soc.* **137**, 667 (1990).
68. N. E. B. Cowern, H. F. F. Jos, K. T. F. Janssen, and A. J. H. Wachters, *Mat. Res. Soc. Symp. Proc.* **163**, 605 (1990).
69. T. O. Sedgwick, Meeting of the Electrochem. Soc., Las Vegas, NV (1985).
70. R. B. Fair, in *Microelectronics Processing* (D. W. Hess and K. F. Jensen, eds.), American Chemical Society, Washington, DC (1989), p. 265.
71. Y. M. Kim, G. Q. Lo, H. Konoshita, and D. L. Kwong, *J. Electrochem Soc.* **138**, 1122 (1991).
72. A. E. Michel, *Nucl. Instrum. Meth. Phys. Res. B.* **37/38**, 379 (1989).
73. O. W. Holland, *Appl. Phys. Lett.* **54**, 798 (1989).
74. A. E. Michel, W. Rausch, P. A. Ronsheim, and R. H. Kastl, *Appl. Phys. Lett.* **50**, 416 (1987).
75. A. E. Michel, W. Rausch, and P. A. Ronsheim, *Appl. Phys. Lett.* **51**, 487 (1987).
76. R. Angelucci, F. Cembali, P. Negrini, M. Servidori, and S. Solmi, *J. Electrochem. Soc.* **134**, 3130 (1987).
77. R. B. Fair, in *Impurity Doping Processes in Silicon* (F. F. Y. Yang, ed.), North Holland (1981), p. 315.
78. J. Huang, D. Fan, R. J. Jaccodine, and F. Stevie, in *VLSI Science and Technology/1987* (S. Broydo and C. M. Osburn, eds.), Vol. 87-11, The Electrochemical Society, Pennington, NJ (1987), p. 340.
79. J. C. C. Tsai, D. G. Schimmel, R. B. Fair, and W. Maszara, *J. Electrochem. Soc.* **143**, 1508 (1987).
80. B. Rogers, Ph.D. Dissertation, Duke University, Durham, NC (1988).
81. J. A. Van Vechten, in *Lattice Defects in Semiconductors 1974* (F. A. Huntley, ed.), Institute of Physics, Bristol, UK (1975), p. 212.
82. C. D. Thurmond, *J. Electrochem. Soc.* **122**, 1133 (1975).
83. W. Frank, in *Lattice Defects in Semiconductors 1979* (F. A. Huntley, ed.), Institute of Physics, Bristol, UK (1975), p. 23.
84. S. D. Brotherton, J. R. Ayers, J. B. Clegg, and B. J. Goldsmith, *Mater. Res. Soc. Symposium Proc.* **104**, 161 (1988).
85. B. L. Crowder, J. F. Ziegler, and G. W. Cole, in *Ion Implantation in Semiconductors and Other Materials*, Plenum, New York (1975), p. 257.
86. D. R. Myers and R. C. Wilson, *Rad. Effects* **47**, 91 (1980).
87. T. M. Liu and W. G. Oldham, *IEEE Electron Dev. Lett.* **EDL-4**, 59 (1984).
88. A. E. Michel, R. H. Kastl, S. R. Mader, B. J. Masters, and J. A. Gardner, *Appl. Phys. Lett.* **44**, 404 (1984).
89. T. E. Seidel, *IEEE Electron Dev. Lett.* **EDL-4**, 353 (1983).
90. R. Angelucci, P. Negrini, and S. Solmi, *Appl. Phys. Lett.* **49**, 1468 (1986).
91. C. I. Drowley, J. Adkission, D. Peters, and S. Chiang, *Mat. Res. Soc. Symp. Proc.* **35**, 375 (1985).
92. Y. Kim, T. Y. Tan, H. Z. Massoud, and R. B. Fair, in *Process Physics and Modeling in Semiconductor Technology* (G. R. Srinivasan, J. D. Plummer, and S. T. Pantelides, eds.), Vol. 91-4, Electrochemical Society, Pennington, NJ (1991), p. 304.

93. A. M. Mazzone, *Phys. Stat. Sol. A* **95**, 149 (1986).
94. Y. Kim, H. Z. Massoud, S. Chevacharveukul, and R. B. Fair, in *Semiconductor Silicon 1990* (H. R. Huff, K. G. Barraclough, and J. Chikawa, eds.), Vol. 90-7, Electrochemical Society, Pennington, NJ (1990), p. 437.
95. J. R. Pfiester and P. B. Griffin, *Appl. Phys. Lett.* **54**, 471 (1988).
96. P. Fahey, Materials Research Society meeting, San Diego, CA (1989).
97. M. Orlowski, *Appl. Phys. Lett.* **53**, 1323 (1988).
98. M. Servidori, S. Solmi, P. Zaumseil, U. Winter, and M. Anderle, *J. Appl. Phys.* **65**, 98 (1989).
99. R. B. Fair, in *Semiconductor Silicon 1990* (H. R. Huff, K. G. Barraclough, and J. Chikawa, eds.), Vol. 90-7, Electrochemical Society, Pennington, NJ (1990), p. 429.
100. T. O. Sedwich, *Nucl. Instr. Meth. Phys. Res. B* **37/38**, 760 (1989).
101. A. E. Michel, in *Process Physics and Modeling in Semiconductor Technology* (G. R. Srinivascan, J. D. Plummer, and S. T. Pantelides, eds.), Vol. 91-4, Electrochemical Society, Pennington (1991), p. 242.
102. H. Jiang, P. Smith, D. Griffis, A-G. Xiao, C. M. Osburn, G. McGuire, and G. A. Rozgonyi, in *Proceedings of the Materials Research*, Electrochemical Society, Los Angeles (May 1989).
103. R. T. Hodgson, V. Deline, S. M. Mader, F. F. Morehead, and J. Gelpsey, in *Proceedings of the Materials Research Society Symposium* (J. C. C. Fan and N. M. Johnson, eds.), Mat. Res. Soc., North-Holland (1984), p. 253.
104. R. Kalish, T. O. Sedgwick, S. Mader, and S. Shatas, *Appl. Phys. Lett.* **44**, 107 (1984).
105. R. Kwor, D. L. Kwong, C. C. Ho, B. Y. Tsaur, and S. Baumann, *J. Electrochem. Soc.* **132**, 1201 (1985).
106. R. Galloni, R. Rizzoli, and A. Nylandsted Larsen, in *Proceedings of the Materials Research Society* (unpublished), Europe Spring meeting, Strasbourg (1985).
107. J. Naryan, O. W. Holland, R. E. Eby, J. J. Wortman, V. Ozguz, and G. A. Rozgonyi, *Appl. Phys. Lett.* **43**, 957 (1983).
108. A. Nylandsted Larsen and V. E. Borisenko, *Appl. Phys. A.* **33**, 51 (1984).
109. V. E. Borisenko and A. Nylandsted Larsen, *Appl. Phys. Lett.* **43**, 582 (1983).
110. S. R. Wilson, W. M. Paulson, R. B. Gregory, A. H. Hamdi, and F. D. McDaniel, *J. Appl. Phys.* **55**, 4162 (1984).
111. T. O. Sedgwick, S. A. Cohen, G. S. Oehrlein, V. R. Deline, R. Kalish, and S. Shatas, in *VLSI Science and Technology/1984* (K. E. Bean and G. A. Rozgonyi, eds.), Electrochemical Society, Pennington (1984).
112. F. F. Morehead and R. T. Hodgson, in *Proceedings of the Materials Research Society Symposium* (D. K. Biegelsen, G. A. Rozgonyi, and C. V. Shank, eds.), Mar. Res. Soc., Pittsburgh, PA (1985), p. 341.
113. R. B. Fair, *Nucl. Instr. Meth. Phys. Res. B* **37/38**, 371 (1989).
114. J. C. C. Tasi, D. G. Schimmel, R. B. Fair, and W. Maszara, *J. Electrochem Soc.* **134**, 1508 (1987).
115. Y. Kim, Ph.D. Thesis, Duke University, Durham, N.C. (1990).
116. Y. Kim., H. Z. Massoud, U. Gosele, and R. B. Fair, in *Process Physics and Modelling in Semiconductor Technology* (G. R. Srinevasan, J. D. Plummer, and S. T. Pantelides, eds.), Vol. 91-4, Electrochemical Society, Pennington, NJ (1991).
117. U. M. Gosele, in *Proceedings of the Materials Research Society Symposium*, Vol. 2, Mat. Res. Soc., North-Holland, Amsterdam (1981), p. 55.
118. M. Schrems, T. Brabec, M. Budel, H. Poetzl, E. Guerrero, D. Huber, and P. Pongsatz, in *Defect Control in Semiconductors,* Inf. Conf. Sci. and Tech. (1989).
119. K. Nishi and D. A. Antoniadis, *Appl. Phys. Lett.* **46**, 516 (1986).

120. U. Gosele, in *Semiconductor Silicon 1986* (H. R. Huff, T. Abe, and B. Kolbesen, eds.), Electrochemical Society, Pennington (1986), p. 541.
121. M. T. Robinson and I. M. Torrens, *Phys. Rev. B* **9**, 5009 (1974).
122. T. Saito, H. Yamakawa, S. Komiya, H. J. Kang, and R. Shimuzu, *Nucl. Instr. Meth. B* **21**, 456 (1987).
123. T. L. Crandle and B. J. Mulvaney, *IEEE Electron Dev. Lett.* **11**, 42 (1990).
124. L. C. Kimberling and D. V. Lang, *Conf. Series-Inst. Phys.* **23**, 589 (1975).
125. G. D. Watkins, J. R. Troxwell, and A. P. Chatterjee, *Conf. Series—Inst. Phys.* **46**, 16 (1979).
126. V. T. Gusets and V. V. Titov, *Physics and Semiconductor Technology* **3**, 3-10 (1969).
127. V. R. Yurkots, *Solid State Phys.* **3**, 563 (1961).
128. Zhang Tonghe, unpublished.
129. R. A. McMahon, D. G. Hasko, H. Ahmed, W. M. Stoble, and D. J. Godfrey, in *Energy Beam-Solid Interactions and Transient Thermal Processing 1984* (D. K. Biegelsen, G. A. Rozgonyi, and C. V. Shank, eds.), Vol. 35, Materials Research Society, Pittsburgh, PA (1984), p. 347.
130. R. B. Fair, *Ibid.*, p. 381.
131. J. J. Wortman, unpublished.
132. N. E. B. Cowern, H. F. F. Jos, and K. T. F. Janssen, *Mat. Sci. Eng. B* **4**, 101 (1989).
133. T. Y. Tan, *Phil. Mag.* **44**, 101 (1981).
134. N. E. B. Cowern, K. T. F. Janssen, and H. F. F. Jos, *J. Appl. Phys.* **68**, 6191 (1990).
135. E. Tannenbaum, *Sol. State Electron* **2**, 123 (1961).
136. M. Finetti, P. Negrini, S. Solmi, and D. Nobile, *J. Electrochem. Soc.* **128**, 1313 (1981).
137. D. Nobili, A. Armigliato, M. Finetti, and S. Solmi, *J. Appl. Phys.* **53**, 1484 (1982).
138. D. L. Kendall and D. B. De Vries, in *Semiconductor Silicon* (R. R. Haberecht and E. L. Kern, eds.), Electrochemical Society, New York (1969), p. 97.
139. R. B. Fair and J. C. C. Tsai, *J. Electrochem. Soc.* **122**, 1689 (1975).
140. N. Natsuaki, M. Tamura, and T. Tokuyama, *J. Appl. Phys.* **51**, 3373 (1980).
141. J. Kato and Y. Ono, *J. Electrochem. Soc.* **132**, 1730 (1985).
142. S. Iwamatsu and I. Kato, *J. Vac. Soc. Jpn.* **25**, 735 (1982).
143. A. Lietoela, R. B. Gold, J. F. Gibbons, and T. W. Sigmon, *J. Appl. Phys.* **52**, 230 (1981).
144. K. Kugimiya and G. Fuse, *Jpn. J. Appl. Phys.* **21**, L16 (1982).
145. F. Cembali, R. Galloni, and Z. Zignagi, *Rad. Eff.* **26**, 161 (1975).
146. M. Miyao, N. Yoshihiro, T. Tokuyama, and T. Mitsuishi, *J. Appl. Phys.* **49**, 2573 (1978).
147. D. K. Sadana, J. Washburn, and C. W. Magee, *J. Appl. Phys.* **54**, 3479 (1983).
148. D. Nobili in *Aggregation Phenomena of Point Defects in Silicon* (E. Sirtl and J. Goorissen, eds.), Vol. 83-4, Electrochemical Society, Pennington, NJ (1982), p. 189.
149. M. Y. Tsai, F. F. Morehead, J. E. E. Baglin, and A. E. Michel, *J. Appl. Phys.* **51**, 3230 (1980).
150. R. B. Fair and G. R. Weber, *J. Appl. Phys.* **44**, 273 (1973).
151. R. B. Fair, in *Semiconductor Silicon 1981* (H. R. Huff, R. J. Kriegler, and Y. Takeishi, eds.), Electrochemical Society, Pennington, NJ (1981), p. 963.
152. Th. Wickert and M. Swanson, *J. Appl. Phys.* **66**, 3026 (1989).
153. M. Orlowski, R. Subrahmanyan, and G. Huffman, *J. Appl. Phys.* **71**, 164 (1992).

7 Silicides

C. M. Osburn
MCNC
Center for Microelectronic Systems Technologies
Research Triangle Park
North Carolina
and
Department of Electrical and Computer Engineering
North Carolina State University
Raleigh, North Carolina

I. Introduction	228
A. Utility of Silicides in ULSI Devices	228
B. Material Choices	237
C. Rapid Thermal Processing of Silicides	239
II. Formation of Silicides	240
A. Silicide Formation Techniques	240
B. Kinetics of Metal–Silicon Reactions and Dopant Redistribution—The Case for RTA	251
C. Effect of Formation Atmosphere	257
D. Issues for RTA Formation of Silicides	259
III. Properties of Silicides and Silicided Junctions	264
A. Conductivity	264
B. Stability	264
C. Dopant Redistribution	273
D. Junction Leakage and Damage Removal	276
IV. Applications of Silicides Formed by RTA and Process/Device Considerations	282
A. Silicided Gates	282
B. Silicided Shallow Junctions	283
C. Junctions Formed by Diffusion through Silicide	287
D. Local Interconnections	289
E. Process Modeling with Silicides	291
V. Summary	292
References	292

I. Introduction

Metal silicides play an important role in VLSI/ULSI device technology, and several reviews [1–10] summarize both the silicide material properties and their device applications. Rapid thermal processing has been widely used for the implementation of metal silicides into devices. Rapid thermal annealing (RTA) has proven to be superior to conventional furnace techniques for silicide formation because of the limited thermal stability of many candidate silicide materials and because of their extreme sensitivity to trace impurities in the annealing atmosphere. It has been essential in reducing the overall thermal budget both during and after silicide formation.

A. Utility of Silicides in ULSI Devices

Silicides were first used on silicon devices as a contact, specifically a Schottky barrier contact in bipolar devices [11–12]. This application made use of the fortuitous metal semiconductor barrier height of PtSi, which allowed the formation of a Schottky barrier diode in parallel with the collector-base junction of an *npn* bipolar device, as shown in Fig. 1. A similar application, namely as a diffusion barrier to retard aluminum spiking of junctions while providing low contact resistance, successfully employed Pd_2Si [13, 14] as illustrated in Fig. 2. In the late 1970s, when patterned feature sizes reached about 1 µm dimensions, the "polycide"

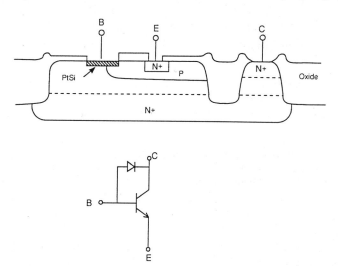

FIGURE 1. Schottky barrier diode in an *npn* bipolar device made with PtSi.

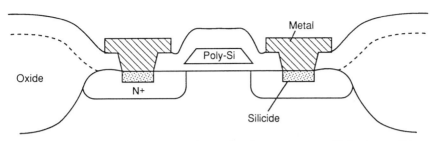

FIGURE 2. Metal silicide used for low-resistance junction contacts to avoid aluminum spiking.

technology was developed, in which polycrystalline silicon (poly-Si) gate electrodes were clad with an overlayer of metal silicide, typically WSi_2, $MoSi_2$, $TaSi_2$, or $TiSi_2$ [15-29]. In this application, the silicide was most frequently codeposited on top of the poly-Si, and the composite gate electrode was reactive ion-etched in a multistep process. The polycide process, as illustrated in Fig. 3, is used to reduce the wiring delay associated with the high resistance–capacitance (RC) product of long poly-Si lines. Figures 4 and 5 show the performance improvement that is obtained by

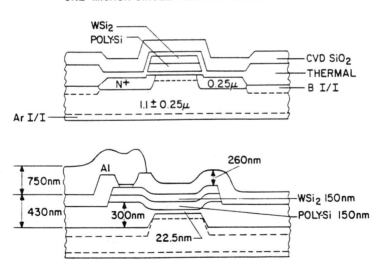

FIGURE 3. The "polycide" process in which metal silicide is clad to the poly-Si regions of MOSFET devices. (From M. Y. Tsai, H. H. Chao, L. M. Ephrath, B. L. Crowder, A. Cramer, R. S. Bennett, C. J. Lucchese, and M. R. Wordeman, One-Micron Polycide (WSi_2 on Poly-Si) MOSFET Technology, *J. Electrochem. Soc.* **128**, 2207, 1981. This figure is reprinted by permission from the publisher, The Electrochemical Society, Inc.)

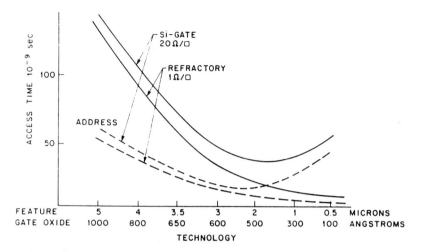

FIGURE 4. SRAM access time as a function of technology feature size, comparing conventional poly-Si gates with refractory silicide–clad gates. (From P. B. Ghate and C. R. Fuller, Multilevel Interconnections for VLSI, *Semiconductor Silicon, 1981*, The Electrochemical Society **81**, 680, 1981.)

using polycide gates rather than poly-Si. Figure 4 gives the access time of a 4K bit static memory as a function of technology feature size [28]. Down to about 2 μm dimensions, the total and address access times decrease with decreasing feature size as a result of device performance improvements; however, below 2 μm the access times increase for a poly-Si gate technology because of its high resistance. In contrast the performance of a refractory gate, having a sheet resistance of only 1 ohm per square, continues to improve as dimensions are scaled into the submicrometer regime. Figure 5 shows that the dynamic memory access time decreases dramatically as the sheet resistance of the poly-Si wordline is reduced with polycide technology [26]. The benefit of the low-resistance polycide gate is most pronounced when more bits are added onto the wordline; for a wordline containing 256 bits, the wordline risetime is reduced by almost an order of magnitude when polycide is substituted for a conventional poly-Si gate. Since the source/drain ion implantation and its subsequent anneal are performed after the silicide is in place for the polycide process, the silicide material has to be stable during high-temperature annealing. In many cases these silicide layers are subject to oxidation to provide sidewall protection to the gate electrode. Thus they must be stable in oxidizing atmospheres.

A further reduction in device dimensions below 1 μm led to the need to also lower the sheet resistance of diffusion regions and to reduce the junction contact resistances that are in series with the active device. This

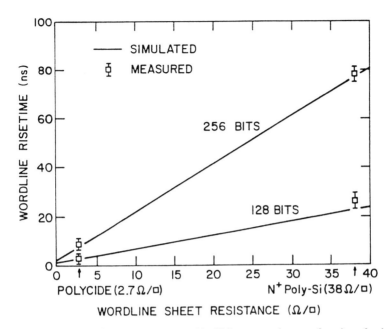

FIGURE 5. DRAM performance as measured by bitline access time as a function of polycide sheet resistance. (From M. Y. Tsai, H. H. Chao, L. M. Ephrath, B. L. Crowder, A. Cramer, R. S. Bennett, C. J. Lucchese, and M. R. Wordeman, One-Micron Polycide (WSi_2 on Poly-Si) MOSFET Technology, *J. Electrochem. Soc.* **128**, 2207, 1981. This figure is reprinted by permission from the publisher. The Electrochemical Society, Inc.)

problem led to the self-aligned silicide (salicide) technology in the early 1980s [30–68]. This technology, shown in Fig. 6, forms a silicide on the source, drain, and gate regions. This process requires the use of a sidewall spacer on the poly-Si gate [69–72] and achieves its self-alignment by reacting metal with exposed silicon regions. In contrast to the silicides used in the polycide process, the silicide in the salicide process is not subjected to high-temperature anneals, other than borophosphosilicate glass (BPSG) reflow, and is thus not required to have as much high-temperature stability. In fact, the metals used in this process must readily react with silicon at reasonable temperatures without further diffusing the junctions. The silicides of Co, Ti, Pd, Pt, and Ni (NiSi) have been most widely considered for this application.

Figure 7 shows the components of sheet resistance on an MOS device [73]. The salicide process reduces both the diffusion resistance (R_{sh} in the figure) by cladding the junction with a high-conductivity overlayer and the contact resistance (R_{co}) by increasing the effective contact area. As an example, consider the geometry of a device whose effective channel length is 0.5 μm and width is 1.5 μm. For a minimum layout geometry shown in

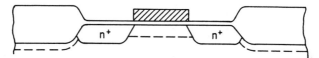

FORM STANDARD DEVICE UP TO DIFFUSIONS

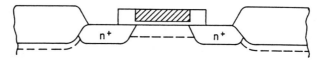

PROTECT SIDEWALLS OF POLY-Si WITH OXIDE

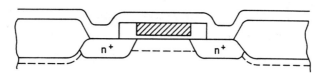

DEPOSIT BLANKET METAL

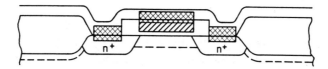

REACT TO FORM METAL SILICIDE OVER SILICON REGIONS

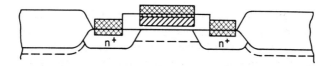

REMOVE ALL FREE METAL

FIGURE 6. The salicide process in which metal silicide is self-aligned to gate and source/drain regions of MOSFET devices. (From C. M. Osburn, M. Y. Tsai, S. Roberts, C. J. Lucchese, and C. Y. Ting, High Conductivity Diffusions and Gate Regions Using a Self-Aligned Silicide Technology, *Proc. 1st Int. Symp. on VLSI Science and Technology*, The Electrochemical Society, **82-1**, 213, 1982.)

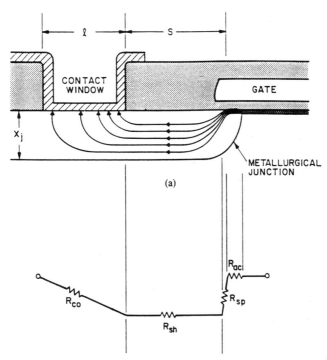

FIGURE 7. Components of sheet resistance on a MOSFET device. (From K. K. Ng and W. T. Lynch, The Impact of Intrinsic Series Resistance on MOSFET Scaling, *IEEE Trans. Electron. Dev.* **ED-34**(3), 503, © 1987 IEEE.) In addition to the channel resistance, the total device resistance includes contact resistance R_{co}, diffusion sheet resistance R_{sh}, spreading resistance at the junction edge R_{sp}, and accumulation layer resistance, R_{ac}.

Fig. 8, the diffusion area is 2.25 μm^2, yet the total contact area is only 0.25 μm^2. With a 10^{-6} Ω-cm^2 contact resistivity, a conventional contact would contribute 400 Ω to R_{co} at both the source and the drain contacts. With a salicided contact, the overall resistance would be increased by only 45 Ω at each end of the device. With this efficient layout, there is less than one square of diffusion in series with the device. For a typical n^+ junction sheet resistance of 100 Ω/square, these diffusions would add an additional 2×100 Ω = 200 Ω to the external resistance, which drops to only 10 Ω if the junctions are clad with a 5 Ω/square silicide. Figure 9 compares current–voltage characteristics of devices with and without silicided contacts [10]. At the maximum applied voltage, the silicided device provides about 20% more current drive capability. This additional current is available to charge a capacitive load in 20% less time, thereby improving circuit performance by this amount. In this example, the primary benefit

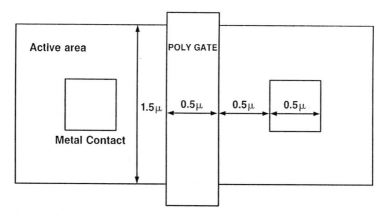

FIGURE 8. Typical layout geometry of $0.5 \times 1.5\,\mu m$ MOSFET device. (From Ref. 10.)

of the silicide is in lowered contact resistance rather than in lowered diffusion resistance. In other layout configurations, where there are more squares of diffusion between the contact and the device, the reduced diffusion resistance due to the presence of a silicide would provide an additional benefit. Just as the polycide process reduces the RC time constant of poly-Si features used as interconnections, the salicide process allows the diffusion layer to be used for interconnections without incurring an inordinate performance penalty.

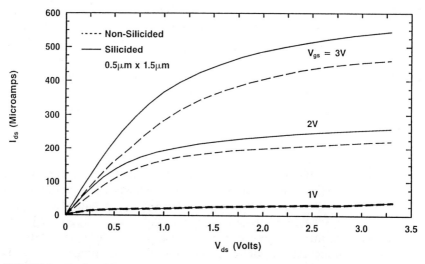

FIGURE 9. Device current–voltage (I–V) characteristics of transistors shown in Fig. 8 with and without salicide. (From Ref. 10.)

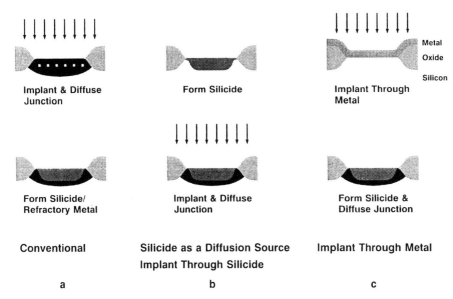

FIGURE 10. Comparison of silicide as a diffusion source (SADS), implant through metal (ITM), and implant through silicide (ITS) processes. (From Ref. 10.)

Several variations of the salicide process form the junction after the silicide or metal is in place. Three common processes use implantation of dopant through the metal (ITM) [67, 74–89], implant through silicide (ITS) [78, 79, 87, 90–99], and silicide as a diffusion source (SADS) [75–77, 91, 94, 97, 99–114]. These processes, compared in Fig. 10, are distinguished by whether the dopant is implanted into metal or silicide and by the depth of the implant with respect to the silicide thickness. All three employ a thermal cycle after implantation to diffuse dopant, form silicide, remove implantation damage, or a combination of the above.

Silicides have also been tested in conjunction with emitter contacts in bipolar devices. However, the use of Pd_2Si as an emitter contact apparently increases the hole injection and thereby leads to lower gain devices than those contacted by aluminum [115].

State-of-the-art bipolar devices often employ n^+ poly-Si contacts to the emitter and p^+ poly-Si for the base contact. Silicide can be self-aligned to the two types of poly-Si to provide low resistance interconnects and to bridge the two dopant types [116]. Silicides have also been considered for permeable base transistor applications [117].

Because the silicided features in MOS technology are self-aligned to the gate *or* to the diffusion area, with a gate spacer oxide or isolation oxide between the two regions, the salicide process still requires the use of metal

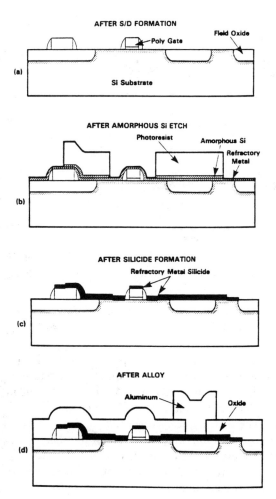

FIGURE 11. Use of metal silicide to form local interconnections across isolation oxide regions. (From D. C. Chen, S. S. Wong, P. P. Voorde, P. Merchant, T. R. Cass, J. Amano, and K.-Y. Chiu, A New Device Interconnect Scheme for Sub-Micron VLSI, *IEDM Tech. Dig.* **118**, © 1984 IEEE.)

to connect poly-Si features to diffusion regions. However, local interconnections between gates and diffusions can be achieved with an additional silicide level that requires one additional masking level (see Fig. 11). In this implementation of local interconnection technology, an amorphous silicon layer is deposited and patterned on top of the metal film [118-121]. This silicon layer then provides the material to form slicide over oxide regions. A variant of this process [122] used the reaction of Ti and N_2 during rapid thermal annealing to form a conductive layer of TiN that can be subsequently patterned to form local interconnections. Since the properties, functions, and processing conditions to form TiN are so similar to those for silicides, the remainder of this chapter will consider titanium nitride as well as the silicides. The local interconnection technology has not been as successful with $CoSi_2$ as it has been with $TiSi_2$. Since cobalt does not form a nitride, the amorphous silicon approach has been necessary. Unfortunately, the presence of small voids at the $CoSi_2/SiO_2$ interface [123] has hampered the development of a satisfactory process.

These applications and the requirements for the materials needed for them will be described in more detail in Section IV.

B. MATERIAL CHOICES

Several desirable material properties make silicides useful in integrated circuit applications. Table I summarizes some of these attributes. Several different silicides have been considered and chosen for each device application as described earlier. A partial summary of these materials and their applications is given in Table II. These silicides have been chosen because they have a low sheet resistance, a desirable barrier height (mid-gap for simultaneaous ohmic contacts to n^+ and p^+ junctions or a high barrier for Schottky diodes), adequate thermal stability to withstand subsequent processing, and acceptably low stress. Table III summarizes these properties for some of the silicides [1, 3, 18, 121, 124–127]. They can be divided into two groups: the refractory metal silicides of groups IV-A, V-A, and VI-A

Table I Desired properties of silicides for integrated circuits.

Low sheet resistivity
Low contact resistivity
Easy to form; high processing yield
Minimal consumption of silicon
No reaction with SiO_2 or lateral silicide growth
Easy to pattern or etch
Stable in oxidizing ambients; oxidizable
Mechanically stable; good adherence; low stress
Smooth surface and interface
Stable throughout processing at high temperatures
Minimal reaction with aluminum or other metallization
Should not contaminate devices, wafers, or processing equipment
Good device characteristics and lifetimes
Minimal junction (silicon) penetration

Table II Metal silicides used in semiconductor devices.

Material	Application	Formation Technique	References
PtSi	Schottky barriers	Thermal reaction of metal	11, 12
PtSi	Gate & junction cladding	Thermal reaction of metal	30, 31, 64
PtSi	Diffusion source	Thermal reaction of metal	99, 104
Pd_2Si	Diffusion barrier	Thermal reaction of metal	13, 14
Pd_2Si	Bipolar emitter contact	Thermal reaction of metal	115
Pd_2Si	Diffusion source	Thermal reaction of metal	104
WSi_2	Gate cladding	Codeposition	15, 16, 20-23, 25-27
WSi_2	Gate & junction cladding (Diffusion souce)	Ion beam mixing	31, 84, 98, 111
WSi_2	Gate & junction cladding	Selective CVD	31
$TaSi_2$	Gate cladding	Codeposition	17-19
$TaSi_2$	Diffusion source	Ion beam mixing	110
$MoSi_2$	Gate cladding	Codeposition	1
$MoSi_2$	Gate & junction cladding (Diffusion source)	Ion beam mixing	31, 39, 74, 77-83, 135
$TiSi_2$	Gate cladding	Codeposition	1, 129
$TiSi_2$	Gate & junction cladding	Thermal reaction of metal	30-31, 33-40, 44-54, 58, 61-62, 66
$TiSi_2$	Gate & junction cladding	Ion beam mixing	80, 81, 83
$TiSi_2$	Local interconnection	Thermal reaction of metal	118-121
$TiSi_2$	Diffusion source	Thermal reaction of metal	3, 90-96, 104, 111
$CoSi_2$	Gate & junction cladding	Thermal reaction of metal	31, 33, 42-43, 55, 57, 63-65, 67
$CoSi_2$	Gate & junction cladding	Ion beam mixing	77, 133
$CoSi_2$	Diffusion source	Thermal reaction of metal	3, 55, 97, 104-106, 117
NiSi	Gate & junction cladding	Thermal reaction of metal	31
NiSi	Diffusion source	Thermal reaction of metal	104
$TiN/TiSi_2$	Diffusion barrier	Thermal reaction of metal	85, 141, 146, 156, 167-169
$TiN/TiSi_2$	Local interconnection	Thermal reaction of metal	122

[Ti, Mo, W, Ta, Zr, V, ...) and the near noble metal silicides of the group VIII elements (Pt, Pd, Co, Ni, ...). While the refractory metal silicides have generally superior thermal stability and can be oxidized, typically these metals do not readily react with silicon, so that their silicides must be formed by codeposition. Titanium is a prominent counterexample to this since it does readily react with silicon; on the other hand, its thermal stability is only moderate. The refractory metal silicides have been the material of choice for polycide applications. The near noble metal silicides as well as $TiSi_2$ are readily formed by metal reaction with silicon, and they have higher conductivities; thus these are the material used for salicide applications even though their thermal stability is limited.

Table III Properties of metal silicides.

Silicide	Resistivity ($\mu\Omega$-cm)	Φ_b (eV)	$T^*_{stability}$ (°C)	Stress (10^9 dyn/cm^2)	References
PtSi	28–35	0.87	~800	5	1, 124
Pd$_2$Si	30–35	0.74	~700	5	1, 124
WSi$_2$	~30	0.65	>1000		1
TaSi$_2$	35–55	0.59	>1000	18	1, 125–126
MoSi$_2$	~40	0.55	>1000	10	1, 125–126
TiSi$_2$	13–16	0.6	~900	15–25	1, 3, 18, 59
CoSi$_2$	18–20	0.64	~1000	8–10	1, 3, 59, 104, 121, 127
ZrSi$_2$	35–40	0.55			1
NiSi	12–15	0.7	~700	~1	1, 18
TiN	40	0.49			127

C. Rapid Thermal Processing of Silicides

Although the historical development of silicide technology predated the widespread availability of rapid thermal processing equipment, silicides and rapid thermal processing have become almost synonymous today. The success of each technology owes a lot to the other. Silicide technology has greatly benefited from the atmospheric control available in RTP systems. As will be discussed in more detail later, the development of silicide processes with conventional furnace processes required an inordinate amount of attention to the mechanics of avoiding oxygen during annealing to the point where the manufacturability of self-aligned silicides was seriously questioned. Short, high-temperature anneals have allowed the nucleation and subsequent growth of the desired, high-conductivity silicide phase without shorting gates to junctions by the extensive lateral overgrowth of the silicide. On the other hand, the reaction of thin metal layers to form silicide occurs rapidly at moderate temperatures, thereby providing a ready-made application for RTP equipment. Silicide processes have been relatively tolerant of the thermal gradients and temperature uncertainties that characterized the early generations of RTA equipment.

Rapid thermal annealing has several applications in silicide technology. First, it can be used to spatially homogenize metal and silicon codeposits that are often not deposited uniformly throughout the film. Second, it can be used to react codeposited films into the desired, stoichiometric phase [128]. Third, and probably the most widespread application of RTA in silicide formation, is its use to directly react metal, often titanium or cobalt, with a patterned silicon substrate to form self-aligned silicides as described earlier [50, 51, 53–59, 65, 67, 82, 129–166]. The silicides of both titanium and cobalt form intermediate phases before the high-conductivity disilicide is achieved. As will be discussed in more detail later, the use of RTA provides more flexibility in choosing formation conditions that consider both nucleation and growth of the desired phase. Rapid thermal processing can also be used to react films, such as $TiSi_2$, with an atmosphere of nitrogen to form a nitride layer such as TiN, on top [140, 146, 156, 167–169]. These nitride layers are often superior diffusion barriers to aluminum penetration.

Because of both the redistribution of dopants from the silicon into the silicide during annealing and the limited thermal stability of many of the silicide materials (e.g., $TiSi_2$ and $CoSi_2$) RTA has been exceedingly beneficial in reducing the thermal cycles that occur after silicidation, particularly the BPSG glass reflow step [170–171]. By reducing the subsequent thermal budget, the silicide integrity is preserved. Thus RTA is indeed an enabling technology for the use of silicides.

The processes that form shallow junctions after silicide formation also heavily rely on RTA to minimize the junction depth and to preserve the silicide integrity. Silicides are used as diffusion sources to achieve the shallowest possible junctions to allow dimensional scaling of device feature sizes. The advantages of using RTA for this application are essentially the same as for using RTA for conventional shallow junction fabrication: first, it allows the process engineer to controllably reduce the total thermal budget to minimize diffusion; second, its high-temperature capability maximizes the dopant solubility so that low resistance contacts can be made to junctions.

II. Formation of Silicides

A. Silicide Formation Techniques

Some of the more common silicide formation techniques are given in Table II. These include (1) codeposition of metal and silicon either from two independent targets or from a hot-pressed alloy followed by sintering, (2) thermal reaction of metal with silicon, (3) ion beam–induced mixing, (4) formation of buried layers by ion implantation of metal (mesotaxy), (5) chemical vapor deposition, and (6) molecular beam or solid phase epitaxy (MBE, SPE). The first two are the most commonly used techniques, where codeposition is used for polycides, and metal reaction is employed for the salicide technology. The first four formation techniques, namely codeposition, metal reaction, ion beam mixing, and buried layer formation can employ rapid thermal annealing as part of the process. Even CVD of silicides can use rapid thermal processing to advantage.

1. Codeposition of Metal and Silicon

The reaction of the more refractory metals (W, Ta, Mo, etc.) with silicon typically requires a high temperature and is easily inhibited by the presence of native oxide on silicon. Thus they are usually codeposited by electron beam evaporation, sputtering or chemical vapor deposition. The as-deposited films are usually intentionally or unintentionally nonstoichiometric. These deposited from dual targets (i.e., metal and silicon) are often spatially nonuniform in composition. The films are then subjected to annealing that (1) allows for diffusion of the components to homogenize the composition, (2) crystallizes the film into the desired high-conductivity silicide phase, and (3) permits sufficient diffusion, using the silicon substrate as a source or sink for silicon, to give a single-phase stoichiometric film on top of the substrate.

Codeposited films are generally amorphous when deposited and have a higher sheet resistance than annealed films. Figure 12 illustrates the decreasing sheet resistance and increasing X-ray diffraction intensity,

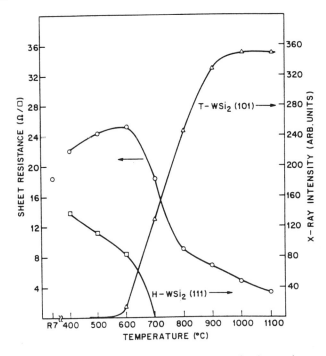

FIGURE 12. The effect of 30-min annealing temperature on the sheet resistance and X-ray diffraction intensity, for both the hexagonal and tetragonal phases of WSi_2, ion 250 nm, cosputtered $WSi_{2.2}$ films. (From Ref. 172.)

associated with growing tetragonal grains, of cosputtered $WSi_{2.2}$ as the sintering temperature is increased [172]. Similar behavior, with the exception of the low-temperature maximum in resistivity, is seen with most of the other codeposited silicides. Figure 13 compares the before and after sintering resistivities of Ti/Si films as a function of the Ti/Si composition [173] for films on oxide (Fig. 13a) and on poly-Si (Fig. 13b). Clearly different behavior is seen depending on whether the substrate is SiO_2 or silicon. Sintering lowers the sheet resistance of films on both substrates; however, lower sheet resistances are seen in films deposited on silicon than in those deposited on oxide. Furthermore the initial Ti/Si ratio has less of an effect on the final resistance on films deposited on silicon due to the ability of the substrate to provide extra silicon to silicon-deficient alloys and to nucleate the deposition of excess silicon in silicon-rich alloys. Nevertheless, a minimum in the resistivity versus metal-to-silicon ratio curve is seen for the silicides of titanium [173], tungsten [126], cobalt [174], and nickel [175], but not molybdenum [125]. Interestingly in Fig. 13, the minimum resistivity of $TiSi_x$ films deposited on oxide does not correspond to stoichiometric $TiSi_2$.

2. Thermal Reaction of Metal and Silicon

The silicides of Ti, Co, Pt, Pd, and Ni are usually formed by the direct reaction of metal with single-crystal, polycrystal, or amorphous silicon. In this case, metal is evaporated or sputtered onto silicon, and the resulting stack is then annealed to form the desired silicide. For patterned structures, as illustrated in Fig. 14, the metal reacts only where it is in contact with silicon, and a selective etchant is used to remove unwanted metal from the oxide regions. Some of the etchants for this purpose are listed in Table IV.

One of the serious issues associated with self-aligning a silicide to patterned oxides in this way is the lateral growth of the silicide (as shown

Table IV Processing conditions for self-aligned silicide formation.

Silicide	Typical First Anneal	Metal Etchant	Typical Second Anneal
$TiSi_2$	120″ 600 °C	$NH_4OH:H_2O_2$; piranha	10″ 800 °C
$CoSi_2$	10″ 550 °C	$3HCl:H_2O_2$; piranha	10″ 700 °C
NiSi	450–550 °C	HNO_3	NA
PtSi	500 °C	Aqua regia	NA
Pd_2Si	300–400 °C	$KI:I_2$	NA

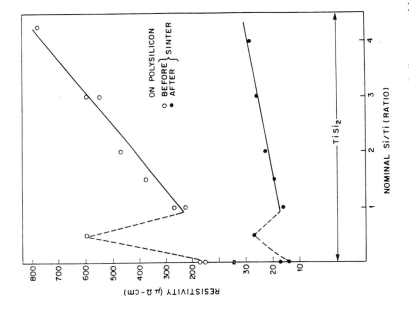

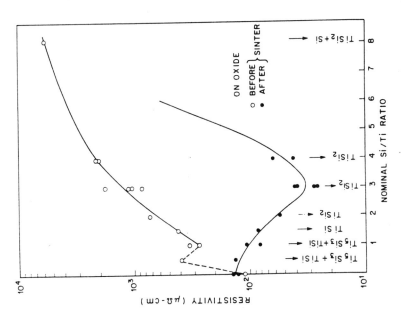

FIGURE 13. The effect of annealing (30 min, 900 °C) on the resistivity of cosputtered Ti-Si alloys as a function of alloy compositions. Left: alloy deposited on oxide; right: alloy deposited on silicon. (From Ref. 173.)

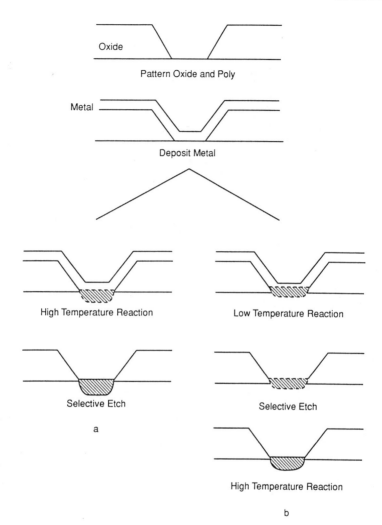

FIGURE 14. Thermal reaction of metal with silicon to form self-aligned silicide: (a) single-step annealing and (b) two-step annealing process.

in Fig. 15) away from the oxide window regions. This overgrowth occurs as a result of lateral diffusion of silicon from the substrate through either metal or formed silicide to the metal where it reacts to form additional silicide. Since silicon is the predominantly diffusing species in $TiSi_2$ formation [176–177], the lateral overgrowth problem is expected to be more severe with titanium than with cobalt, where (at least for Co_2Si and $CoSi_2$ formation) the metal is the diffusing species [178–180]. Since grain

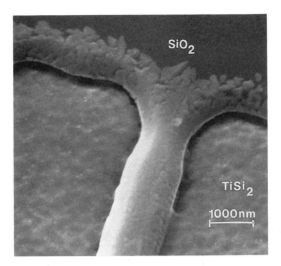

FIGURE 15. Lateral overgrowth of TiSi$_2$ onto oxide regions after a 10-sec, 900 °C single-step RTA. (From T. Brat, C. M. Osburn, T. Finstad, J. Liu, and B. Ellington, Self-Aligned Ti Silicide Formed by Rapid Thermal Annealing Effects, *J. Electrochem. Soc.* **133**, 1451, 1986. This figure is reprinted by permission from the publisher, The Electrochemical Society, Inc.)

boundary diffusion is faster than bulk diffusion, the forming silicide front, both vertically as well as laterally, is not necessarily planar. Silicon, for instance, diffuses along grain boundaries in titanium silicide and in titanium so that the silicide phase can start around the edges of the grains and move inwards. Thus TiSi$_2$ formation is observed on top of titanium films before the underlying metal is totally converted to silicide. Figure 16a, schematically illustrates the silicon diffusion in TiSi$_2$ to form silicide around the grains and at the surface. Figure 16b shows a cross-sectional TEM of cobalt silicide formation where enhanced metal diffusion along the silicide grain boundary initially results in the formation of a thicker silicide in the grain boundary region. Although a void or spike associated with excess silicon consumption is expected at the feature edge whenever lateral overgrowth occurs, surprisingly this has not been much of an issue, at least with titanium [181]. In the cases where metal diffusion predominates, a thicker silicide is expected at the window edge with a thinning of the nearby metal over the oxide. Such behavior is indeed seen [182] with thinner cobalt films; however, at more common silicide thicknesses (~ 75 nm) these edge effects are minor.

Lateral growth of silicide can short out adjacent features, particularly junctions and gates that are separated by only a narrow spacer as depicted

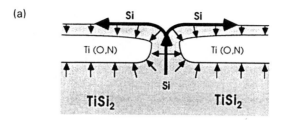

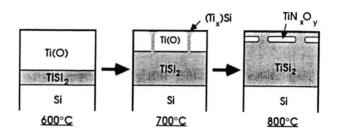

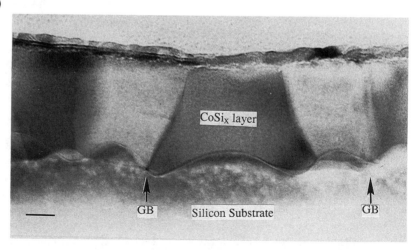

FIGURE 16. Effect of grain boundary diffusion on polycrystalline silicide formation: (a) schematic illustration of TiSi$_2$ where Si is the diffusing species (from Ref. 3); (b) XTEM of cobalt silicide where metal diffusion along the grain boundaries results in a thicker silicide film. (From Ref. 279.)

in Fig. 6. This lateral overgrowth can be minimized by using an anneal-etch-anneal sequence, portrayed in Fig. 14b, in which a first, low-temperature anneal is used to allow only a modest amount of silicon/metal inter-diffusion. Often only a subsilicide (e.g., Co_2Si, $CoSi$, or $TiSi$) is formed over exposed silicon during this step. The etch step is then able to cleanly remove the unreacted metal over the oxide regions. Finally a higher temperature anneal is used to form the desired, high-conductivity silicide phase. Table IV gives some typical conditions that might be employed in the two-step annealing process. Figure 17a shows the window for the initial RTA time and temperature to anneal titanium films over a MOSFET device having oxide spacers separating the poly-Si gate from the junctions. Those annealing conditions within the acceptable window avoid lateral bridging of the junction and gate [183]. Figure 17b shows plots of the spacer bridging yield for $CoSi_2$ formed from 38 nm of cobalt as a function of the initial, 30-sec RTA temperature [184].

A second technique used with titanium to minimize lateral overgrowth is to perform the initial annealing step in a nitrogen atmosphere [34, 140, 167-9, 185]. The nitrogen is believed to fill the grain boundaries of the titanium metal over the oxide regions to retard the subsequent lateral diffusion of silicon. Over the silicon regions, this nitrogen "stuffing" actually results in a layer of TiN (or more commonly TiON) where the nitrogen (and oxygen) from the metal snowplow to the outer surface during silicidation to form a surface layer with titanium.

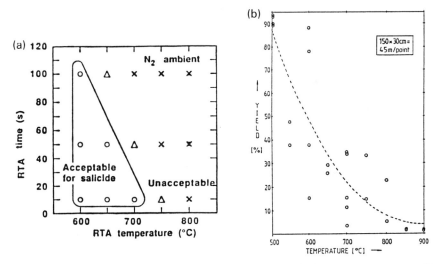

FIGURE 17. Process windows giving the acceptable range of RTA temperatures to avoid lateral bridging of spacers in (a) $TiSi_2$ (from Ref. 183) and (b) $CoSi_2$. (From Ref. 184.)

The use of a two-step annealing process also minimizes the reaction of metal with oxide that can occur at high temperatures. Titanium, for example, is not thermodynamically stable on SiO_2. At least in an argon annealing atmosphere, it will reduce the oxide to form Ti_5Si_3 [129, 140, 187]. After a nitrogen anneal, layers of $TiN/TiO/Ti_5Si_3$ are observed [3]. While this effect is beneficial in reducing any native oxide and thereby promoting the reaction of titanium with silicon, it can be deleterious in the isolation areas where pinholes could form where the oxide is attacked. As will be discussed later, limiting the temperature of the initial reaction step limits the amount of this oxide that is reduced. Apparently annealing in nitrogen also results in less oxide consumption than in argon [3]. Cobalt, on the other hand, is thermodynamically stable on oxide. However, although cobalt does not react directly with SiO_2, it does agglomerate [3, 33, 129, 182, 186], and trace quantities of cobalt are incorporated into the oxide at rapid thermal annealing temperatures above 500 °C [188]. Figure 18 is a

FIGURE 18. SEM micrograph showing cobalt agglomeration on oxide regions of patterned wafers after hot-stage annealing around 500 °C. (From Ref. 182.)

SEM micrograph showing discontinuous cobalt globules left on oxide lines, compared to the relatively featureless, dark regions where the cobalt reacted with the silicon substrate in oxide openings. Interestingly, agglomerated cobalt on oxide has been reported to cause morphological changes in the oxide that replicate the shape of the cobalt globules [3].

Rapid thermal annealing has been widely used for the direct thermal reaction of metal with silicon since the early 1980s. The kinetics of the reaction is sufficiently fast, so that all the necessary phase transformations can occur in a very short period of time (seconds) even without needing to raise the process temperatures over those used with furnace annealing. Furthermore, the ability of RTA equipment to preserve the purity of the annealing atmosphere has resulted in silicides that are lower in resistance and are more uniform in thickness. Sections B and C describe in more detail the kinetics of the reactions and the effect of impurities in the annealing atmosphere during RTA.

3. Ion Beam-Induced Mixing

Because of the difficulty associated with directly reacting many of the refractory metals with silicon, presumably because of the presence of its native oxide, ion bombardment has been used to provide some intermixing of the metal and silicon prior to thermal annealing [78, 81-84, 88-89, 189-192]. In the ion beam-induced mixing process, typically a heavy ion, like Ge or As, is implanted with an energy such that the peak of the dose occurs at or deeper than the original metal/silicon interface. Dopant species can also be used for the mixing; if they are implanted deep enough, they can simultaneously form a junction beneath the silicide. This process is often referred to as implantation through metal (ITM). After the intermixing and disruption of any native oxide, furnace annealing or, more typically, RTA is used to promote the silicidation reaction. One of the major issues associated with ITM is the knock-on of metal into the silicon where it can create crystal damage or generation-recombination centers that adversely affect junction leakage.

4. Buried Ion Implantation (Mesotaxy)

Direct implantation of metal into silicon or poly-Si has recently been used in an attempt to improve control of the silicide formation process and to prove a larger grained, smoother silicide film that is more thermally stable than those formed by other methods [193-201]. Buried silicide layers can be formed within a silicon substrate by implanting with a very high dose of energetic metallic ions. The implantions can be carried out at an elevated temperature in order to dynamically anneal the damage generated by the collision cascades of the implanted metal ions and to form a silicide layer. Depending on the metal implantation energy, the technique can also be used to form a surface silicide layer. As an alternative to *in situ* heating during implantation, postimplantation annealing can be used to homogenize the metal and the silicon and to form the silicide. Since considerable damage is produced during the ion implantation step, furnace annealing is generally used in this process rather than RTA. This mesotaxy technique has produced epitaxial layers on both $\langle 111 \rangle$ and $\langle 100 \rangle$ silicon [200–201].

5. Chemical Vapor Deposition

Chemical vapor deposition (CVD) has been extensively explored for silicide formation [202–214], particularly for WSi_2 layers on poly-Si. The chemical reactions typically employ the decomposition of a metal halide and silane:

$$WF_6 + 2SiH_4 \rightarrow WSi_2 + 6HF + H_2 \qquad (7.1)$$

$$MoF_5 + 2SiH_4 \rightarrow MoSi_2 + 5HF + 3/2H_2 \qquad (7.2)$$

$$TiCl_4 + 2SiH_4 \rightarrow TiSi_2 + 4HCl + 2H_2 \qquad (7.3)$$

There are several issues associated with the CVD of silicides. First of all, in many of these cases it is possible to get direct deposition of metal; for instance, the silane reduction of WF_6 is the process most commonly used to get selective tungsten deposition. Thus the CVD process must be carefully designed to obtain only a single-phase silicide deposition. Second, a disproportionation reaction can occur that can etch the substrate:

$$4WF_6 + Si \rightarrow 4WF_5 + SiF_4 \qquad (7.4)$$

$$4TiCl_4 + Si \rightarrow 4TiCl_3 + SiCl_4 \qquad (7.5)$$

Finally, other reactions can occur that consume silicon from the substrate to form silicide or metal deposits:

$$TiCl_4 + 2Si + 2H_2 \rightarrow 4TiSi_2 + 4HCl \qquad (7.6)$$

$$2WF_6 + 3Si \rightarrow 2W + 3SiF_4 \qquad (7.7)$$

The occurrence of the displacement reaction has been used as the basis for selectively depositing silicide on silicon. However, the process window needed to get a selective deposition without also etching extra substrate material appears to be narrow [213].

Silicides

6. Molecular Beam and Solid Phase Epitaxy

Molecular beam epitaxy has been used to form epitaxial silicides [215-219], especially $CoSi_2$ on $\langle 111 \rangle$ Si. By heating the substrate under ultrahigh vacuum conditions, it is possible to epitaxially deposit films. Epitaxial films can also be formed by reacting a very thin cobalt layer with the underlying silicon followed by evaporation and reaction of additional Co.

Recently the solid phase epitaxial growth of $CoSi_2$ on $\langle 100 \rangle$ Si has been reported [220-221] from the deposition of Co/Ti alloys and/or layers and subsequent furnace or rapid thermal annealing. Apparently the cobalt diffuses through a titanium layer, thereby providing a source for uniform epitaxial growth on the substrate, and the resulting films exhibit superior thermal stability.

B. Kinetics of Metal-Silicon Reactions and Dopant Redistribution—The Case for RTA

The kinetics of the metal-silicon reaction is such that rapid thermal annealing can be used as a direct replacement for many of the alternative furnace processes. Figures 19 and 20 compare the sheet resistivities for both RTA (10 sec) and furnace (30 min) annealing as a function of temperature for titanium or cobalt films, respectively. The temperature dependencies are nearly identical for both types of annealing: they both exhibit a maximum in resistivity at low annealing temperatures, and both produce a sharp drop in resistivity above a critical annealing temperature. The magnitude of the resistivity maximum and the temperatures at which both the maximum resistivity and the sharp drop-off occur are nearly the same for RTA and furnace annealing.

The kinetics of silicide growth has been studied in most of the systems of interest (see Refs. 222-236 for a more complete listing), particularly for thicker layers where the reactions do not go to completion during the shortest thermal cycles. These reactions are typically either diffusion controlled or limited by nucleation of the growing phase. Oxide or polymer layers between the metal and the silicon can reduce the growth rate. Likewise, the presence of dopants or other impurities either within the silicon substrate or in the deposited metal can inhibit the silicidation reaction.

Diffusion-controlled kinetics is observed in most silicides, and the growth rate increases as the square root of time. Table V summarizes some of the experimentally measured reaction diffusivities. Figure 21 gives an Arrhenius plot of the silicide growth rate parameter showing that a single activation energy describes the $TiSi_2$ growth kinetics over a wide range of annealing

Table V Kinetics of silicide formation.

Silicide	Starting Material	Reaction Diffusivity	
		D_0 (cm^2/s)	E_a (eV)
TiSi$_2$	Ti/Si		1.8 [5, 225]
TiSi$_2$	Ti/α-Si	3.5×10^{-3}	1.8 [224]
Co$_2$Si	Co/Si	5×10^{-3}	1.5 [223]
CoSi	Co$_2$Si/Si	10^{-1}	1.9 [223]
CoSi$_2$	CoSi/Si	$2\text{-}20 \times 10^3$	2.6–2.8 [149, 180, 200, 226–227]
CoSi$_2$	CoSi/Si	2×10^{-3}	1.78 [228], thin film + RTA
CoSi$_2$	CoSi/α-Si		2.3 [201]
Pt$_2$Si	Pt/Si		1.5 [223]
PtSi	Pt$_2$Si/Si		1.5 [223]
Ni$_2$Si	Ni/Si		1.5, 1.3 [229, 230]
NiSi	Ni$_2$Si/Si		1.4 [230]

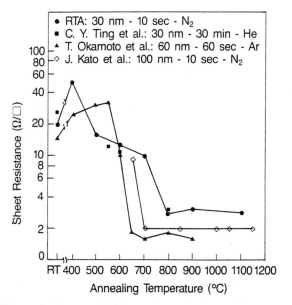

FIGURE 19. Sheet resistance as a function of furnace (30 min) or RTA (10 and 60 sec) annealing temperature for 30–100 nm titanium films on silicon. (From T. Brat, C. M. Osburn, T. Finstad, J. Liu, and B. Ellington, Self-Aligned Ti Silicide Formed by Rapid Thermal Annealing Effects. *J. Electrochem. Soc.* **133**, 1451, 1986. This figure is reprinted by permission from the publisher, The Electrochemical Society, Inc.)

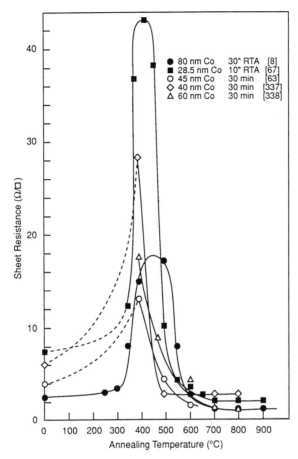

FIGURE 20. Sheet resistance as a function of furnace (30 min) or RTA (10 or 30 sec) annealing temperature for cobalt films on silicon. Data were taken from references 8, 63, 67, 337, and 338.

temperatures and times ranging from seconds to hours [5]. Because of the rapid silicide growth rates at moderate temperatures (500–800 °C), the thin films typically employed in VLSI technology (< 100 nm thick) can easily be formed in a few seconds of RTA. Extended RTA time and or furnace anneals are not necessary to fully react these films. To the contrary, such extended annealing may contribute to agglomeration of the silicide and, thus, can be deleterious. As shown earlier in Figs. 19 and 20, the sheet resistance versus annealing temperature characteristics for $TiSi_2$ and $CoSi_2$ are more or less identical over a wide temperature range for both conventional furnace and rapid thermal annealing.

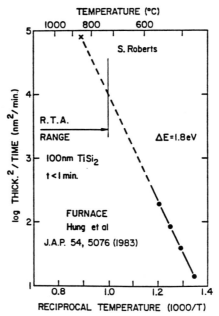

FIGURE 21. Kinetics of silicide formation in the furnace and the RTA regimes. (From Ref. 6.)

Nucleation-controlled kinetics favors the use of rapid thermal annealing for silicide growth. Both $CoSi_2$ and the C54 phase of $TiSi_2$ are believed to be limited by their nucleation. Unfortunately, nucleation-controlled processes are believed to result in rougher films [5]. By forming many nuclei at high temperature with RTA, it presumably is possible to eliminate this limiting factor, and at least one study [141] has reported smoother films of $CoSi_2$ when using RTA.

Only at temperatures near that required for phase transformation or for thicker films does annealing time becomes important. In these cases the annealing time becomes important during the first step of a two-step annealing process, particularly in the formation of $CoSi_2$ where two intermediate phases, Co_2Si and $CoSi$, are formed. Figure 22 illustrates this temperature region, namely 400–500 °C, where there is a pronounced effect of annealing time on the film sheet resistance. Since the resistivity of Co_2Si is higher than that of cobalt metal, the sheet resistance actually increases with annealing time at the lowest temperature (<400 °C) as seen by the boxes in the curve. At 450 °C (circles), the resistance increases for short annealing times, but decreases with additional annealing, reflecting the formation of the $CoSi$ phase. At higher temperatures the resistance monotonically decreases as $CoSi$ and $CoSi_2$ progressively are formed.

Silicides

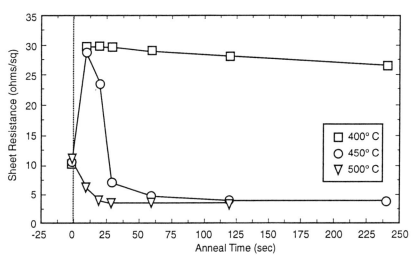

FIGURE 22. Dependence of $CoSi_2$ resistance on RTA time [courtesy of M. Kellam].

The presence of native oxide on the silicon surface [237–239] and dopants within the silicon [240–248] are two important factors that can affect silicidation kinetics. One of the advantages of using titanium is its ability to reduce the native oxide. Thus native oxides only add a little incubation time but, otherwise, apparently do not impact the silicide growth rate [237]. Cobalt, on the other hand, does not reduce SiO_2, and relatively thin oxides (2 nm) can completely impede its silicidation [3]; however, dilute HF dips prior to metallization have proven to be adequate to ensure a good reaction. Both oxygen and carbon, from the substrate surface as well as from within the deposited metal, are observed to "snowplow" ahead of the growing silicide layer and ultimately are left as an oxygen- or carbon-rich layer, often as an oxide, along the top surface. Figure 23 illustrates this snowplow effect during the formation of titanium and cobalt disilicide; oxygen that was originally in the metal film or at the metal/silicon interface results in a surface oxide or oxygen-rich layer. In cases where metal is the predominately diffusing species (e.g., Co), the metal is believed to diffuse through the oxide, leaving the oxygen at the top; diffusion of metal through the oxide also explains why a native oxide does not totally inhibit silicidation in those systems. For systems where silicon is the diffusing species, the origin of the snowplow is more complex. Apparently the segregation coefficients of O and C are such that they diffuse ahead of the growing silicide phase so that they remain in the metal phase. In both systems, the concentration of the snowplowed oxygen can exceed the solubility limit in the metal so that an oxide phase is formed. Since titanium silicide can grow downward from

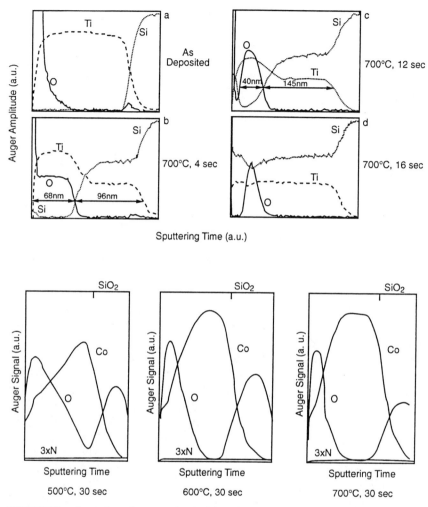

FIGURE 23. Snowplow of oxygen ahead of the silicide during RTA formation of (top): $TiSi_2$ (from Ref. 153), (bottom) $CoSi_2$ (from Ref. 149).

the metal top surface and upward from the substrate, the snowplow of oxygen ahead of the growing silicide can actually result in a buried oxygen-rich layer, as shown in Fig. 23 after 12 or 16 sec RTA at 700 °C, or even in a layer of oxide precipitates. Interestingly, more pronounced RTAs (i.e., higher temperature or longer times), can actually result in a decreased oxygen content in the film, presumably because of volatilization into the ultrahigh purity vapor phase.

Silicides

There have been many reports about the effect of dopants on silicide growth and phase transformation kinetics [10, 147, 240–248]. Especially for $TiSi_2$, a wide range of sheet resistances has been reported on differently doped starting substrates. The substrate doping effect seems to be most pronounced for $TiSi_2$ growth on n^+ silicon. Naem [242] ascribes much of the effect to unactivated arsenic, where concentrations above the solubility limit have the profound influence of completely stopping the silicidation reaction.

The presence of polymer on the starting silicon surface is another factor that might inhibit the silicidation. In a typical salicide application, a sidewall oxide spacer is formed by selective reactive ion etching of oxide. In order to achieve selectivity (i.e., to etch oxide but not silicon), a fluorine deficient etching gas is usually employed. Such etching is known to leave CF_x polymer on the silicon substrate. This polymer can be several nanometers thick and if not removed—for instance in an oxygen plasma—can block any subsequent formation of silicide. Even plasma removal of this polymer can create two additional problems: the carbon and fluorine in the polymer can be knocked-on into the substrate where they can also interfere with silicidation kinetics, and extra care must be taken to remove all of the oxide grown on the silicon during the oxygen plasma strip. Any oxygen left after the strip can be knocked into the silicon by the source/drain implant where it has been observed to retard silicidation [241].

For $TiSi_2$ the transformation from the high-resistivity C49 to the low-resistivity C54 phase is another kinetic limitation and has been the cause of considerable study [249–255], because its rate depends on impurities, dopant, substrate crystallinity, and feature size [255]. Since the transformation is nucleation controlled, it is impeded in very narrow lines when nuclei are not present, but is enhanced when the recrystallization of an amorphous silicon substrate accompanies the silicide formation reaction.

C. Effect of Formation Atmosphere

Annealing of metals in atmospheres containing even sub parts per million of oxidizing impurities such as H_2O is known to result in oxidation of the metal at the exposed surface that competes with silicidation at the metal–silicon interface. Table VI shows how the resistivity of titanium silicide depends on the oxygen concentration in a diffusion furnace. The x-ray photoelectron spectroscopy (XPS) profiles in Fig. 24 show the presence of a considerable amount of oxygen in the film when the formation atmosphere contains very much oxygen. This sensitivity to impurities during the silicide formation anneal is one reason why RTA has become so important for silicide

Table VI Effect of oxygen impurities on silicide sheet resistance [263].

Gas	Impurities	Silicide Resistance
Nitrogen	0.1 ppm Oxygen	2.6 Ω/□
Nitrogen	1 ppm Oxygen	3.4 Ω/□
Nitrogen	10 ppm Oxygen	100 Ω/□
Argon	<0.1 ppm Oxygen	1.8 Ω/□

formation. The impurity levels present in VLSI gases, N_2 or Ar, using leak-free piping are adequately low for this application. Nevertheless, because of backstreaming of air from the outlet end, the oxygen levels in a typical diffusion furnace tube are quite high, and much work [35, 256–263] has focused on measuring and reducing the oxygen contamination in such systems. Figure 25, for example, shows the concentration of backstreamed oxygen in a conventional diffusion furnace as a function of position in the furnace tube and of flow rate of the input gas. At low flow rates and positions nearest the outlet, the contamination levels can be very high (greater than 1 ppm) and results in the oxidation of the metal rather than its silicidation.

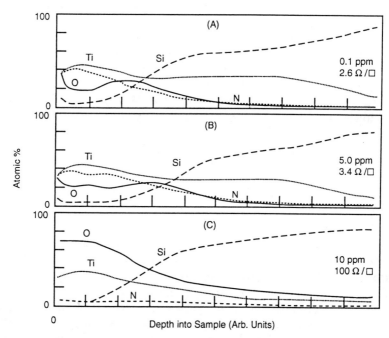

FIGURE 24. XPS profiles of $TiSi_2$ films formed in N_2 having 0.1, 1.0, and 10 ppm of O_2 impurities. The reaction was for 30 minutes at 800 °C. (From Ref. 263.)

Silicides

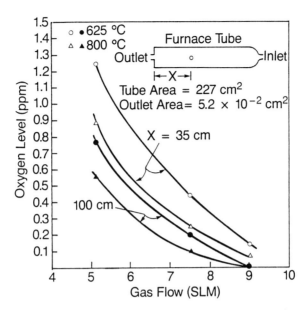

FIGURE 25. Concentration of backstreamed oxygen in a typical diffusion furnace geometry as a function of flow rate, position in the furnace, and temperature. (From Ref. 263.)

Even when using rapid thermal annealing, the gas purity has an effect on the silicide resistivity. Impurity levels greater than about 1 ppm of oxygen in nitrogen or argon degrade the resistivity; however, these purity levels are readily attainable with state-of-the-art gas distribution and RTA systems.

Capping layers on top of the metal [264–266] have been used by several groups to provide a less ambient-sensitive structure. With this scheme, a metal (possibly sacrificial) is deposited on top of the desired silicide source. For example, molybdenum and amorphous silicon layers on titanium and on cobalt have been tested. After annealing, the top metal remains intact on top of a good, low-resistivity silicide. While such capping makes the use of furnace annealing more practical, the extra processing complexity detracts from its attractiveness.

D. Issues for RTA Formation of Silicides

The theoretical amount of silicon consumed and the amount of silicide formed by the reaction of metal with silicon can be calculated from the theoretical densities as shown in Table VII. Nevertheless, because of lateral diffusion of either metal or silicon, edge effects and/or excess silicon

Table VII Silicide formation and silicon consumption based on theoretical densities*

Silicide	Metal Thickness (nm)	Silicide Formed (nm)	Silicon Consumed (nm)
$TiSi_2$	1.00	2.50	2.24
$CoSi_2$	1.00	3.49	3.63
NiSi	1.00	2.22	1.84
Pd_2Si	1.00	1.42	0.68
PtSi	1.00	1.98	1.32

*Data from Nicolet and Lau [2].

consumption are observed. For instance, during the formation of $TiSi_2$ on patterned substrates, silicon diffuses laterally from openings in oxide windows into the titanium on top of oxide; in this process there can be a significant loss of silicon from the opening. When argon atmospheres are used, considerable lateral overgrowth occurs and the silicon consumption is almost twice the value that would be calculated from Table VII. Even when nitrogen annealing is used to suppress the lateral growth of silicide, the silicon consumption can be 25% greater than that expected. To avoid vertical junction penetration with this excess silicon consumption requires a large buffer between the junction and the "expected" bottom of the silicide. To ensure high-yield, low-leakage junctions, the buffer needs to be as much as half the junction depth for $TiSi_2$. Even when metal is the predominant diffusing species, as with Co, a thicker silicide can form along window edges also giving rise to deeper than expected penetration of the silicide.

One of the issues associated with the formation of silicides using RTA is the generic problem of accurate temperature measurement and control. Since the silicide growth rates are quite rapid at the temperatures used in the second annealing step, precise temperature control is generally not needed at that step. However, since lateral overgrowth is sensitive to the silicide phase that is formed, particularly in the Co–Si system, as shown earlier in Fig. 17b, it can be anticipated that much attention needs to be paid to temperature measurements and control during the first step anneal. Figure 22 illustrates the dramatic effect of temperature differences of only 50 °C on the phase that is formed, as measured by sheet resistance. The increased resistance seen after 400 and 450 °C annealing in the figure represents the formation of Co_2Si, while the decreased resistance at 500 °C comes from the CoSi phase. At 450 °C the data show the formation of Co_2Si for short annealing times followed by that of CoSi at longer times. Indeed, sheet resistance versus time data, taken by different investigators using different RTA tools, show the same qualitative trends as shown in Fig. 22, only at temperatures that are offset by 50–75 °C. Further complicating the temperature control problem are the differences in absorption of lamp radiation between silicides and silicon. It has been shown that differences in radiation absorption between oxide and silicon can result in substantial lateral thermal gradients, especially between chip and kerf or edge areas [267]. These gradients are expected to be exacerbated when silicides are present during RTA.

Another serious issue in silicide formation is the difficulty encountered in producing ultrathin silicide layers from evaporated metal films. This problem has two components. The first component is the occurrence of a nonzero intercept in a plot of film thickness versus sheet resistance, as seen in Fig. 26. This offset is seen with both metal and silicide films, at least for

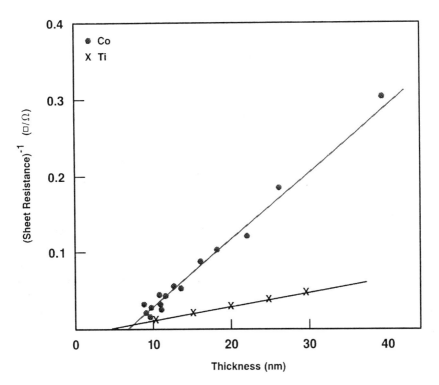

FIGURE 26. Relationship between deposited metal thickness and silicide resistivity for cobalt and titanium. (From Ref. 10.)

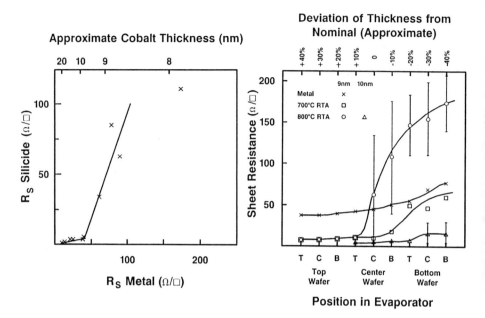

FIGURE 27. Nonlinear relationship between metal thickness as measured by resistivity (left) or position in evaporator (right) and silicide resistance. (From Ref. 10.)

Ti and Co, and can be interpreted as a critical film thickness that does not contribute to the overall conductivity, presumably because the film is primarily a nonconducting oxide rather than metal or silicide. The offsets are almost 10 nm in these metals, making it exceedingly difficult to obtain conducting silicide layers that are below about 30 nm in thickness. The second component of the problem is the lack of uniformity or control of the resistivity of very thin films. The uniformity problem is illustrated in Fig. 27 by the large error bars describing the range of sheet resistance of thin silicide. Even though different wafers in the same evaporation may differ in average thickness by only 10–20%, their resistivities may differ by as much as a factor of two. This uniformity issue is believed to be due in part to the offset just described and in part to the agglomeration of very thin films. The ultrathin silicides are roughened at their formation temperature and thus exhibit a wide range of resistance.

As shown earlier in Table III, high stress levels can be generated by silicide formation [3, 6, 268–271]. Since silicided junctions tend to be quite shallow and because the stress is concentrated at film edges where extended defects might be formed, care must be taken in choosing the silicide thickness and processing temperature. Stress is generated by the thermal

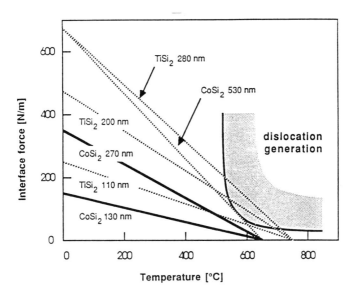

FIGURE 28. Allowable process window to avoid dislocation generation along silicide edges. (From L. Van den hove, J. Vanhellemont, R. Wolters, W. Classen, R. DeKeersmaecker, and G. Declerk, Correlation Between Stress and Film-Edge Induced Defect Generation for $TiSi_2$ and $CoSi_2$ Formed by Metal-Si Reaction, *Proc. of First Int. Symp. on Advanced Materials for ULSI, Electrochem. Soc.* **88-19**, 165, 1988.)

expansion mismatch between silicon and the substrate at temperatures below which stress relaxation can occur. Dislocation loops lying on the $\langle 111 \rangle$ plane along $\langle 110 \rangle$-oriented feature edges are observed if the silicide is too thick. Van de hove et al. [270] observed stress relaxation down to about 750 °C for $TiSi_2$ and 630 °C in $CoSi_2$. Above those temperatures the stress is zero; at lower temperatures, the force exerted on a window edge is given by the difference in thermal expansion coefficient between the silicide and the silicon and by the silicide thickness. Figure 28 shows the resultant interfacial force as a function of temperature for different thicknesses of each of these silicides. The shaded region represents process regions where dislocation generation in the silicon occurs. Fortunately for less than about 500 nm of $CoSi_2$, the problem is minimized; however, defects can be seen along the edges of $TiSi_2$ that is only 100 nm thick. Part of the superiority of $CoSi_2$ is due to the fact that thermal stress builds up only below 630 °C for $CoSi_2$ versus 750 °C for $TiSi_2$. In this regard furnace processing offers an advantage over rapid thermal processing. For slow cooling, stress is relieved down to 500 °C for $CoSi_2$ whereas it builds up starting at 630 °C for RTA [3].

III. Properties of Silicides and Silicided Junctions

A. CONDUCTIVITY

The conductivity of silicide films is more complicated than the often cited tables, including Table III, of their resistivities would imply. In addition to its bulk resistivity, the resistance of a polycrystalline silicide film is affected by impurities such as oxygen and carbon (as shown earlier), dopants, including those from the substrate as well as those implanted into the silicide, degree of agglomeration, the silicide phase that is present (e.g., C49 or C54 $TiSi_2$), and presumably grain size as well as ion implantation damage within the silicide. Furthermore, as shown earlier in Fig. 26, the film resistance (at least of polycrystalline $CoSi_2$ and $TiSi_2$) does not scale with its thickness because there is a fixed thickness that apparently does not contribute to the conductance. Indeed, the use of sheet resistance as a monitor of the silicide thickness can be very misleading. The resistivity of most silicides varies linearly with temperature. However, the residual resistivities and temperature coefficients of resistivity have been seen to be functions of dopant [181, 272–273], processing conditions, and other impurities in the silicide as evidenced by the wide range of reported values.

B. STABILITY

The thermal stability of metal silicides is one of the limiting factors that must be considered when choosing a material [90, 251–255, 274–288]. Even though they have a higher resistivity, the refractory metal silicides are typically able to withstand the higher thermal budget required for junction annealing. Therefore they are used for polycide applications where the silicide is in place before the diffusions are formed. The silicides of the near-noble metals, on the other hand, have the lowest resistivity but are not able to withstand the temperatures associated with most postimplantation anneals. Thus their applications are usually restricted to salicide applications, where the junction is formed prior to metal deposition and silicide formation. The use of the near-noble metal silicides as diffusion sources strains their stability limits and requires that the thermal budget associated with the solid-source diffusion be minimized.

At high temperatures silicides undergo a thermal grooving process [289–290] until they have agglomerated into islands. Figure 29 shows an example of this agglomeration in which a 9-nm thick cobalt film is reacted with silicon at 800 °C for 10 sec. For the polycrystalline silicides of $TiSi_2$ or $CoSi_2$, the initial films have a peak-to-peak roughness of about one-third

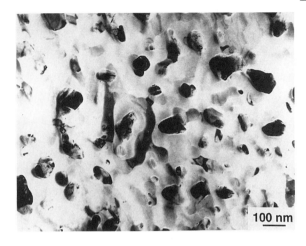

FIGURE 29. Cross-sectional TEM micrographs of $CoSi_2$ films showing agglomeration. (From Ref. 10.)

of the silicide thickness. This roughness increases with further annealing until the film becomes discontinuous. Thermal grooving occurs at the intersection of three regions (silicide-ambient in Fig. 30a, silicide-ambient-silicon in Fig. 30b, and silicide-silicon-oxide in Fig. 30c). In the grooving process, the angles between the grain boundaries change at the triple points in order to balance the forces (represented by the surface tension vectors in the figure) while minimizing the system free energy (interfacial and bulk). The equilibrium angles between grains at triple points, and hence the amount of growing, depend on the values of the interfacial energies and on the film thickness.

The silicide resistivity provides a convenient measure of the amount of agglomeration [104, 277–278]. Figure 31 shows the annealing and temperature dependence of the sheet resistance of $TiSi_2$ and $CoSi_2$ films; for a fixed annealing time the resistance dramatically increases at higher temperatures. Figure 32 gives the sheet resistance of $CoSi_2$ as a function of annealing time and illustrates that the amount of degradation depends on both the silicide thickness as well as the annealing atmosphere. Other silicides exhibit the

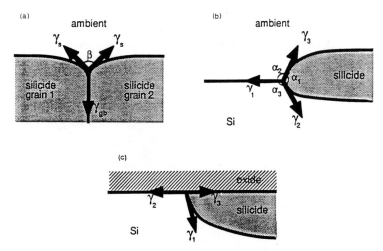

FIGURE 30. Model for thermal grooving. (From Ref. 3.)

same kind of qualitative behavior. Jiang *et al.* [104] found it useful to define an agglomeration time constant t_{degrad} as the time it takes for the film resistivity to degrade (increase) by 30% from its initial value. By plotting the experimentally measured degradation time as a function of reciprocal annealing temperature in Fig. 33, an activation energy for agglomeration can be extracted. For cobalt silicide on single crystal silicon, the resistivity degradation has a thermal activation energy of about 5 eV. Different activation energies are observed for other silicides or for poly-Si substrates. Table VIII gives some of the empirical constants that describe the thickness and temperature dependence of the stability of several silicides of interest.

Table VIII Thermal degradation of metal silicides. Parameters for the equation $\tau_{\text{degradation}} = (d - d_0)^2/D_0 \exp[(-E_a/kT)]$.

Silicide	Degradation Rate Parameters				
	d_0 (nm)	D_0 (N$_2$) (cm^2/s)	E_a (N$_2$) (eV)	D_0 (Ar) (cm^2/s)	E_a (Ar) (eV)
CoSi$_2$	11	1.3×10^6	4.8	6.1×10^8	5.3
TiSi$_2$	11 est	~50	3.5		
PtSi	11 est	~10	3.1		
Pd$_2$Si		τ = 10 sec at 850 °C			
NiSi		τ = 10 sec at 800 °C			

Silicides

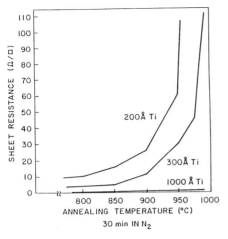 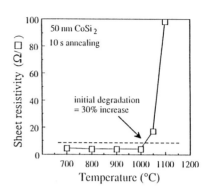

FIGURE 31. Temperature dependence of the resistivity of furnace-annealed TiSi$_2$ films (left) and 50 nm CoSi$_2$ after 10-sec RTA [104] (on right). (These figures are reprinted by permission from the publisher, The Electrochemical Society, Inc. Left from C. V. Ting, F. M. d'Heurle, S. S. Iyer, and P. M. Fryer, High. Temp. Process Limitation on TiSi$_2$, *J. Electrochem. Soc.* **133**(12), 2621, 1986. Right from H. Jiang, C. M. Osburn, Z.-G. Xiac, G. McGuire, and G. A. Rozgonyi, Ultra Shallow Junction Formation using Diffusion for Silicides: III. Diffusion into Silicon, Thermal Stability of Silicides and Junction Integrity, *J. Electrochem. Soc.* **139**(1), 211, 1992.)

The thermal stability of silicides depends on several parameters including film thickness, initial distribution of grain sizes, the presence or absence of capping layers, and the annealing atmosphere. Figure 34 shows the thickness dependence of the degradation time constant for conventionally formed cobalt silicide [104]. The degradation occurs at shorter times for thinner silicide films. The data in the figure fit the empirical equation

$$(d - d_0)^2 = D_{degrad} \cdot t_{degrad} \tag{7.8}$$

One very interesting feature in Fig. 34 and Eq. 7.8 is the nonzero intercept d_0. Silicides less than about 30 nm thick are not able to withstand any appreciable annealing beyond their formation anneals before they have markedly agglomerated. Indeed, as pointed out earlier with Fig. 27, there appears to be a minimum metal thickness (~8 nm for Co and ~7 nm for Ti) below which it is not possible to get the normally expected silicide conductivity. The constant of proportionality between the degradation time and the square of the corrected thickness, D_{degrad}, has the units of cm^2/sec and can be considered as a "degradation diffusivity".

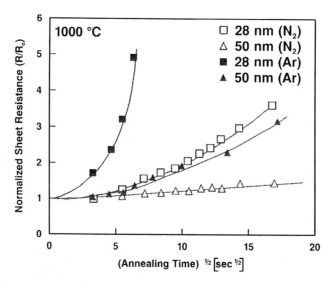

FIGURE 32. Time dependence of the resistivity of 28 and 50 nm CoSi$_2$ annealed at 1000 °C, showing improved stability of films annealed in nitrogen.

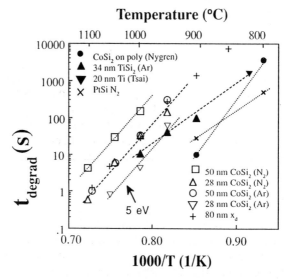

FIGURE 33. Arrhenius plot of the degradation time constant for different thicknesses of cobalt and titanium silicide. (From H. Jiang, C. M. Osburn, Z.-G. Xiac, G. McGuire, and G. A. Rozgonyi, Ultra Shallow Junction Formation using Diffusion for Silicides: III. Diffusion into Silicon, Thermal Stability of Silicides and Junction Integrity, *J. Electrochem. Soc.* **139**(1), 211, 1992. This figure is reprinted by permission from the publisher, The Electrochemical Society, Inc.)

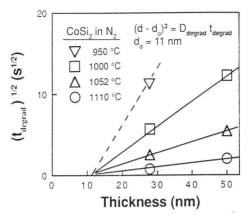

FIGURE 34. Thickness dependence of the degradation time constant for $CoSi_2$ films. (From H. Jiang, C. M. Osburn, Z.-G. Xiac, G. McGuire, and G. A. Rozgonyi, Ultra Shallow Junction Formation using Diffusion for Silicides: III. Diffusion into Silicon, Thermal Stability of Silicides and Junction Integrity, *J. Electrochem. Soc.* **139**(1), 211, 1992. This figure is reprinted by permission from the publisher, The Electrochemical Society, Inc.)

Since silicide agglomeration occurs at grain boundaries, it is not at all surprising that its stability depends on the starting grain size. Larger grain, and presumably single-crystal, silicide films are more stable in that they agglomerate into fewer islands than fine-grained films. This observation has led to increased efforts to produce smoother, larger grained films using techniques like sputter precleaning of the substrate before metal deposition [280] or ion beam mixing of the metal and silicon before RTA. Single crystal films of $CoSi_2$ grown by SPE from Co/Ti layers [220–221] have superior stability to conventional polycrystalline films. The grain size effect may also partially explain the increased stability of thicker films, because the grain size typically scales with film thickness. While larger grained films produce fewer islands after high-temperature annealing, the islands themselves contain more material so that they can form thicker globules. Thus larger grained films exhibit greater surface and interface roughness after extreme agglomeration than fine-grained films, and the degradation in film resistance can be more severe because the probability of interconnecting a small number of large islands can actually be less than that of interconnecting many small islands.

Reaction of metal with amorphous silicon results in smoother, finer grained silicides that remain smooth up to higher temperatures [251–253]. Apparently both cobalt and titanium silicides nucleate more readily when they react with amorphous silicon than when they are reacted with a single crystal substrate, even though the silicidation reaction competes kinetically with the amorphous silicon recrystallization at about the same temperature. As a consequence of this enhanced nucleation rate the final silicide grains, which are formed on an initially amorphous substrate, are smaller and do not significantly add to the film roughness after high-temperature annealing.

The stability of silicides on poly-Si is even more limited than that on single crystal substrates [254, 274–277, 282–284, 288]. In addition to the thermal grooving process, silicides on polycrystalline substrates are affected by grain growth of the poly-Si. Silicide overlayers provide a medium into which the poly-Si can dissolve, transport through, and nucleate as large grains on the outer surface. This additional mechanism greatly detracts from the thermal stability of the system. Cobalt silicide, particularly, is not very stable on poly-Si electrodes and limits the allowable post-silicidation thermal cycle. In this system a reversal of the layering of $CoSi_2$ on poly-Si/SiO_2 occurs after annealing to give poly-Si/$CoSi_2$/SiO_2 [292]. Some device designers have chosen to use other materials for the gate cladding (e.g., TiN) rather than $CoSi_2$ to circumvent its mediocre stability [291].

The annealing atmosphere also impacts the silicide stability [3, 104, 278–279, 282]. Both $TiSi_2$ and $CoSi_2$ are more stable when they are annealed in nitrogen rather than in argon. This result is not surprising for $TiSi_2$, because it reacts with nitrogen to form TiN. However, the reason for the improvement in $CoSi_2$ is much less certain since nitrogen is not believed to react with it. Nevertheless the improvement can be quite dramatic, as seen in Fig. 35. Figure 35 compares D_{degrad} values obtained from data like those shown in Fig. 34 for $CoSi_2$ annealed in Ar or in N_2. The degradation diffusivity of silicide annealed in argon is over half an order of magnitude higher than that of films annealed in nitrogen.

Capping of the silicide film with a layer of CVD oxide or nitride [129, 170, 278, 293] has a complex effect on the stability of the silicide. First of all, the cap impacts the surface free energy on the top of the silicide in Fig. 30. Second, it mechanically constrains the silicide and thus provides a barrier to agglomeration. Nevertheless, capped films of $TiSi_2$ and $CoSi_2$ also exhibit agglomeration. One study [293] found that nitride-capped films were intermediate in stability to uncapped films that were annealed in Ar and those annealed in N_2. In this instance, the stability of nitride-capped films is explained by a combination of improved mechanical stability over straight Ar-annealed silicide but degraded stability compared to those annealed in N_2 as a result of an impermeable cap.

Because agglomeration is a three-dimensional process, there is a feature size dependence of the film resistivity, where narrower lines of silicide are more affected by annealing than wider lines. Figure 36 shows a cross-sectional TEM micrograph of $TiSi_2$ on a narrow poly-Si line after annealing

Silicides

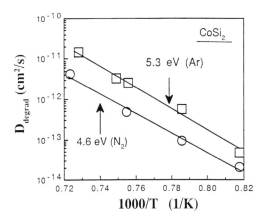

FIGURE 35. Effect of annealing atmosphere on the thermal stability of $CoSi_2$ films (see also Figs. 32, 33). (From H. Jiang, C. M. Osburn, Z.-G. Xiac, G. McGuire, and G. A. Rozgonyi, Ultra Shallow Junction Formation using Diffusion for Silicides: III. Diffusion into Silicon, Thermal Stability of Silicides and Junction Integrity, *J. Electrochem. Soc.* **139**(1), 211, 1992). This figure is reprinted by permission from the publisher, The Electrochemical Society, Inc.)

at 900 °C for 30 min. The silicide tends to agglomerate on the edges of the poly-Si lines, somewhat depleting the center [283–284, 288]. Another manifestation of a linewidth dependence of silicide resistivity is illustrated in Fig. 37 for $TiSi_2$ features on single crystal silicon [255]. The top graph shows that narrow lines require higher temperatures to transform to a low sheet resistance structure. This phenomena is explained by the relative lack of nuclei for the C49 to C54 phase transformation. The phase transformation process is characterized by a relatively slow nucleation rate and a fast growth rate for the C54 phase. The probability of forming a nucleation

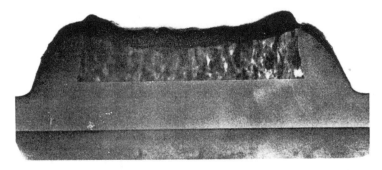

FIGURE 36. Cross-sectional TEM micrograph of $TiSi_2$ agglomeration on a narrow poly-Si line. (From Ref. 284.)

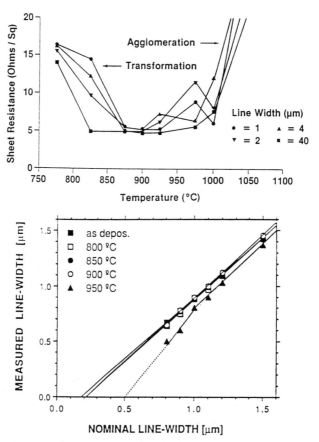

FIGURE 37. Linewidth dependence of the formation and agglomeration of $TiSi_2$ films. The top figure shows the need for higher annealing temperatures for the transformation from the C49 to the C54 phase in narrower lines. (From J. Lasky, J. Nakos, O, Cain, and P. Geiss, Comparison of Transformation to Low-Resistivity Phase and Agglomeration of TiS_2 and $CoSi_2$, *IEEE Trans. Electron. Dev.* **ED-38**, 262, © 1991 IEEE.) The bottom figure shows increasing linewidth bias (difference between mask dimension and final electrically equivalent dimension) for decreasing feature sizes after high-temperature annealing. (From Ref. 283.)

site within a narrow line is diminished. Thus the narrowest lines may be predominately the high-resistivity C49 phase, while the wide lines are the high-conductivity C54 phase. The bottom graph in Fig. 37 shows that the apparent bias between the mask dimension and the electrically measured linewidth increases as the line gets smaller. This behavior, at high annealing temperatures, illustrates that narrow lines of $TiSi_2$ have poorer thermal stability than wide lines.

C. Dopant Redistribution

The redistribution of dopants in silicides [294–299] has an important effect on silicided devices and provides restrictions for their utility. The rapid diffusion of some dopants in some silicides makes it difficult to silicide a preexisting junction without depleting the surface dopant and thereby increasing the contact resistance. On the other hand, the relatively slow diffusion of other dopants in other silicides makes it difficult to use them as diffusion sources.

Some examples of dopant redistribution following silicidation of n^+ (arsenic) and p^+ (boron) junctions using RTA are shown in Figs. 38 and 39 for titanium and cobalt silicide respectively. These junctions are formed by implantation and anneal before silicidation. The dopant profiles of the as-formed junctions are barely affected by the lowest silicide annealing temperatures shown in the figure. However, after high-temperature annealing (i.e., after 10 sec at 900 °C) dopant is depleted at the silicon–silicide interface, and the dopant diffuses to the silicide surface. The contact resistivity of the silicide–silicon contact depends on both the silicide barrier

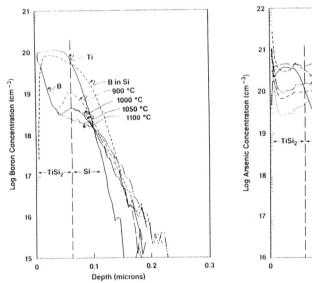
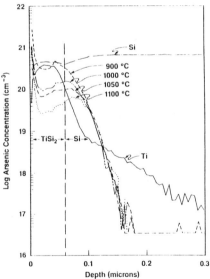

FIGURE 38. Dopant redistribution with TiSi$_2$. (From C. M. Osburn, T. Brat, D. Sharma, N. Parikh, W.-K. Chu, D. Griffs, S. Corcoran, and S. Lin, The effects of Titanium Silicide Formation on Dopant Redistribution, *Proc. First Intl. Symp. on ULSI Sci. and Tech.*, The Electrochemical Society **135(6)**, 1900, 1988. This figure is reprinted by permission from the publisher, The Electrochemical Society, Inc.)

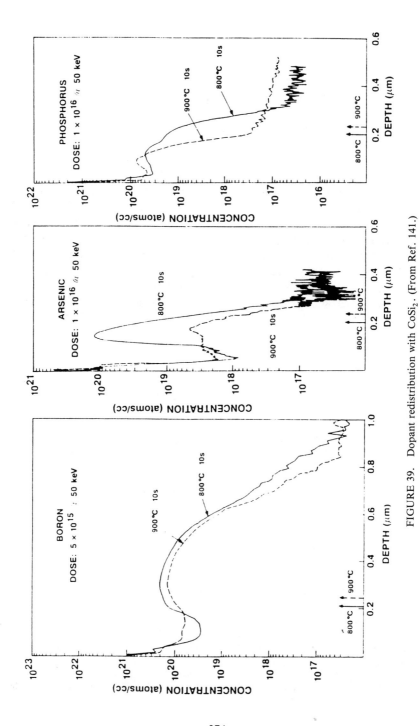

FIGURE 39. Dopant redistribution with $CoSi_2$. (From Ref. 141.)

Table IX Silicide contact resisitivity during subsequent thermal processing [3], measured on $3 \times 3 \, \mu m^2$ contacts. (Results are in Ω-cm^2.)

Treatment	CoSi$_2$		TiSi$_2$	
	As	B	As	B
Reference	2.1×10^{-7}	1.1×10^{-7}	1.8×10^{-7}	2.2×10^{-7}
800 °C, 30 min	3.1×10^{-7}	7.6×10^{-7}	1.9×10^{-7}	7.5×10^{-7}
900 °C, 30 min	2.2×10^{-7}	3.3×10^{-6}	9.3×10^{-7}	5.2 ± 10^{-6}
900 °C, 30 sec	—	7.4×10^{-7}	1.3×10^{-7}	1.0 ± 10^{-6}
1000 °C, 30 sec	6.7×10^{-7}	4.6×10^{-6}	6.1×10^{-7}	1.1×10^{-5}
1100 °C, 30 sec	3.3×10^{-6}	9.5×10^{-6}	3.2×10^{-6}	1.6×10^{-5}

height and the silicon surface doping. Thus a depletion of the surface doping due to silicidation increases the contact resistivity. Table IX shows the effect of annealing on the contact resistivity of CoSi$_2$ and TiSi$_2$ [3]. Higher temperatures and prolonged times result in more dopant depletion from the substrate and redistribution into the silicide, thereby giving higher contact resistance. In the extreme case, the increased contact resistivity may be larger than the decrease in external series resistance because of the lowered diffusion sheet resistance and the larger effective contact area, so that there is no net advantage of siliciding the device. Indeed there have been several instances where silicided devices actually had higher series resistance and lower current drive than conventional devices. Dopant redistribution in silicides on poly-Si can also present a problem when sufficient dopant is depleted to actually change the poly-Si work function and hence the device threshold voltage [47, 298].

Two techniques can be used to minimize the dopant depletion at a silicide contact: minimize the thermal cycle or add additional dopant. Since most of the postsilicide thermal cycle comes from the glass reflow step that is used to smooth out the CVD oxide over devices to reduce steps for the metallization, some optimization of the BPSG process is helpful in reducing the thermal cycle; however, engineering tradeoffs or compromises between contact resistance and planarization are required in many applications [50]. The second approach, namely that of adding additional dopant, is especially suited to NMOS technology but entails additional masking if it is to be used in CMOS. Taur *et al.* [300–302] used this approach to get excellent characteristics on n^+ contacts.

Dopant redistribution in silicides plays an even more important role when the silicide is used as a diffusion source. For the SADS process, dopant must redistribute (diffuse) through the silicide to the silicon–silicide interface, where it may segregate, prior to its diffusion into the substrate. Unfortunately, dopants also diffuse to the silicide surface where they can evaporate, thereby reducing the total source. While a capping layer of silicon nitride or oxide prevents evaporation, some of the dopants can diffuse into these layers and be lost to the process anyway.

Dopant redistribution in the silicon, as well as junction formation from a doped silicide source, depends on several phenomena in the silicide: segregation of dopant into the silicide bulk or grain boundaries, compound formation with the dopants [303–307], diffusion of dopants through the silicide via bulk or grain boundary mechanisms [303–313], surface segregation or compound formation, and dopant evaporation from a free silicide surface or diffusion into a capping layer. There has been an extensive amount of work done to quantify each of these phenomena, and some of the results are given in Table X. These effects can be very large; for instance, 99% of the arsenic dopant can evaporate from a $TiSi_2$ or $CoSi_2$ layer that is subjected to rapid thermal annealing. As more and better data become available on these effects and as these data are incorporated into process models (see Section IV E for instance) the ability to predict dopant profiles in complex silicides/silicon structures will improve.

The lateral diffusion of dopants along silicide lines remains as a serious issue. Especially in CMOS where silicides contact both n^+ and p^+ junctions with a local interconnect technology or where both conductivity types of poly-Si are employed, dopants can diffuse laterally along the silicide over the polysilicon and counterdope regions having the opposite conductivity type [43, 55, 314–318]. This counterdoping of the poly-Si leads to shifts in device threshold voltages that depend on the proximity of the gate to a source of the opposite dopant species. Sometimes a barrier layer (e.g., TiN) can be used on top of the poly-Si to block the outdiffusion and subsequent lateral motion of dopant. In other cases, sufficient space must be provided between oppositely doped regions.

D. Junction Leakage and Damage Removal

The reverse leakage current is one of the most important characteristics of silicided junctions, because high leakage produces excessive power dissipation and rapid loss of stored charge in logic or dynamic RAM applications. One of the key issues in siliciding preexisting junctions has been the need to allow a buffer between the bottom of the silicide and the metallurgical

Table X Parameters affecting dopant redistribution in silicon/silicide structures.

Parameter	Silicide		
	$CoSi_2$	$TiSi_2$	PtSi
Diffusion			
Boron (lattice)	$0.09\,e^{-2.1/kT}$ [312]	Very low [308]	
Boron (grain B)	$0.004\,e^{-2.0/kT}$ [310]	Slow [103]	Slow [103]
Boron (composite)	$4.9\,e^{-2.09/kT}$ [313]		
Arsenic (lattice)	Low [310]	$5 \times 10^{-6}\,e^{-1.8/kT}$ [308]	
Arsenic (grain B)		Very fast [311]	Slow [103]
Arsenic (grain B)		$\sim 10^{-3}$ at 600 °C	
Arsenic (composite)	$10^3\,e^{-2.91/kT}$ [313]	$4.8\,e^{-2.13kT}$ [313]	
Phosphorus (lattice)		$4 \times 10^{-5}\,e^{-2.0/kT}$ [308]	
Phosphorus (composite)		$392\,e^{-2.64/kT}$ [313]	
Silicon		$2 \times 10^{-3}\,e^{-1.8/kT}$ [224]	
Solid solubility			
Arsenic		$1.06 \times 10^{25}\,e^{-1.0/kT}$ [311]	
Compound formation			
Boron		TiB_2 [199, 303–306]	
Arsenic		TiAs [97, 246, 308]	
Evaporation coefficient			
Boron	$7.6 \times 10^9\,e^{-4.4/kT}$ [103]		
Arsenic	$7.0\,e^{-2.0/kT}$ [103]	$140\,e^{-2.3/kT}$ [103]	Comparable [103]
Surface segregation			
Boron	1.6×10^{14} [103]	Observed [92]	
Arsenic	ND [103]		

junction. For $TiSi_2$ this buffer has needed to be as thick as the silicide itself, making it very difficult to simultaneously have low silicide resistivity and shallow junctions. The need for this generous buffer is probably explained by a combination of silicide roughness, which can amount to about one-third of the silicide thickness, and of the excess silicon consumption due to lateral motion of silicon from the substrate into titanium regions over oxide, as previously described. Because this lateral diffusion is minimized with $CoSi_2$ formation, much less of a buffer is needed to get low-leakage diodes. One study reported low leakage even when the cobalt silicide consumed up to 90% of the junction [64, 186]; however, other work suggests that the buffer should be at least half of the silicide thickness [3]. When a sufficient buffer is maintained, low-leakage diodes, having a narrow statistical spread of leakage currents, can be produced beneath $TiSi_2$, $CoSi_2$, PtSi, and Pd_2Si. Figure 40 illustrates low-leakage, tightly distributed reverse junction characteristics of n^+ and p^+ junctions as

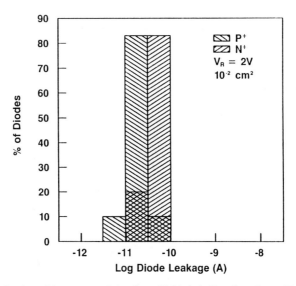

FIGURE 40. Leakage histograms of titanium silicided shallow junctions. (From Ref. 10.) The junctions were 70–90 nm deep, formed by 10 sec, 950 °C RTA of 8 keV BF_2 or 15 keV As implants, and silicided from 10 nm of titanium.

shallow as 70 nm that are silicided with 10 nm of $TiSi_2$. Arrhenius plots of their leakage ($\log I_r/T^3$ vs $1/T$) yield an activation energy equal to the silicon band gap (1.1 eV), which is characteristic for carrier diffusion to the junction rather than the generation of leakage currents via defect states within the depletion region.

In contrast to the widely reported, low-leakage data on silicided shallow junctions [10], a large variation in leakage is seen in devices where the junction is formed after the metal or silicide is deposited. Only a few ITM diodes exhibit good leakage ($< 10 \, nA/cm^2$), while SADS and ITS diodes generally exhibit low leakage [10], but careful optimization of the process conditions is necessary to minimize junction leakage. Otherwise, unacceptable devices can result from this process also.

Diode leakage data for n^+ and p^+ SADS diodes after 10 sec, 800 °C RTA are given in Fig. 41. Immediately after ion implantation, the leakage current is higher than in unimplanted, control Schottky barrier diodes, presumably because of the radiation damage in the perimeter oxide that is created during the implantation process and is eliminated by a 400 °C anneal. Progressively higher annealing temperatures then cause a dramatic reduction in reverse leakage currents as the Schottky diodes become junction diodes with good leakage characteristics. Only a relatively modest annealing temperature (600–800 °C) is needed to produce good diodes. The projected

Silicides

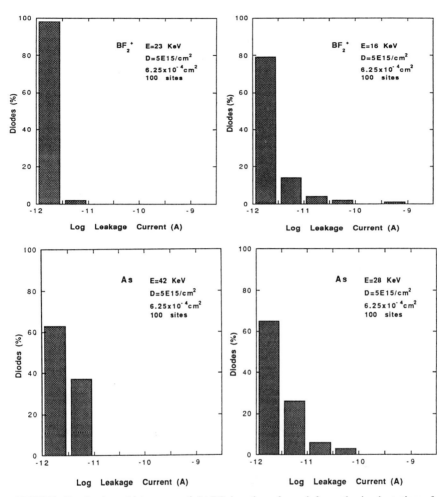

FIGURE 41. Leakage histograms of SADS junctions formed from the implantation of arsenic or BF_2 into 60 nm of $CoSi_2$ and subsequently annealed for 10 sec at 800 °C [courtesy of Q. F. Wang].

dopant motion can be computed from the conventionally accepted values of boron or arsenic diffusivity or from the enhanced values reported by Jiang et al. [319]. Plots of this dopant motion are shown in Fig. 42 for both values of diffusivity; as can readily be seen, the junction motion for 10-sec RTA at 900 °C and below is expected to be in the 0.1 to 10 nm range. This very low leakage with such little dopant motion is indeed encouraging. Increased leakage is observed at the higher annealing temperatures, where silicide agglomeration may occur.

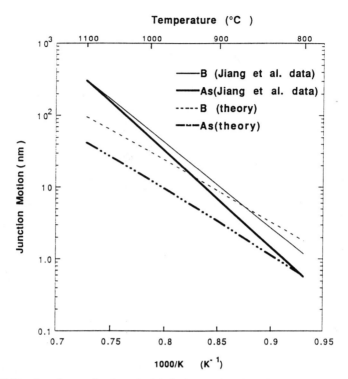

FIGURE 42. Junction motion from SADS diodes as a function of 10-sec RTA temperature. (From Ref. 319.)

The possible presence of deep energy levels within the silicon band gap, associated with the transition metal located beneath the silicide itself, has been a concern for salicide technology. Such deep levels would dramatically increase the magnitude of the leakage current and decrease the activation energy of the leakage. Both the solid solubility and the diffusivity of these metals are appreciable at typical postsilicidation temperatures. Deep level transient spectroscopy (DLTS) has indeed detected Co at concentrations up to $10^{11}/cm^3$ after 1100 °C RTA [3]. Fortunately this concentration is four orders of magnitude below the cobalt solubility limit in silicon at this temperature, and it is not enough to noticeably increase the diode leakage. One study [156] observed a DLTS peak associated with $1-5 \times 10^{14}$ Ti/cm^3 at 0.30 eV above the valance band, and another [320] concluded that titanium had diffused past the junction to introduce trap levels that were responsible for SRH (Shockley-Read-Hall) generation currents; however, the diode leakages seen in that instance were higher than those seen elsewhere and the activation energies for leakage were below those seen in other studies.

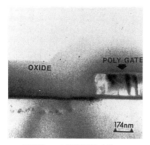 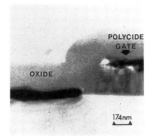

Without Ti Silicide   **With Ti Silicide**

FIGURE 43. Cross-sectional TEMs showing defect removal in silicided shallow junctions.

One somewhat unexpected benefit of siliciding shallow junctions with $TiSi_2$ is the enhanced removal of the junction ion-implantation damage [131, 321–327]. Ion implantation typically creates interstitials that coalesce during annealing to form interstitial dislocation loops. The cross-sectional TEMs of Fig. 43 compare the end-of-range loop densities of control and silicided junctions where, in this case, almost complete damage removal is seen with the silicide.

The reduction in number and size of these interstitial loops is attributed to the injection of vacancies at the silicide–silicon interface during silicide formation. These vacancies, which are created as the silicon from the substrate moves into the metal or silicide, diffuse into the substrate where they annihilate the interstitials composing the dislocation loops left by the dopant implantation step. The reduction of implantation damage is particularly dramatic because it occurs during the silicidation and temperatures, such as 10 sec at 800 °C, that are well below those at which any thermal annealing of defects occurs. Since the vacancy injection relies on silicon diffusion into the forming silicide, defect reduction is expected to occur only for those silicides, or phases, where silicon, and not metal, is the primary moving species. To date, this reduction has only been seen for $TiSi_2$. On the basis of this injection of point defects argument, however, one might expect to observe increased damage beneath silicides that are formed by metal diffusion and therefore would inject interstitials. Such an enhancement in damage has not been observed; however, most other studies have focused on $CoSi_2$, which has multiple phases, each formed with different moving species. It is possible that vacancy injection from one phase cancels interstitial injection from another.

Injection of point defects by silicidation reactions might then be expected to affect the diffusion of dopants, a process that is critically dependent on point defect concentrations. Indeed, enhanced diffusion in silicon has been reported in several systems but not in others [104, 328–330]. More comprehensive studies are underway to systematically quantify the enhancement factors and to determine the lifetimes of the injected excess point defects. Not only might this enhanced diffusion affect junction depths, depending on the postsilicidation annealing conditions, but it is also expected to influence dopant redistribution into the silicide and thereby the contact resistance.

IV. Applications of Silicides Formed by RTA and Process/Device Considerations

A. SILICIDED GATES

Because cladding of poly-Si gates with silicides (i.e., polycide technology) is now a relatively mature process, most of the materials and process issues associated with it have been satisfactorily addressed. Four areas have received a considerable amount of attention: (1) control of the silicide deposition to obtain a low-resistance, stoichiometric silicide; (2) dielectric breakdown of thin oxides underneath polycide gates; (3) etching of silicide/poly-Si stacked structures so as to obtain relatively vertical walls, without undercutting the poly-Si during the overetch required to clear the gate over oxide steps; and (4) modest changes in the threshold voltage of polycide gate transistors compared to poly-Si gated ones.

Polycide processes typically employ deposited silicides, and the importance of controlling the film composition to minimize its resistance was discussed earlier. The composition of the deposited film has also been seen to have an impact on gate oxide breakdown [22]. For instance if a tungsten-rich silicide film is initially deposited, the excess tungsten reacts and consumes some of the underlying poly-Si gate. If an adequate buffer of poly-Si is not left behind, the gate oxide is degraded. Thus maintaining good gate oxide integrity restricts the amount by which the poly-Si thickness can be scaled down in a polycide process. In this regard, thermal reaction of metal with the poly-Si is not as preferred as codeposition of silicide. Furthermore, the more limited thermal stability of those silicides formed by reaction with polysilicon such as $TiSi_2$ and $CoSi_2$ (compared to the more refractory silicides of W, Mo, and Ta) make them less able to withstand the annealing times and temperatures often associated with junction formation.

Curiously, polycide-gate transistors typically have threshold voltages that are 10–40 mV different (typically more positive) than poly-Si devices [50]. Because the threshold voltage of a device depends on the metal–semiconductor work function, oxide charge, and doping level in the underlying silicon, it has not been absolutely possible to unambiguously explain the effect. Dopant redistribution out of the poly-Si and into the silicide, thereby changing the metal–semiconductor work function difference, is largely believed to be responsible for the threshold changes. Nevetheless, fluorine motion through the poly-Si gate and into the gate oxide has been observed during CVD of WSi_2 using WF_6. While the impact of fluorine on oxide charge is unclear, it is known to enhance the diffusion of boron from the gate, through the oxide, and into the substrate. The effect of the silicide itself on boron diffusion through the gate oxide is only beginning to be studied [331].

B. SILICIDED SHALLOW JUNCTIONS

1. Device Characteristics

As described earlier, the advantage of siliciding junctions is to reduce both the series resistance associated with shallow diffusions and the high contact resistance associated with small contact areas. For a half-micrometer device technology, illustrated in Figs. 8 and 9, the presence of a silicide reduces the parasitic series resistance to allow a 20% higher current drive capability and improved circuit performance. The parasitic resistances of polysilicon and diffusion interconnects are projected to increase as technologies are further scaled down, because short channel effects force reductions in thermal budget and junction depths. In addition, the parasitic resistance associated with metal-to-silicon contacts increases as the square of the scaling parameter and begins to dominate the total parasitic resistance for submicrometer dimensions.

Minimum size devices, namely those with the shortest channel length and narrowest widths with minimum diffusion spacing between the gate edge and the metal contact, should show the smallest degradation because of parasitic series resistance and provide the most conservative estimate of the beneficial properties of the silicide. Modeling of these devices using the PISCES computer program provides insight into the applications of silicides to shallow junction technology [10]. One observation is that most of the drive current loss in nonsilicided devices is caused by the voltage drop across the resistance at the source end of the device, which reduces the effective applied gate potential. A series voltage drop in the drain end of the device does not have a large effect on the current when the device is in the saturation region. The effect of the series resistance in nonsilicided devices is greatest in the triode region where the resistance of both the source and the drain are important [73]. Another important observation for minimum geometry silicided devices is that the largest contributor to the parasitic voltage drop is the metal-to-silicon contact resistance, not the high sheet resistance of the diffusion. The silicide acts to distribute this resistance over the entire drain area, significantly lowering the voltage drops across the contact.

Thus for minimum-geometry devices, the contact resistance is typically more critical than the diffusion resistance in establishing the device performance. In this case, it is especially important to ensure that dopant depletion and redistribution into the silicide are minimized so that the contact resistivity remains low; the silicide sheet resistance is a secondary concern. However, the sheet resistance does play a limiting role in devices that are connected with many squares of diffusion interconnect and in very wide devices that have a low channel resistance. Process choices that reduce the silicide sheet resistivity while increasing the contact resistivity may not lead to the optimum reduction in parasitic resistance effects.

2. Contact Resistance

As shown earlier, the primary benefit of salicide technology is in its reduced metal-to-diffusion contact resistance. This decreased resistance occurs because of an enlarged diffusion contact area. Because the metal-to-silicide contact resistivity ($\sim 10^{-8}$ Ω-cm^2) is about two orders of magnitude lower that the silicide-to-diffusion resistivity, the contact resistance of a silicided junction is decreased roughly by the ratio of diffusion area to contact area, or about 4–10 in a typical layout, compared to a conventional structure. Since the Fermi levels of $CoSi_2$ and $TiSi_2$ are very near the center of the Si band gap, both make excellent choices for simultaneously contacting both n^+ and p^+ junctions. Other silicides, like PtSi, that make good Schottky barrier diodes because of their barrier height make even lower contact resistance to one type of junction (e.g., n^+) but generally have unacceptably high resistance to the complementary junction.

Dopant depletion from the silicon surface into the silicide is the biggest hurdle to achieving low contact resistivity in practice. Special care must be taken to ensure a high surface dopant concentration. The first step involves the implantation of a sufficiently high dose to begin with, as shown in Fig. 44. The figure shows that the contact resistance of both p^+ and n^+ junctions to either $TiSi_2$ or $CoSi_2$ continuously decrease for increasing junction implantation dose up to 10^{16} ions/cm^2, where good, low-contact resistivities ($\sim 10^{-7}$ Ω-cm^2) are observed. Because they do not suffer from the same dopant redistribution, nonsilicided junctions typically do not require such a high implantation dose to achieve a good contact resistance. The second step in reducing the contact resistance involves minimizing the

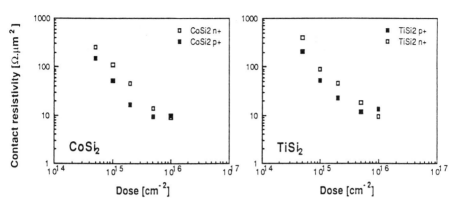

FIGURE 44. Contact resistivity as a function of implantation dose for $CoSi_2$ and $TiSi_2$ contacts to n^+ and p^+ junctions. (From Ref. 3.)

postsilicidation annealing to minimize dopant redistribution. Finally, the implantation of additional dopant into the silicide can be performed to maintain a high surface concentration.

3. Hot-Electron Stability of Silicided Devices

The stability of devices subjected to hot-electron stressing is a key concern for current and future generations of technology. Siliciding the diffusion regions of devices has two effects on this stability. First, the reduction in external series resistance has the adverse effect of reducing the external voltage drop. The resulting higher voltage across the device increases the magnitude of the resultant hot-electron instability. For example, Haken [41] reported a 10% decrease in transistor current drive in silicided transistors after hot-electron stressing under conditions in which no degradation was observed in conventional devices. Second, the presence of the high-conductivity silicide on the junction causes a subtle shift in the current flow path and in the electric field patterns near the drain. The magnitude of these shifts is expected to depend on the design and implementation of the lightly doped drain (LDD) implant and sidewall spacer structure. An additional consideration is that silicidation can modify the lateral diffusion profiles of the LDD and threshold-adjust implants and, thus, have a pronounced impact on the electric field profiles [54, 90].

Figure 45 compares the hot electron–limited lifetimes of silicided and unsilicided transistors as a function of applied drain voltage. The devices shown had a channel length of 0.95 μm and a 0.2 μm deep junction, which was optically silicided with 70 nm of $CoSi_2$. The end of the useful device

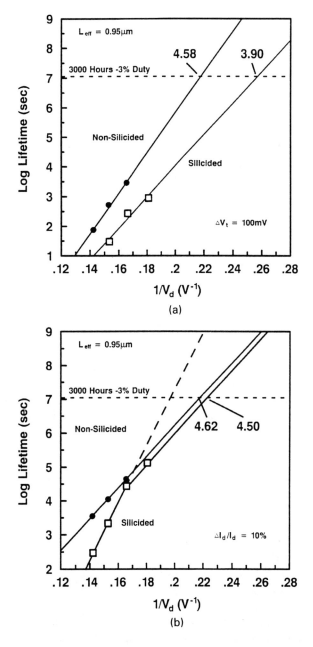

FIGURE 45. Hot electron lifetimes for silicided and unsilicided one micron devices (a) based on $\Delta V_t = 100$ mV, and (b) based on $\Delta I_d/I_d = 10\%$. (From Ref. 10.)

life is based on 100 mV of allowable threshold voltage shift in Fig. 45a or 10% drop in drive current at the end of life in Fig. 45b [10]. Even though device simulations indicate that the series voltage drop associated with diffusion and contacts is only 0.2 V, equivalent degradation in transistor threshold occurs at about 0.7 V lower stress voltage in silicided devices than in unsilicided ones. Equivalent changes in current drive occurred at 0.2–0.5 V lower voltages in silicided devices. Extrapolation of the data to 10 years' operation at 3% duty cycle shows that the unsilicided devices can safely operate up to 4.6 V. Depending on which end-of-life criterion is used, the silicided devices can only withstand 3.9–4.5 V. Thus, by reducing the parasitic series resistance, silicidation increases the active voltage across the channel and thereby enhances the hot electron–induced degradation. However, the data also indicate that the difference between silicided and unsilicided devices is more complex than would be expected merely by the difference in external voltage drop.

4. Layout and Latch-up in Siliced Devices

The silicide process has a beneficial impact on device layout ground rules and packing density, particularly on the rules related to butting n-well and substrate contacts. Butting contacts occur when an n^+ and p^+ diffusion implant share the same active area without a field oxide barrier between them. Because the n^+-to-p^+ diffusion contact may not be ohmic, a metal contact connecting the two regions is normally required. The placement of the extra metal contact increases the size of the active area and the n^+ select mask because of the alignment of the contact to n^+ region. If the silicide is used to strap the n^+ and p^+ regions, the high area cost of placing a metal contact may be saved.

Another benefit of siliciding CMOS devices is an improvement in the latch-up voltage obtained by better strapping device junctions to the well contact [44].

C. Junctions Formed by Diffusion through Silicide

Formation of silicide prior to junction formation for the SADS, ITS, or ITM processes, illustrated in Fig. 10, is an alternative to the conventional process for siliciding shallow junctions; thus, the device applications of both techniques are similar. However, because the implantation depths are reduced by the stopping power of the higher atomic weight surface metals or silicide film, these techniques offer the potential for reducing the dopant implantation depth. Furthermore, if the implant depth is less than the film thickness the end-of-range ion implantation damage is confined within the film. This allows the use of a lower thermal budget, because there is no need either to remove implant damage or to diffuse the junction beyond a damage region. Both of these factors lead to shallower junctions than can be formed with conventional techniques. Thus these silicide processes are especially useful for making devices having the smallest feature sizes and shallowest junctions.

1. Dopant Diffusion from Silicides

Control of dopant diffusion from the silicide into the silicon is one of the key issues associated with the SADS process. For the silicides of most interest, the diffusion of at least one of the common dopants is relatively slow within the silicide. For instance, the bulk diffusion of B in $TiSi_2$ or As in $CoSi_2$ is relatively slow at realistic processing temperatures. Furthermore, the formation of compounds such as TiB_2 can occur in several of the silicide/dopant systems of potential interest [199, 303–306]. Thus grain boundary diffusion is responsible for the transport of most dopant atoms from within the silicide–silicon interface; however, to maintain a high interfacial doping concentration and thereby low contact resistance, it is necessary to employ high implantation doses of dopant. SADS and ITS junctions typically use between 5×10^{15} and 10^{16} dopant atoms/cm^2, which is a factor of 2–10 higher than would be used for a conventionally formed junction of a comparable depth.

Since enhanced diffusion of dopant within the silicon has been observed [104] when a silicide source is employed, the thermal budget must be limited in order to restrict the junction motion to a few tens of nanometers. The activation energy for this enhanced dopant diffusion is actually higher than for conventional diffusion, so that the junction depth can be most easily controlled by reducing the drive-in temperature. The expected dopant motion for boron and arsenic junctions after 10-sec rapid thermal annealing is illustrated in Fig. 42 where it is noted that junction motion of only a few nanometers is possible at annealing temperatures of 700–900 °C. Such little dopant motion allows the total junction depth to be only slightly greater than the silicide thickness.

2. Device Characteristics

Good, low-leakage diodes (~ 1 nA/cm^2, 1 pA/cm) have been made with SADS processing for $CoSi_2$ [3, 10, 77, 97, 104–106], $TiSi_2$ [3, 90–96, 104–105], and PtSi [99, 104]. As long as the implantation depth is kept moderate, the implant through silicide process also results in good electrical characteristics. As described earlier, an interesting progression occurs as SADS junctions are annealed at progressively higher temperatures. Initially the forward and reverse characteristics are those of silicide–silicon Schottky barrier diodes. Very low annealing temperatures (600–700 °C) are then needed to obtain junction diode characteristics. The lowest leakage and tightest distribution of reverse currents are obtained at intermediate temperatures. Higher temperature annealing degrades the junction leakage because of agglomeration of the silicide. Diode leakage is more sensitive to this agglomeration than is the silicide resistivity. As a result, poor junction characteristics are observed at lower temperatures than those at which significant increases in silicide sheet resistance occur.

The implant-through-metal process has been more likely to produce leaky junctions, presumably because of knocked-on metal. The good diodes produced by SADS and ITS have ideality factors of about 1.0 and thermal activation energies of the reverse leakage current of about 1.0 eV, signifying the absence of generation centers within the junction depletion region.

These encouraging results from junction diodes have led the way towards using SADS and ITS processes in full MOS devices.

D. LOCAL INTERCONNECTIONS

The local interconnection technology shown earlier in Fig. 11 provides several important advantages for device technology. The first advantage is that it allows the silicide layer to cross over oxide isolation regions to connect gates with diffusion areas or to connect nearby diffusions or gates to each other. Without the topological constraints inherent in the salicide process, the local interconnection technology provides another wiring level that has capabilities similar to a layer of tungsten wiring or even another level of aluminum. The higher silicide resistivity and smaller thickness restrict its use to making nearby connections, but nevertheless very significant chip area savings accrue from the use of local interconnects, especially in selected applications like SRAMs.

A second advantage is that local interconnections form self-aligned contacts to gate and diffusion regions. Other interconnections (metal or tungsten) require a contact hole mask. Borders around the contact hole are typically required for both the layer above and the layer below the contact level to allow for mask alignment as well as lithography and etching biases and tolerances. It is not at all uncommon to require that the contact land be four times larger than the contact hole itself. The local interconnect contact land can be even smaller than the minimum lithographic feature size.

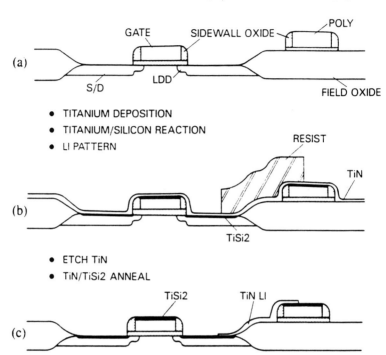

FIGURE 46. TiN option for the formation of local interconnections. (From T. E. Tang, C.-C. Wei, R. A. Haken, T. C. Holloway, L. R. Hite, and T. G. W. Blake, Titanium Nitride Local Interconnect Technology for ULSI, *IEEE Trans. Electron. Dev.* **ED-34**, 682, © 1987 IEEE.)

In addition to the obvious area savings of these contacts, the chip circuit's performance is enhanced by the reduced stray capacitance associated with the smaller area and by the shortened wiring lengths that occur when the layout is compacted.

Another advantage, closely related to that of the self-aligned contacts, is the ability to make metal-to-diffusion contacts over the thick isolation oxide. This capability can be used to minimize the diffusion area itself and thereby its capacitance, both factors which improve circuit performance.

A final advantage of the local interconnection technology occurs with the TiN option. This option, depicted in Fig. 46, forms a thick TiN layer on top of the $TiSi_2$. Since TiN is a far superior metallurgical barrier to Al, the need for a barrier in addition to the $TiSi_2$ is eliminated with a savings in process complexity and cost.

Silicides

At present the local interconnection technology is largely restricted to $TiSi_2$ because of problems associated with formation of other silicides on SiO_2. However, one can hope that new processes will be developed that will overcome this limitation.

E. Process Modeling with Silicides

One measure of the maturity of any VLSL process technology is the availability and accuracy of process models for the technology. Silicide formation via RTA is described in several process modeling programs: PREDICT, developed at the MCNC Center for Microelectronics [332], SUPREM as partially modified by Moynagh *et al.* [333], and programs developed at Delft University [334] or elsewhere [335–336].

Some of the phenomena associated with silicided shallow junction formation are given in Table XI along with the methods used by some of these models to account for them. As can be seen from the table, the PREDICT process model is quite comprehensive in its ability to simulate a wide variety of effects. Despite the good agreement between simulations and experimental data, much work remains to fully verify the accuracy of all these models over a wide range of conditions. Currently the process models themselves are probably more sophisticated than the state of the physical data that have been taken to date. Work still needs to be done to develop a better physical understanding of several silicide processes in order

Table XI Features of silicide process models.

	Model		
Feature	PREDICT [332]	Delft Univ. [334]	Mod SUPREM [333]
Formation kinetics	Yes	Yes	Yes
Silicide thickness	$1.95d_{Ti}$ (Ar); $1.36d_{Ti}$ (N$_2$)	$1.88(d_{Ti} - 20$ nm)	Yes
Silicide thickness	$3.2d_{Co}$		
Dopant redistribution into silicide	Yes	No	Yes
Segregation	Yes	No	Yes
Evaporation	Yes	No	Yes
Surface accumulation	No	No	No
I/I into silicide	Multiplier of Si R_p, ΔR_p	No	Yes
Diffusion in silicide	Rapid	No	10^3–10^4 Si value
Diffusion in silicon	Solid Source	No	8–100 × Si value
Agglomeration	No	No	No
Oxidation of silicide	No	No	

to improve the accuracy of models. For instance, none of the current models accounts for agglomeration and its effects on dopant profiles as measured by secondary ionization mass spectrometry.

V. Summary

Silicides have found widespread acceptance in semiconductor technology. In most applications that require silicides that are self-aligned to diffusion regions, rapid thermal annealing has become the preferred method to form the silicide via reaction of metal with exposed silicon. Rapid thermal annealing provides superior control over impurities in the atmosphere during silicide formation, restricts junction motion, and minimizes agglomeration of the silicide as compared to conventional annealing.

Advances in silicide technology have closely followed advances and the acceptance of RTA technology, and the two are expected to remain closely linked in the future.

References

1. S. P. Murarka, *Silicides for VLSI Applications*, Academic Press (1983).
2. M. A. Nicolet and S. S. Lau, Formation and Characterization of Transition-Metal Silicides, in *VLSI Electronics*, Vol. 6 (N. G. Einspruch and G. B. Larrabee, eds.), Academic Press (1983).
3. L. Van den hove and R. F. DeKeersmaecker, Silicidation by Rapid Thermal Processing, in *Reduced Thermal Processing for VLSI* (R. A. Levy, ed.), Plenum Press, New York (1989).
4. S. Wolf and R. N. Tauber, *Silicon Processing for the VLSI Era: Volume 1—Process Technology*, Chapter 11, Lattice Press, Sunset Beach, CA (1986).
5. F. M. d'Heurle, R. T. Hodgson, and C. Y. Ting, Silicides and Rapid Thermal Annealing, *Mat. Res. Soc. Symp. Proc.* **52**, 261 (1986).
6. F. M. d'Heurle, Material Properties of Silicides for VLSI Technology, in *Solid State Devices* (P. Balk and O. G. Folberth, eds.), Elsevier, Amsterdam, p. 213 (1986).
7. S. P. Murarka and M. C. Peckarar, *Electronic Materials Science and Technology*, Chapter 6, Academic Press, San Diego, CA (1989).
8. L. Van den hove, *Advanced Interconnection and Contact Schemes Based on $TiSi_2$ and $CoSi_2$: Relevant Materials Issues and Technological Implementation*, Ph.D. Thesis, Katholieke Universiteit Leuven, June (1988).
9. C. M. Osburn, Formation of Silicided, Ultra-Shallow Junctions using Low Thermal Budget Processing, *J. Electron. Mater.* **19**(1), 67 (1989).
10. C. M. Osburn, Q.-F. Wang, M. Kellam, C. Canovai, P. L. Smith, G. E. McGuire, Z. G. Xiao, and G. A. Rozgonyi, Incorporation of Metal Silicides and Refractory Metals in VLSI Technology, *Appl. Surf. Sci.* **53**, (1991).
11. M. P. Lepselter and J. M. Andrews, Ohmic Contacts to Silicon, in *Ohmic Contacts to Semiconductors* (B. Schwartz, ed.), The Electrochemical Society, Princeton, NJ, p. 159 (1969).

12. M. P. Lepselter and S. M. Sze, *Bell Syst. Tech. J.* **47**, 195 (1968).
13. W. V. T. Rusch and C. A. Burrus, *Solid State Electron.* **11**, 517 (1968).
14. C. J. Kircher, Metallurgical Properties and Electrical Characteristics of Palladium Silicide-Silicon Contacts, *Solid State Electron.* **14**, 507 (1971).
15. S. Zirinsky and B. L. Crowder, Refractory Silicides for High Temperature Compatible IC Conductor Lines, *J. Electrochem. Soc.* **124**, 463 (1977).
16. B. L. Crowder and S. Zirinsky, One Micron MOSFET VLSI Technology: Part VII-Metal Silicide Interconnection Technology—A Future Perspective, *IEEE Trans. Electron Dev.* **ED-26**, 369 (1979).
17. S. P. Murarka, Refractory Silicides for Low Resistivity Gates and Interconnections, *IEDM Tech. Dig.* 454 (1979).
18. S. P. Murarka, Refractory Silicides for Integrated Circuits, *J. Vac. Sci. Technol.* **17**, 775 (1980).
19. A. K. Sinha, W. S. Linden Berger, D. B. Fraser, S. P. Murarka, and E. N. Fuls, MOS Compatibility of High-Conductivity $TaSi_2/n^+$ Poly-Si Gates, *IEEE Trans. Electron Dev.* **ED-27**, 1425 (1980).
20. H. J. Geipel, N. Hsieh, M. Ishaq, C. Koburger, and F. White, Composite Silicide Gate Electrodes—Interconnections for VLSI Device Technologies, *IEEE Trans. Electron Dev.* **ED-27**, 1417 (1980).
21. C. Koburger, M. Ishaq, and H. J. Geipel, Electrical Properties of Composite Silicide Gate Electrodes, Extended Abstract, ECS Spring Meeting, St. Louis, MO, 162 (1980).
22. C. Koburger, M. Ishaq, and H. J. Geipel, Electrical Properties of Composite Evaporated Silicide/Polysilicon Electrodes, *J. Electrochem. Soc.* **129**, 1307 (1982).
23. M. Y. Tsai, F. M. d'Huerle, C. S. Peterson, L. Ng, and C. J. Lucchese, Annealing Studies of WSi_2 on Poly-Silicon, *Electron Material Conf.*, Ithaca, New York (1980).
24. M. Morimo, M. Sugimoto, K. Terada, K. Takahasha, T. Ishijima, M. Muta, and S. Suzuki, High Speed 4Kb Static RAM With Silicide Wafer Wiring, *12th Conference on Solid State Devices*, Tokyo, Japan, Digest of Tech. Papers (1980).
25. M. Y. Tsai, F. M. d'Huerle, C. S. Petersson, and R. W. Johnson, Properties of WSi_2 Film on Poly-Si, *J. Appl. Phys.* **52**, 5350 (1981).
26. M. Y. Tsai, H. H. Chao, L. M. Ephrath, B. L. Crowder, A. Cramer, R. S. Bennett, C. J. Lucchese, and M. R. Wordeman, One-Micron Polycide (WSi_2 on Poly-Si)MOSFET Technology, *J. Electrochem. Soc.* **128**, 2207 (1981).
27. H. H. Chao, R. H. Dennard, M. Y. Tsai, M. R. Wordeman, and A. Cramer, A 34 μm^2 DRAM Cell Fabricated with a 1 μ Single Level Polycide FET Technology, *IEEE J. Solid State Circuits* **SC-16**, 499 (1981).
28. P. B. Ghate and C. R. Fuller, Multilevel Interconnections for VLSI, *Semiconductor Silicon, 1981*, The Electrochemical Society **81**, 680 (1981).
29. N. Kobayashi, S. Iwata, N. Yamamutu, Refractory Metals for Silicides, *IEDM Tech. Dig.* 122 (1984).
30. T. Shibata, K. Hieda, M. Sato, M. Konaka, R. L. M. Dang, and H. Iizuka, An Optimally Designed Process for Submicron MOSFET's, *IEDM Tech. Dig.* 647 (1981).
31. C. M. Osburn, M. Y. Tsai, S. Roberts, C. J. Lucchese, and C. Y. Ting, High Conductivity Diffusions and Gate Regions using a Self-Aligned Silicide Technology, *Proc. 1st Int. Symp. on VLSI Science and Technology*, Electrochemical Society **82-1**, 213(1982).
32. C. Y. Ting, S. S. Iyer, C. M. Osburn, G. J. Hu, and A. M. Schweighart, The use of $TiSi_2$ in a Self-Aligned Silicide Technology, *Proc. 1st Inter. Symp. on VLSI Science and Technol.*, Electrochem. Soc. Proc. **82**(7), 224 (1982).
33. C. J. Lucchese, Use of Cobalt Silicide in VLSI, *Proc. 1st Inter. Symp. on VLSI Science and Technol.*, Electrochem. Soc. Proc. **82**(7), 232 (1982).

34. C. K. Lau, Y. C. See, D. B. Scott, J. M. Perna, and R. D. Davies, Titanium Disilicide Self-Aligned Source-Drain + Gate Technology, *IEDM Tech. Dig.* 714 (1982).
35. F. Runovc, H. Norstrom, R. Buchta, P. Woklund, and S. Petersson, Titanium Disilicide in MOS Technology, *Physica Scripta* **26**, 108 (1982).
36. T. Yamaguchi, S. Morimoto, G. H. Kawamoto, H. K. Park, and G. C. Eiden, High-Speed Latchup-Free 0.5 μm Channel CMOS using Self-Aligned $TiSi_2$ and Deep Trench Isolation Technologies, *IEDM Tech. Dig.* 522 (1983).
37. K. Maex, R. DeKeersmaecker, P. F. A. Alkemade, F. H. P. M. Habraken, and W. F. van der Weg, $TiSi_2$ Formation using Rapid Thermal Processing and Related Impurity Redistribution, *Proc. MRS Europe Conference*, 315, June (1984).
38. K. Tsukamoto, T. Okamoto, M. Shimizu, T. Matsukawa, and H. Nakata, Self-Aligned Titanium Silicidation of Submicron MOS Devices by Rapid Lamp Annealing, *IEDM Tech. Dig.* 130 (1984).
39. C. Y. Ting, Silicide for Contacts and Interconnects, *IEDM Tech. Dig.* 110 (1984).
40. R. A. Haken, Application of the Self-Aligned Titanium Silicide Process to VLSI NMOS and CMOS Technologies, *Proc. Conference on Refractory Metal Silicides*, San Juan Batista, CA (1985).
41. R. A. Haken, Application of the Self-Aligned Titanium Silicide Process to Very Large-Scale Integrated N-metal-oxide-semiconductor and Complementary Metal-oxide-semiconductor Technologies, *J. Vac. Sci. Technol.* **B 3**(6), 1657 (1985).
42. L. Van den hove, R. Wolters, K. Maex, R. F. DeKeersmaecker, and G. J. Declerck, A Self-Aligned $CoSi_2$ Interconnection and Contract Technology for VLSI Applications, *IEEE Trans. Electron Dev.* **ED34**(3), 554 (1987).
43. S. J. Hillenius, R. Liu, G. E. Georgiou, R. L. Field, D. S. Williams, A. Komblit, D. M. Boulin, R. L. Johnston, and W. T. Lynch, A Symmetric Submicron CMOS Technology, *IEDM Tech. Dig.* 252 (1986).
44. F. S. Lai, L. K. Wang, Y. Taur, Y. C. Sun, K. E. Petrillo, S. M. Chicotka, E. J. Petrillo, M. R. Polcari, T. J. Bucelot, and D. S. Zicherman, A Highly Latchup-Immune 1 μm CMOS Technology Fabricated with 1 MeV Ion Implantation and Self-Aligned $TiSi_2$, *IEDM Tech. Dig.* 513 (1985).
45. D. C. Chen, T. R. Cass, J. E. Turner, P. Merchant, and K. Y. Chiu, The Impact of $TiSi_2$ on Shallow Junctions, *IEDM Tech. Dig.* 411 (1985).
46. D. C. Chen, T. R. Cass, J. E. Turner, P. Merchant, and K. Y. Chiu, $TiSi_2$ Thickness Limitations for Use with Shallow Junctions and SWAMI or LOCOS Isolation, *IEEE Trans. Electron Devices* **ED-33**, 1463 (1986).
47. M. E. Alperin, T. C. Holloway, R. A. Haken, C. D. Gosmeyer, R. V. Karnaugh, and W. D. Parmantie, Development of the Self-Aligned Titanium Silicide Process for VLSI Applications, *IEEE Solid State Circuits* **SC-20**(1), 61 (1985).
48. J. Amano, K. Mauka, M. P. Scott, and J. E. Turner, Junction Leakage in Titanium Self-Aligned Silicide Devices, *Appl. Phys. Lett.* **49**(12), 737 (1986).
49. P. Merchant, Impurity and Dopant Effects on $TiSi_2$ Formation, *Proc. Conference on Refractory Metal Silicides*, San Juan Batista, CA (1985).
50. D. Levy, P. Delpech, M. Paoli, C. Masurel, M. Vernet, N. Brun, J.-P. Jeanne, J.-P. Gonshond, M. Ada-Hanifi, M. Haond, T. T. D'Ouville, and H. Mingam, Optimization of a Self-Aligned Titanium Silicide Process for Submicron Technology, *IEEE Trans. Semicond. Manuf.* **3**(4), 168 (1990).
51. N. S. Parekh, H. Roede, A. A. Bos, A. G. M. Jonkers, and R. D. J. Verhaar, Characterization and Implementation of Self-Aligned $TiSi_2$ in Submicrometer CMOS Technology, *IEEE Trans. Electron. Dev.* **ED-38**(1), 88 (1991).

52. D. B. Scott, R. A. Chapman, C. Wei, S. S. Mahant-Shetti, R. A. Haken, and T. C. Holloway, *IEEE Trans. Electron Dev.* **ED-34**, 562 (1987).
53. R. P. Kramer, MEGA Project: Some Aspects of ULSI Technology, *Proc. Second Inter. Symp. on ULSI Science and Technol.*, The Electrochemical Society **89-9**, 27 (1989).
54. C. Y. Lu and J. M. Sung, Reverse Short Channel Effects on Threshold Voltage in Submicron Salicide Devices, *IEEE Electron Dev. Lett.* **10**(10), 446 (1989).
55. S. J. Hillenius, H. I. Cong, J. Leebowitz, J. M. Andrews, R. L. Field, L. Manchanda, W. S. Lindenberger, D. M. Boulin, and W. T. Lynch, A Self Aligned $CoSi_2$ Source/Drain/Gate Multi-Gigahertz Symmetric CMOS Technology, *Proc. Second Inter. Symp on ULSI Science and Technol.*, The Electrochemical Society, **89-9**, 51 (1989).
56. M.-L. Chen, S. J. Hillenius, W. Juengling, T. S. Yang, A. Kornblit, W. S. Lindenberger, J. A. Swinderski, and D. P. Favreau, Self-Aligned Silicided Inverse-T Gate LDD Devices for Sub-Half Micron CMOS Technology, *IEDM Tech. Dig.* 829 (1990).
57. E. K. Broadbent, R. F. Irani, A. E. Morgan, and P. Maillot, Application of Self-Aligned $CoSi_2$ Interconnection in Submicrometer CMOS Transistors, *IEEE Trans. Dev.* **ED-36**(11), 2440 (1989).
58. C. Arena, S. Deleonibus, G. Guegan, P. Laporte, F. Martin, and J. L. Pelloie, 1 μm MOS Devices with Self-Aligned Titanium Silicide and CVD Tungsten As First Metallization Level, *Solid State Devices*, Elsevier Science Publishers, North-Holland, p. 41 (1988).
59. C. S. Wei, G. Raghavan, M. Lawrence, A. Dass, M. Frost, T. Brat, and D. B. Fraser, Comparison of Cobalt and Titanium Silicides for Salicide Process and Shallow Junction Formation, *VMIC Conference*, p. 241 (1989).
60. H. Okabayashi, M. Morimoto, and E. Nagasawa, Low-Resistance MOS Technology using Self-Aligned Refractory Silicidation, *IEEE Trans. Electron Dev.* **ED-31**, 1329 (1984).
61. C. Y. Ting and S. S. Iyer, The Use of $TiSi_2$ for Self Aligned Silicide (Salicide) Technology, *Proc. IEEE V-MIC* 307 (1985).
62. C. K. Lau, Characterization of the Self-Aligned $TiSi_2$ Process, *Electrochem. Soc. Extended Abstracts* **83**(1), 569 (1983).
63. S. P. Murarka, D. B. Fraser, A. K. Sinha, H. J. Levinstein, E. J. Lloyd, R. Liu, D. S. Williams, and S. J. Hillenius, Self-Aligned Cobalt Disilicide for Gate and Interconnection and Contacts to Shallow Junctions, *IEEE Trans. Electron Dev.* **ED34**, (102), 108 (1987).
64. E. K. Broadbent, M. Delfino, A. E. Morgan, D. K. Sadana, and P. Maillot, Self-Aligned Silicided (PtSi and $CoSi_2$) Ultra-Shallow P^+N Junctions, *IEEE Electron Dev. Lett.* **EDL-8**, 318 (1987).
65. L. Van den hove, R. Wolters, M. Geyselaers, R. De Keersmaecker, G. Declerck, Key Issues to the Self-Aligned Formation of $CoSi_2$ in a Salicide Process, *Mat. Res. Soc. Symp. Proc.* **100**, 99 (1988).
66. D. Moy, S. Basavaiah, H. Protschka, L. K. Wang, F. d'Heurle, J. Wetzel, S. Brodsky, and R. Volant, Use of Thin Titanium Salicides for Submicron ULSI CMOS, *Proceedings of the First International Conference on ULSI*, Electrochem. Soc. **87-11**, 381 (1987).
67. A. E. Morgan, E. K. Broadbent, M. Delfino, B. Coulman, and D. K. Sadana, Characterization of a Self-Aligned Cobalt Silicide Process, *J. Electrochem. Soc.* **134**(4), 925 (1987).
68. A. Guldan, V. Schiller, A. Steffan, and P. Balk, Formation and Properties of $TiSi_2$ Films, *Thin Solid Films* **100**, 1 (1983).
69. J. Riseman, Method for Forming an Insulator Between Layers of Conductive Material, U.S. Patent 4234362 (1980).
70. S. Ogura, P. Tsang, W. Walker, D. Critchlow, and J. Shepard, Design and Characteristics of the Lightly Doped Drain-Source (LDD) Insulated Gate Field-Effect Transistor, *IEEE Trans. Electron Dev.* **ED-27**, 1359 (1980).

71. P. J. Tsang, J. F. Shephard, J. Lechaton, and S. Ogura, Characterization of Sidewall Spacers Formed by Anisotropic RIE, RNP553 ECS Spring Meeting, Minneapolis, MN (1981).
72. P. J. Tsang, S. Ogura, W. W. Walker, J. F. Shephard, and D. Critchlow, Fabrication of High Performance FET's with Oxide Sidewall Spacer Technology, *IEEE Trans.* **ED-29**, 590 (1982).
73. K. K. Ng and W. T. Lynch, The Impact of Intrinsic Series Resistance on MOSFET Scaling, *IEEE Trans. Electron Dev.* **ED-34**(3), 503 (1987).
74. E. Nagasawa, M. Morimoto, H. Okabayashi, Mo-Silicided Low Resistance Shallow Junctions, Dig. Tech. Papers, *IEEE Symp. VLSI Technol.*, p. 26 (1982).
75. R. Liu and T. P. H. F. Wendling, Leakage Mechanisms for Shallow Silicided Junctions, Workshop on Refractory Metals and Silicides for VLSI IV, San Juan Bautista, CA, May (1986).
76. R. Liu, D. S. Williams, and W. T. Lynch, Mechanism for Process-Induced Leakage in Shallow Silicided Junctions, *IEDM Technical Dig.*, 58 (1986).
77. R. Liu, D. S. Williams, and W. T. Lynch, A Study of the Leakage Mechanisms of Silicided n^+/p Junctions, *J. Appl. Phys.* **63**(6), 1990 (1988).
78. C. Dehm, G. Valyi, J. Gyulai, and H. Ryssel, Ion-Beam Mixed $MoSi_2$ Layers: Formation and Contact Properties, *Proc. ESSDERC*, Berlin, 253 (1989).
79. R. Angelucci, M. Merli, L. Don, G. Pizzochero, S. Solmi, and R. Canteri, Shallow Junctions Fabrication By Using Molybdenum Silicide and Rapid Thermal Annealing, *Proc. ESSDERC*, Berlin, 237 (1989).
80. K. Maex and R. DeKeersmaecker, Rapid Thermal Processing for Simultaneous Annealing of Shallow Implanted Junctions and Formation of their $TiSi_2$ Contacts, *Physicia* **129B**, 192 (1985).
81. D. L. Kwong and N. S. Alvi, Electrical Characterization of Ti-Silicided Shallow Junctions Formed by Ion-Beam Mixing and Rapid Thermal Annealing, *J. Appl. Phys.* **60**(2), 688 (1986).
82. D. L. Kwong, D. C. Meyers, and N. S. Alvi, Simultaneous Formation of Silicide Ohmic Contacts and Shallow P^+-N Junctions by Ion-Beam Mixing and Rapid Thermal Annealing, *IEEE Electron Dev. Lett.* **EDL-6**(5), 244 (1985).
83. E. Nagasawa, H. Okabayashi, and M. Morimoto, Mo- and Ti-Silicided Low-Resistance Shallow Junctions Formed Using the Ion Implantation Through Metal Technique, *IEEE Trans. Electron Dev.* **ED-34**, 581 (1987).
84. B. Y. Tsaur, C. K. Chen, C. H. Anderson, Jr., and D. L. Kwong, Selective Tungsten Silicide Formation by Ion Beam Mixing and Rapid Thermal Annealing, *J. Appl. Phys.* **57**(6), 1890 (1985).
85. M. Delfino, E. K. Broadbent, A. E. Morgan, B. J. Burrow, and M. H. Norcott, Formation of $TiN/TiSi_2/p^+$-Si/n-Si by Rapid Thermal Annealing (RTA) Silicon Implanted with Boron Through Titanium, *IEEE Electron Dev. Lett.* **EDL-6**, 591 (1985).
86. B. Shah, Temperature Dependent Current-Voltage Characteristics of $TiSi(2)/n+p$ Silicon Shallow Junctions, M.S. Thesis, N.J. Institute of Technology (1990).
87. S. Solmi, R. Angelucci, and M. Merli, Shallow Junctions for ULSI Technology, *European Trans. on Telecommunications and Related Technologies* **1**(2), 159 (1990).
88. T. Gessner, M. Rennau, S. Schubert, and E. Vetter, Mo-Silicided Shallow Junctions Formed using the ITM Technique and the Influence of Mo Recoil Atoms, *European Workshop on Refractory Metals and Silicides*, p. B.3.1, Leuven, Belgium, April (1989).
89. T. Gessner, R. Reich, W. Unger, and W. Wolke, The Influence of Rapid Thermal Processing on the Properties of $MoSi_2$ Layers Formed by Using the Ion Implantation Through Metal Technique, *Thin Solid Films* **177**, 225 (1989).

90. C. Y. Lu, J. M. J. Sung, R. Liu, N. S. Tasi, R. Singh, S. J. Hillenius, and H. C. Kirsch, Process Limitation and Device Design Tradeoffs of Self-Aligned TiSi$_2$ Junction Formation in Submicrometer CMOS Devices, *IEEE Trans. Electron Dev.* **ED-38**, 246 (1991).
91. D. L. Kwong, Y. H. Ku, S. K. Lee, and E. Lewis, Silicided Shallow Junction Formation by Ion Implantation of Impurity Ions into Silicide Layers and Subsequent Drive-in, *J. Appl. Phys.* **61**(11), 5084 (1987).
92. Y.-H. Ku, Rapid Thermal Processing and Self-Aligned Silicide Technology for VLSI Application, Ph.D. Thesis, University of Texas, Austin, TX, May (1988).
93. L. Rubin, D. Hoffman, D. Ma, and N. Herbots, Shallow-Junction Diode Formation by Implantation of Arsenic and Boron through Titanium-Silicide Films and Rapid Thermal Annealing, *IEEE Trans. Electron Dev.* **ED-37**, 183 (1990).
94. L. M. Rubin, N. Herbots, D. Hoffman, and D. Ma, Integrated Processing of Silicided Shallow Junctions using Rapid Thermal Annealing Prior to Dopant Activation, *MRS Symp. Proc.* **146**, 191 (1989).
95. Y. H. Ku, S. K. Lee, and D. L. Kwong, Shallow, Silicided P + /N Junction Formation and Dopant Diffusion In SiO$_2$/TiSi$_2$/Si Structure, *Appl. Phys. Lett.* **54**(17), 1684 (1989).
96. T. Yoshida, M. Fukumoto, and T. Ohzone, Self-Aligned Titanium Silicided Junctions Formed By Rapid Thermal Annealing In Vacuum, *J. Electrochem. Soc.* **135**, 481 (1988).
97. V. Probst, P. Lippens, L. Van den hove, K. Maex, H. Schaber, and R. DeKeersmaecker, Shallow Junction Formation Using CoSi$_2$ as a Diffusion Source, *Proc. 17th ESSDERC*, 437, Bologna, Italy, Sept. (1987).
98. F. C. Shone, K. C. Saraswat, and J. D. Plummer, Formation of 0.1 μm N$^+$/P and P$^+$/N Junctions by Doped Silicide Technology, *IEDM Tech. Dig.* 407 (1985).
99. B.-Y. Tsui, J.-Y. Tsai, and M.-C. Chen, Formation of PiSi-contacted p$^+$n Shallow Junctions by BF$_2$ Implantation and Low-Temperature Furnace Annealing, *J. Appl. Phys.* **69**(8), 4354 (1991).
100. H. Jiang, C. M. Osburn, Z.-G. Xiao, G. A. Rozgonyi, G. McGuire, A Study of Ultra Shallow Junctions by Diffusion from Self-Aligned Silicides, *ESSDERC*, Berlin, Sept. (1989).
101. H. Jiang, P. Smith, D. Griffis, Z.-G. Xiao, C. M. Osburn, G. McGuire, and G. A. Rozgonyi, Shallow Implantation into CoSi$_2$ and Diffusion into Silicon, presented at Electrochem. Soc. Annual Meeting, Los Angeles, CA (May, 1989).
102. H. Jiang, C. M. Osburn, P. Smith, Z.-G. Xiao, D. Griffis, G. McGuire, and G. A. Rozgonyi, Ultra Shallow Junction Formation using Diffusion from Silicides: I. Silicide Formation, Dopant Implantation and Depth Profiling, *J. Electrochem. Soc.* **139**(1), 196 (1992).
103. H. Jiang, C. M. Osburn, Z.-G. Xiao, G. McGuire, G. A. Rozgonyi, B. Patnaik, N. Parikh, and M. Swanson, Ultra Shallow Junction Formation Using Diffusion from Silicides: II. Diffusion in Silicides and Evaporation, *J. Electrochem. Soc.* **139**(1), 206 (1992).
104. H. Jiang, C. M. Osburn, Z.-G. Xiao, G. McGuire, and G. A. Rozgonyi, Ultra Shallow Junction Formation Using Diffusion from Silicides: III. Diffusion into Silicon, Thermal Stability of Silicides and Junction Integrity, *J. Electrochem. Soc.* **139**(1), 211 (1992).
105. P. B. Moynagh, C. P. Chew, K. B. Affolter, P. J. Rosser, The Outdiffusion of Boron and Arsenic from Preformed Cobalt Disilicide Layers, *ESSDERC*, Berlin, W. Germany, p. 248 (1989).
106. B. M. Ditchek, M. Tabasky, and E. S. Bulat, Shallow Junction Formation by the Redistribution of Species Implanted into Cobalt Silicide, *MRS Symp. Proc.* **92**, 199 (1987).

107. R. Liu, F. A. Baiocchi, L. M. Heimbrook, J. Kovalchick, D. L. Malm, D. S. Williams, and W. T. Lynch, Formation of Shallow N+/P Junctions with CoSi$_2$, *Proc. of First Int. Symp. on Ultra Large Scale Integration Science and Technology*, The Electrochem. Soc. **87-11**, 446 (1987).
108. L. Van den hove, P. Lippins, K. Maex, L. Hobbs, and R. DeKeersmaecker, Comparison Between CoSi$_2$ and TiSi$_2$ as Dopant Source for Shallow Silicided Junction Formation, European Workshop on Refractory Metals and Silicides, Leuven, Belgium, **A.2.2**, April (1989).
109. D. X. Cao, H. B. Harrison, and G. K. Reeves, Shallow Junction Formation by the Thermal Redistribution of Implanted Arsenic into TiSi$_2$, *MRS Symp. Proc.* **100**, 737 (1988).
110. H. Gierisch, F. Neppl, E. Frenzel, P. Eichinger, and K. Hieber, Dopant Diffusion from Ion-Implanted TsSi$_2$, *Proc. MRS Symp.* Spring, Palo Alto, Ca. (1986).
111. N. Kobayashi, N. Hashimoto, K. Ohyu, T. Kaga, and S. Iwata, Comparison of TiSi$_2$ and WSi$_2$ for Sub-Micron CMOSs, *IEEE Symposium on VLSI Technology*, 49 (1986).
112. J. D. Plummer, Post Diffused Shallow Junctions, Workshop on Refractory Metals and Silicides for VLSI V, San Juan Bautista, CA, May (1987).
113. N. F. Stogdale and K. J. Barlow, Refractory Metal Silicides with Applications to Sub-Micron CMOS Processes, European Workshop on Refractory Metals and Silicides, Leuven, Belgium **A.2.6**, April (1989).
114. K. Maex, L. Van den hove, and R. F. der Keersmaecker, *Thin Solid Films* **140**, 149 (1986).
115. T. H. Ning and R. D. Isaac, *IEDM Tech. Dig.*, 473 (1979).
116. Y. Koh, F. Chien, and M. Vora, Self-Aligned TiSi$_2$ for Bipolar Applications, *Proc. Conference on Refractory Metal Silicides*, San Juan Batista, CA (1985).
117. E. Rosencher, S. Delage, Y. Campidelli, and F. Armaud d'Avitaya, *Electron. Lett.* **20**, 762 (1984).
118. D. C. Chen, S. S. Wong, P. V. Voorde, P. Merchant, T. R. Cass, J. Amano, and K.-Y. Chiu, A New Device Interconnect Scheme for Sub-Micron VLSI, *IEDM Tech. Dig.*, 118 (1984).
119. S. S. Wong, D. C. Chen, P. Merchant, T. R. Cass, J. Amano, K. Y. Chiu, HPSAC—A Silicided Amorphous-Silicon Contact and Interconnect Technology for VLSI, *IEEE Trans. Electron Dev.* **ED-34**(3), 587 (1987).
120. H. J. W. Van Houtum, A. A. Bos, A. G. M. Jonkers, and I. J. M. Raaijmakers, TiSi$_2$ Strap Formation by Ti-Amorphous-Si Reaction, *J. Vac. Sci. Technol. B* **6**(6), 1734 (1988).
121. A. A. Bos, N. S. Parekh, A. G. M. Jonkers, Formation of TiSi$_2$ From Titanium and Amorphous Silicon Layers for Local Interconnect Technology, *Thin Solid Films* **197**, 169 (1991).
122. T. E. Tang, C.-C. Wei, R. A. Haken, T. C. Holloway, L. R. Hite, and T. G. W. Blake, Titanium Nitride Local Interconnect Technology for VLSI, *IEEE Trans. Electron Dev.* **ED-34**, 682 (1987).
123. R. Wolters, L. Van den Hove, The Feasibility of CoSi$_2$, TiW and TiW(N) as Local Interconnection in a Self-Aligned CoSi$_2$ Technology, presented at the 1988 VLSI Multilevel Interconnection Conference (1988).
124. J. Angilello, F. d'Heurle, S. Peterson, and A. Segmuller, Observations of Stresses in Thin Films of Palladium and Platinum Silicides on Silicon, *J. Vac. Sci. Technol.* **17**, 471 (1980).
125. S. P. Murarka, D. B. Fraser, T. F. Retajczyk, Jr., and T. T. Sheng, *J. Appl. Phys.* **51**, 5380 (1980).

126. H. J. Geipel, Jr., N. Hsieh, M. H. Ishaq, C. W. Korburger, and F. R. White, *IEEE J. Solid State Circuits* **SC-15**, 482 (1980).
127. M. Wittmer, Properties and Microelectronic Applications of Thin Films of Refractory Metal Nitrides, *J. Vac. Sci. Technol.* **A3**, 1797 (1985).
128. T. Brat, J. C. S. Wei, J. Poole, D. Hodul, N. Parikh, and C. Wickersham, High Purity Titanium Silicide Films Formed by Sputter Deposition and Rapid Thermal Annealing, *MRS Symp. Proc.* **92**, 191 (1987).
129. A. Reader, Microstructural Defects in Rapid Thermally Processed IC Materials, in *Reduced Thermal Processing for VLSI* (R. A. Levy, ed.), Plenum Press, New York (1989).
130. R. A. Powell, R. Chow, C. Thridandam, R. T. Fulks, I. A. Belch, and J-D. T. Pan, Formation of Titanium Silicide Films by Rapid Thermal Processing, *IEEE Electron Dev. Lett.* **EDL-4**, 380 (1983).
131. K. Maex, R. de Keersmaecker, C. Claeys, J. Vanhellemont, and P. F. A. Alkemade, The Kinetics of Silicide Formation Using Rapid Thermal Processing and Related Defect Behavior, *Proc. of the 5th Inter. Symp. on Silicon Mater. Sci. and Tech.*, The Electrochem. Soc. **86-4**, 346 (1986).
132. K. Maex, R. F. DeKeersmaecker, and P. F. A. Alkemade, Characterization of Doped Si-TiSi Bilayers Formed by Ion Beam Mixing and Rapid Thermal Annealing, *MRS Symp. Proc.* **45**, 153 (1985).
133. P. B. Moynagh, A. A. Brown, and P. J. Rosser, The Outdiffusion of Boron and Arsenic from Pre-Formed Ion-/Beam-Mixed Cobalt Disilicide Layers Using Rapid Thermal Processing, *MRS Symp. Proc.* **146**, 261 (1989).
134. E. A. Maydell-Ondrusz, R. E. Harper, I. H. Wilson, and K. G. Stephens, Formation of $TiSi_2$ by Electron Beam Heating of Arsenic Implanted Titanium Films on Silicon Substrates, *Vacuum* **34(10/11)**, 995 (1984).
135. D. L. Kwong, D. C. Meyers, and N. S. Alvi, Molybdenum Silicide Formation by Ion Beam Mixing and Rapid Thermal Annealing, *Second Int. Symp. on VLSI Science and Technol.*, Electrochem. Soc. **85-5**, 195 (1985).
136. H. J. Bohm, H. Wendt, and H. Oppolzer, K. Masseli, and R. Kassing, Diffusion of B and As from Polycrystalline Silicon During Rapid Optical Annealing, *J. Appl. Phys.* **62(7)**, 2784 (1987).
137. J. Narayan, T. A. Stephenson, T. Brat, D. Fathy, and S. J. Pennycook, Formation of Silicides by Rapid Thermal Annealing over Polycrystalline Silicon, *J. Appl. Phys.* **60(2)**, 631 (1986).
138. N. Natsuaki, K. Ohyu, T. Suzuki, N. Kobayashi, N. Hashimoto, and Y. Wada, Rapid Thermal Annealing Process for Titanium-Silicide Contact Formation, *IEEE 1986 Symposium on VLSI Technology*, 37 (1986).
139. T. Brat, C. M. Osburn, T. Finstad, J. Liu, and B. Ellington, Self-Aligned Ti Silicide Formed by Rapid Thermal Annealing Effects, *J. Electrochem. Soc.* **133**, 1451 (1986).
140. A. E. Morgan, E. K. Broadbent, and A. H. Reader, Formation of Titanium Nitride/Silicide Bilayers by Rapid Thermal Anneal in Nitrogen, *MRS Symp. Proc.* **52**, 279 (1986).
141. M. Tabasky, E. S. Bulat, B. M. Ditchek, M. A. Sullivan, and S. Shatas, Direct Silicidation on Co on Si by Rapid Thermal Annealing, *MRS Symp. Proc.* **52**, 271 (1986).
142. M. Tanielian and Blackstone, *J. Vac. Sci. Technol.* **A3**, 714 (1985).
143. M. Tanielian, R. Lajos, and S. Blackstone, Titanium Silicide by Rapid Thermal Annealing, *J. Electrochem. Soc.* **85-1**, 388 (1985).
144. T. Okamoto, K. Tsukamoto, M. Shimizu, and T. Matsukawa, Titanium Silicidation by Halogen Lamp Annealing, *J. Appl. Phys.* **57(12)**, 5251 (1985).

145. T. Okamoto, M. Shimizu, K. Tsukamoto, and T. Matsukawa, Titanium Silicidation by Halogen Lamp Annealing, *MRS* **35**, 471 (1985).
146. T. Okamoto, K. Tsukamoto, M. Shimizu, Y. Mashiko, and T. Matuskawa, Simultaneous Formation of TiN and TiSi$_2$ by Rapid Lamp Annealing in NH$_3$ Ambient for VLSI Contacts, *Proc. IEEE 1986 Symp. VLSI Technol.*, 51 (1986).
147. T. Okamoto, K. Tsukamoto, M. Shimizu, and T. Matsukawa, Effects and Behavior of Arsenic During Titanium Silicidation by Halogen Lamp Annealing, *J. Appl. Phys.* **61**(9), 4530 (1987).
148. L. Van den hove, K. Maex, N. Saks, R. DeKeersmaecker, and G. Declerck, A Self-Aligned Titanium Silicide Technology Using Rapid Thermal Processing and its Application to NMOS Devices, *ESSDERC*, Aachen, Germany, Sept. (1985).
149. L. Van den hove, R. Wolters, K. Maex, R. De Keersmaecker, and G. Declerck, A Self-Aligned Cobalt Silicide Technology Using Rapid Thermal Processing, *J. Vac. Sci. Technol. B* **4**(6)(11), 1358 (1986).
150. D. Levy, J. P. Ponpon, A. Grob, J. J. Grob, and R. Stuck, Rapid Thermal Annealing and Titanium Silicide Formation, *Appl. Phys. A.* **38**, 23 (1985).
151. C. X. Dexin and H. B. Harrison, Titanium Silicides Formed by Rapid Thermal Vacuum Processing, *J. Appl. Phys.* **63**(6), 2171 (1988).
152. C. S. Wel, J. VanDerSpiegel, and J. Santiago, Incoherent Radiative Processing of Titanium Silicides, *Thin Solid Films* **118**, 155 (1984).
153. R. Pantel, D. Levy, and D. Nicolas, Oxygen Behavior During TiSi$_2$ Formation by Rapid Thermal Annealing, *J. Appl. Phys.* **62**(10), 4319 (1987).
154. E. Ma, M. Natan, B. S. Lim, and M-A. Nicolet, Comparisons of Silicide Formation by Rapid Thermal Annealing and Conventional Furnace Annealing, *MRS Symp. Proc.* **92**, 205 (1987).
155. C. Mallardeau, Y. Morand, and E. Abonneau, Characterization of TiSi$_2$ Ohmic and Schottky Contact Formed by RTA Technology, *J. Electrochem. Soc.* **136**(1), 238 (1989).
156. M. Ada-Hanifi, A. Chantre, D. Levy, J. P. Gonchond, Ph. Delpech, and A. Nouailhat, Leakage Mechanisms of Titanium Silicided n+/p Junctions Fabricated Using Rapid Thermal Processing, *Appl. Phys. Lett.* **58**(12), 1280 (1991).
157. G. Krooshof, F. Habraken, W. van der Weg, L. Van den hove, K. Maex, R. De Keersmaecker, Study of the Rapid Thermal Nitridation and Silicidation of Ti Using Elastic Recoil Detection, Part I and II, *J. Appl. Phys.* **63**(10), 5104 (1988).
158. C. J. Sofield, R. E. Harper, and P. J. Rosser, Boron Redistribution during Transient Thermal Metal Silicide Growth on Si, *MRS Symp. Proc.* **35**, 445 (1984).
159. C. A. Moore, J. J. Rocca, and G. J. Collins, Titanium Disilicide by Wide Area Electron Beam Irradiation, *Appl. Phys. Lett.* **45**(2), 169 (1984).
160. S. Shatas and R. Ramani, Heat-Pulse Rapid Thermal Processing for Annealing Refractory Metal Silicides, presented at the Workshop on Refractory Metal Silicides for VLSI, San Juan Bautista, CA, May (1983).
161. K. Maex, R. F. De Keersmaecker, RTP For Simultaneous Annealing of Shallow Implanted Junction and Formation of Their TiSi$_2$ Contacts, *Physica* **129B**, 192 (1985).
162. L. Van den hove, N. Saks, K. Maex, R. De Keersmaecker, and G. Declerck, A Self-Aligned Titanium Silicide Technology using Rapid Thermal Processing, *Proc. European Solid State Device Research Conference*, Aachen, Germany, 281 (1985).
163. M. Bakli, G. Goltz, M. Vernet, J. Torres, J. Palleau, N. Bourhila, and J. C. Oberlin, Towards Development of a Salicide WSi$_2$ Process using RTA, *Appl. Surf. Science* **38**, 441 (1989).

164. E. Ma, M. Nalin, B. S. Lin, and M. A. Nicolet, Comparisons of Silicide Formation by Rapid Thermal Annealing and Conventional Furnace Annealing, *MRS Symp. Proc.* **92**, 205 (1987).
165. P. J. Rosser and G. J. Tomkins, Rapid Thermal Annealing of $TiSi_2$ for Interconnects, *Mat. Res. Soc. Symp. Proc.* **35**, 457 (1984).
166. A. Wang and J. Lien, Self-Aligned Titanium Polycide Gate and Interconnect Formation Scheme Using Rapid Thermal Annealing, *VSLI Science and Technology*, The Electrochemical Soc. **85-5**, 230 (1985).
167. Y. H. Ku, E. Louis, S. K. Lee, D. K. Shih, and D. L. Kwong, Formation of Shallow Junctions With $TiN_xO_y/TiSi_2$ Ohmic Contacts For Self-Aligned Silicide Technology, Advanced Processing of Semiconductor Devices, *SPIE* **797**, 61 (1987).
168. U. N. Mitra, P. W. Davies, R. K. Shukla and J. S. Multani, "Material Characterization of Selectively Formed Titanium Nitride and Silicide Thin Films," in *Semiconductor Silicon* (H. R. Huff, T. Abe, and B. Kobesen, eds.), The Electrochem. Soc., Pennington, NJ, p. 316 (1986).
169. P. J. Rosser and G. J. Tomkins, Self Aligned Nitridation of $TiSi_2$: A $TiN/TiSi_2$ Contact Structure, *Mat. Res. Soc. Symp. Proc.* **37**, 607 (1985).
170. R. G. M. Penning de Vries and K. Osinski, Enhanced Process Window for BPSG Flow in a Salicide Process Using a LPCVD Nitride Cap Layer, *Proc. ESSDERC*, Berlin p. 45 (1989).
171. R. Burmester, H. Joswig, and A. Mitwalsky, Reduction of Titanium Silicide Degradation During Borophosphosilicate Glass Reflow, *Proc. ESSDERC*, Berlin, p. 231 (1989).
172. S. P. Murarka, M. H. Read, and C. C. Chang, *J. Appl. Phys.* **52**, 7450 (1981).
173. S. P. Murarka and D. B. Fraser, Silicide Formation in Thin Cosputtered (Titanium + Silicon) Films on Polycrystalline Silicon and SiO_2, *J. Appl. Phys.* **51**, 350 (1980).
174. S. P. Murarka and S. Vaidya, Cosputtered Silicides on Silicon, Polycrystalline Silicon, and Silicon Dioxide, *J. Appl. Phys.* **56**, 3404 (1984).
175. S. P. Murarka and D. B. Fraser, Thin Film Interaction Between Titanium and Polycrystalline Silicon, *J. Appl. Phys.* **51**, 342 (1980).
176. W. K. Chu, J. W. Mayer, H. Muller, M.-A. Nicolet, and K. N. Tu, Identification of the Dominant Diffusing Species in Silicide Formation, *Appl. Phys. Lett.* **25**(8), 454 (1974).
177. W. K. Chu, S. S. Lau, J. W. Mayer, H. Muller, and K. N. Tu, Implanted Noble Gas Atoms as Diffusion Markers in Silicide Formation, *Thin Solid Films* **25**, 393 (1975).
178. C. D. Lien, M. Bartur, and M.-A. Nicolet, Marker Experiments for the Moving Species in Silicides During Solid Phase Epitaxy of Evaporated Si, *Mat. Res. Soc. Symp. Proc.* **25**, 51 (1984).
179. G. J. Van Gurp, W. F. van der Weg, and D. Sigurd, Interactions in the Co/Si Thin-Film System. II. Diffusion-Marker Experiments, *J. Appl. Phys.* **49**(7), 4011 (1978).
180. F. M. D'Heurle and C. S. Peterson, Formation of Thin Films of $CoSi_2$: Nucleation and Diffusion Mechanisms, *Thin Solid Films* **128**, 283 (1985).
181. P. L. Smith, C. M. Osburn, D. S. Wen, and G. McGuire, Silicon Consumption During Self-Aligned Titanium Silicide Formation on Shallow Junctions, *MRS Symp. Proc.* **160**, 299 (1990).
182. Q. F. Wang, C. M. Osburn, P. L. Smith, C. A. Canovai, and G. E. McGuire, Thermal Stability of Thin Submicrometer Lines of $CoSi_2$, *J. Electrochem. Soc.* **140(1)**, 200 (1993).
183. V. Q. Ho and D. Poulin, Formation of Self-Aligned $TiSi_2$ for VLSI Contacts and Interconnects, *J. Vac. Sci. Technol.* **A5**, 1396 (1987).
184. R. D. J. Verhaar, A. A. Bos, H. Kraaij, R. A. M. Wolters, K. Maex, and L. Van den hove, Self-Aligned $CoSi_2$ in a Submicron CMOS Process, *Proc. ESSDERC*, 229 Berlin, Springer-Verlag, Heidelberg (1989).

185. S. S. Iyer, C.-Y. Ting, and P. M. Fryer, Ambient Gas Effects on the Reaction of Titanium with Silicon, *J. Electrochem. Soc.* **302**, 2240 (1985).
186. E. K. Broadbent, R. Irani, and A. E. Morgan, Application of Self-Aligned $CoSi_2$ Interconnection in Sub-Micron CMOS Transistors, *Proc. IEEE V-MIC*, 175 (1988).
187. A. E. Morgan, E. K. Broadbent, K. N. Ritz, D. K. Sadana, and B. J. Burrow, *J. Appl. Phys.* **64**, 344 (1988).
188. B. Bhushan and S. P. Murarka, Stability of Cobalt Films on Silicon Dioxide Surfaces in Relation to Self-Aligned Silicide Process, to be published.
189. M. Y. Tsai, C. S. Petersson, F. M. d'Heurle and V. Maniscalco, Refractory Metal Silicide Formation Induced by As^+ Implantation, *Appl. Phys. Lett.* **37**, 295 (1980).
190. F. M. d'Heurle, C. S. Petersson, and M. Y. Tsai, Observation on the Hexagonal Form of $MoSi_2$ and WSi_2 Films Produced by Ion Implantation and on Related Snow-Plow Effects, *J. Appl. Phys.* **51**, 5976 (1980).
191. G. Valyi, H. Ryssel, O. Ganschow, and R. Jede, Ion-Beam Mixed $MoSi_2$ Layers: The Dependence of Composition on Formation Temperature, European Workshop on Refractory Metals and Silicides, p. B.1, Leuven Belgium, April (1989).
192. F. M. d'Heurle, M. Y. Tasi, C. S. Petersson, and B. Stritzker, Comparison of Annealing and Ion Implantation Effects During Solid State Disilicide Formation, *J. Appl. Phys.* **53**, 3069 (1982).
193. M. N. Kozicki and J. M. Robertson, Silicide Formation on Polycrystalline Silicon by Direct Metal Implantation, *J. Electrochem. Soc.* **136**(3), 878 (1989).
194. A. E. White, K. T. Short, R. C. Cynes, J. P. Garno, and J. M. Gibson, *Appl. Phys. Lett.* **50**, 95 (1987).
195. M. N. Kozicki and J. M. Roberston, Inst. Phys. Conf. Ser., Electron Beam Annealing of Co and Cr Implanted Polycrystalline Silicon, **67**, 137 (1983).
196. G. J. Campisi, H. B. Dietrich, M. Delfino, and D. K. Sadana, *Mat. Res. Soc. Symp. Proc.* **54**, 747 (1986).
197. C. Zaring, H. Jiang, M. Ostling, H. J. Whitlow, C. S. Peterson, and T. Phil, Formation of Imbedded $CoSi_2$ Layers by High Energy Implantation and Annealing, *Le Vide les Couches Minces* **42**,55 (1987).
198. M. F. Wu, A. Vertomme, H. Pattyn, G. Langouche, K. Maex, J. Vanhellemont, J. Vanacken, H. Vloeberghs, and Y. Bruynserade, *Nucl. Instrum. Meth. B* **45**, 658 (1990).
199. K. Elst, W. Vandervorst, T. Clarysse, W. Eichammer, and K. Maex, SIMS-SRP Study on the Outdiffusion from Poly and Mono Crystalline Cobalt Silicide, *Proc. First Intl. Workshop on the Measurement and Characterization of Ultra-Shallow Doping Profiles in Semiconductors* **II**, 410 (1991). *J. Vac. Sci. Technol. B* **10**, 524 (1992).
200. K. Maex, J. Vanhellemont, S. Petersson, and A. Lauwers. Formation of Ultrathin Buried $CoSi_2$ Layers by Ion Implantation in (100) Si, European Workshop on Refractory Metals and Silicides, p. 91; *Appl. Surf. Science* **53** (1991).
201. G. Bai, M.-A. Nicolet, and T. Vreeland, Jr., Elastic and Thermal Properties of Mesotaxial $CoSi_2$ Layers on Si, *J. Appl. Phys.* (9), 6451 (1991).
202. K. C. Saraswat, D. L. Brors, J. A. Fair, K. A. Moning, and R. Beyers, Properties of Low-Pressure CVD Tungsten Silicide for MOS VLSI Interconnections, *IEEE Trans. Electron Dev.* **ED-30**, 1497 (1983).
203. T. Ohba, S. Inoue, and M. Maeda, Selective CVD Tungsten Silicide for VLSI Applications, *IEDM Tech. Dig.* 213 (1987).
204. K. Shenai, P. A. Piacente, R. Saia, and B. J. Baliga, Blanket LPCVD Tungsten Silicide Technology for Smart Power Applications, *IEEE Electron Dev. Lett.* **EDL-10**(6), 270 (1989).

205. A. Bouteville, A. Royer, A. Bouamrane, and J. C. Remy, *Le Vide les Couches Minces* **41**, 291 (1986).
206. A. E. Morgan, W. T. Stacy, J. M. DeBlasi, and T.-Y. J. Chen, *J. Vac. Sci. Technol. B* **4**, 723 (1986).
207. G. A. West, K. W. Beeson, and A. Gupta, *J. Vac. Sci. Technol. A* **3**, 2278 (1985).
208. P. K. Tedrow, V. Ilderem, and R. Reif, *Appl. Phys. Lett.* **46**, 189 (1985).
209. F. Pintchovski, U.S. Patent No. 4,619,038 (1986).
210. A. Bouteville, A. Royer, and J. C. Remy, *J. Electrochem. Soc.* **134**, 2080 (1987).
211. G. J. Reynolds, C. B. Cooper III, and P. J. Gaczi, Selective Titanium Disilicide by Low-Pressure Chemical Vapor Deposition, *J. Appl. Phys.* **68**(8), 3212 (1989).
212. V. Iderem and R. Reif, Very Low Pressure Chemical Vapor Deposition Process for Selective Titanium Silicide Films, *Appl. Phys. Lett.* **58**(8), 687 (1988).
213. J. Mercier, J. L. Regoni, and D. Bensahel, Selective TiSi$_2$ Deposition with No Silicon Substrate Consumption by Rapid Thermal Processing in a LPCVD Reactor, *J. Electronic Mater.* **19**(3), 253 (1990).
214. K. Saito, T. Amazawa, and Y. Arita, Ohmic Contact Formation to Shallow Junctions by Selective Titanium Silicide Chemical Vapor Deposition, *Proc. 3rd Symp. on ULSI Science and Technol.* (J. M. Andrews and G. K. Cellar, eds.), The Electrochemical Soc. **91-11**, 276 (1991).
215. S. S. Lau, Z. L. Liau, M.-A. Nicolet, Solid Phase Epitaxy in Silicide-Forming Systems, *Thin Solid Films* **47**, 313 (1977).
216. H. von Kanel, J. Henz, M. Ospelt, J. Hugi, E. Muller, and N. Onda, Epitaxy of Metal Silicides, *Thin Solid Films* **184**, 295 (1990).
217. R. Tung, J. C. Bean, J. M. Gibson, J. M. Poate, and D. C. Jacobson, *Appl. Phys. Lett.* **40**, 684 (1982).
218. K. N. Tu, E. I. Alessandrini, W. K. Chu, H. Krautle, and J. W. Mayer, *Jpn. J. Appl. Phys.* Suppl. **2**, Part I, 669 (1974).
219. S. M. Yalisove, R. T. Tung, D. Loretto, *J. Vac. Sci. Technol. A* **7**, 1472 (1989).
220. M. Lawrence, A. Dass, D. B. Fraser, and S.-S. Wei, Growth of Epitaxial CoSi$_2$ on (100) Si, *Appl. Phys. Lett.* **58**(12), 1308 (1991).
221. S. L. Hsia, T. Y. Tan, P. L. Smith, and G. E. McGuire, Formation of Epitaxial CoSi$_2$ Films on (001) Silicon Using Ti-Co Alloy and Bimetal Source Materials, *J. Appl. Phys.* **70**, 7579 (1991).
222. G. Ottavani, Review of Binary Alloy Formation by Thin Film Interactions, *J. Vac. Sci. Technol.* **16**(5), 112 (1979).
223. S. S. Lau and J. W. Mayer, Interactions in the Co/Si Thin-Film Systems. I. Kinetics, *J. Appl. Phys.* **49**(7), 4005 (1978).
224. L. S. Hung, J. Gyulai, and J. W. Mayer, Kinetics of TiSi$_2$ Formation by Thin Ti Films on Si, *J. Appl. Phys.* **54**(9), 5076 (1983).
225. G. G. Bentini, R. Nipoti, A. Armigliato, M. Berti, A. V. Drigo, and C. Cohen, Growth and Structure of Titanium Silicide Phases Formed by Thin Ti Films on Si Crystals, *J. Appl. Phys.* **57**(2), 270 (1985).
226. A. Appelbaum, R. V. Knoell, and S. P. Murarka, Study of Cobalt-Disilicide Formation From Cobalt Monosilicide, *J. Appl. Phys.* **57**(6), 1880 (1985).
227. C. D. Lien, M. A. Nicolet, and S. S. Lau, *Appl. Phys. A* **34**, 249 (1984).
228. S.-H. Ko and S. P. Murarka, Ellipsometric Measurements of the CoSi$_2$ Formation from Very Thin Cobalt on Silicon, *J. Appl. Phys.* **71**, 4892 (1992).
229. K. N. Tu, W. K. Chu, and J. W. Mayer, *Thin Solid Films* **25**, 403 (1975).
230. D. J. Coe and E. H. Roderick, *J. Phys. D* **19**, 965 (1976).

231. G. Dalmai, J. P. Nys, and X. Wallart, Study of Ti Silicide Growth on Si Heated at Different Temperatures, *Vide Les Couches Minces* **42**, 133 (1987).
232. I. J. M. M. Raaijmakers, A. H. Reader, and J. W. Van Houtum, Nucleation and Growth of Titanium Silicide Studied by In-Situ Annealing in a Transmission Electron Microscope, *J. Appl. Phys.* **61**(7), 2527 (1987).
233. S. T. Kakshmikumar and A. C. Rastogi, The Growth of Titanium Silicides in Thin Film Ti/Si Structures, *J. Vac. Sci. Tech. B* **7**(4), 604 (1989).
234. Y. L. Corcoran and A. H. King, Grain Boundary Diffusion and Growth of Titanium Silicide Layers on Silicon, *J. Electron. Mat.* **19**(11), 1177 (1990).
235. G. J. Van Gurp and C. Langereis, Cobalt Silicide Layers on Si. I. Structure and Growth, *J. Appl. Phys.* **46**(10), 4301 (1975).
236. P. Ruterana, P. Houdy, and P. Boher, A Transmission Electron Microscopy Study of Low-Temperature Reaction at the Co-Si Interface, *J. Appl. Phys.* **68**(3), 1033 (1990).
237. H. J. W. van Houtum and I. J. M. M. Raaijmakers, First Phase Nucleation and Growth of Titanium Disilicide with an Emphasis on the Influence of Oxygen, *Mat. Res. Soc. Symp. Proc.* **54**, 37 (1986).
238. D. Pramanik, A. N. Saxena, O. K. Wu, G. G. Pererson, and M. Tanielian, Influence of the Interfacial Oxide on Titanium Silicide Formation by Rapid Thermal Annealing, *J. Vac. Sci. Technol. B* **2**(4), 775 (1984).
239. H. C. Swart, G. L. P. Berning, and J. DuPlessis, The Influence of Oxygen on Cobalt Silicide Formation, *Thin Solid Films* **189**, 321 (1990).
240. F. Nava, A. D'Amico, and A. Bearzotti, Phase Transformation Induced by Rapid Thermal Annealing In Ti-Si and W-Si Alloys, *J. Vac. Sci. Technol. A* **7**, 3023 (1989).
241. H. Matsui, H. Ohtsuki, M. Ino, and S. Ushio, *Mat. Res. Soc. Symp. Proc.* **54**, 769 (1986).
242. A. Naem, private communication, cited in *Reduced Thermal Processing for VLSI* (R. R. Levy, ed.), Plenum Press, NY, pp. 208–210 (1989).
243. T. P. Chow, W. Katz, and G. Smith, Titanium Silicide Formation on BF_2 Implanted Silicon, *Appl. Phys. Lett.* **46**, 41 (1985).
244. T. P. Chow, W. Katz, R. Goehner, and G. Smith, Titanium Silicide Formation on Boron-Implanted Silicon, *J. Electrochem. Soc.* **132**, 1914 (1985).
245. S. W. Sun, F. Pintchovshi, P. J. Tobin, and R. L. Hance, *Mat. Res. Soc. Symp. Proc.* **92**, 165 (1987).
246. R. Beyers, D. Coulman, and P. Merchant, Titanium Disilicide Formation on Heavily Doped Silicon Substrates, *J. Appl. Phys.* **61**(11), 5110 (1987).
247. P. Revesz, J. Gyimesi, and E. Zsoldos, Growth of Titanium Silicide on Ion-Implanted Silicon, *J. Appl. Phys.* **54**(4), 1860 (1983).
248. H. K. Park, J. Sachitano, M. McPherson, T. Yamaguchi, and G. Lehman, Effects of Ion Implantation Doping on the Formation of $TiSi_2$, *J. Vac. Sci. Technol. A* **2**, 264 (1984).
249. F. d'Heurle, P. Gas, I. Engstrom, S. Nygren, M. Ostling, and C. S. Petersson, The Two Crystalline Structures of $TiSi_2$: Identification and Resistivity, IBM Research Report RC 11151 (1985).
250. R. Beyers and R. Sinclair, Metastable Phase Formation in Titanium-Silicon Thin Films, *J. Appl. Phys.* **57**, 5240 (1985).
251. Z.-G. Xiao, H. Jiang, J. Honeycutt, G. Rozgonyi, C. M. Osburn, and G. McGuire, Influence of Process Sequence on Crystalline Properties of $CoSi_2$ Formed on Shallow Junctions, presented at Electrochem. Soc. Annual Meeting, Los Angeles, CA, May (1989).
252. Z. G. Xiao, H. Jiang, J. Honeycutt, C. M. Osburn, G. McGuire, and G. A. Rozgonyi, $TiSi_2$ Thin Films Formed on Crystalline and Amorphous Silicon, *MRS Symp. Proc.* **181**, 167 (1990).

253. Z.-G. Xiao, H. Jiang, J. Honeycutt, C. M. Osburn, G. McGuire, and G. A. Rozgonyi, Dependence of the Structural Properties of $CoSi_2$ and $TiSi_2$ Layers on the Initial State of the Si Substrate, private communication.
254. S. Vaidya, S. P. Murarka, and T. T. Sheng, Formation and Thermal Stability of $CoSi_2$ on Polycrystalline Si, *J. Appl. Phys.* **58**(2), 971 (1985).
255. J. Lasky, J. Nakos, O. Cain, and P. Geiss, Comparison of Transformation to Low-Resistivity Phase and Agglomeration of $TiSi_2$ and $CoSi_2$, *IEEE Trans. Electron Dev.* **ED-38**, 262 (1991).
256. H. Berger and C. Tollin, A High Purity Inert Gas System That Insures Ti Silicides Reproduceability, *Proc. SEMICON/East Technical Sessions*, 31 (1987).
257. H. Berger and S.-Y. Lin, Effects of Inert Purity on Ti Silicide Films, *Proc. First Inter. Conf. on ULSI Sci. and Technol.*, Electrochem. Soc. **87-11**, 434 (1987).
258. H. Berger, Gas Purity Requirements for Titanium Silicide Metallization, *Semicond. Int.* **10**(10), 137 (1987).
259. R. Sherman and H. Berger, Electron Spectroscopy for Chemical Analysis Results Relating Titanium Silicide Formation to Gas Purity, *J. Vac. Sci. Technol.* **5**(4), 1418 (1987).
260. T. G. Wolfe, R. Roberge, and G. A. Brown, Efect of Gaseous Growth Ambient Impurities on the Electrical Properties of Gate Oxides, *Electrochemical Society Extended Abstracts* **85-2**, 306 (1985).
261. R. Roberge, A. Francis Jr., S. Fisher, and S. Schmitz, Gaseous Impurity Effects in Silicon Epitaxy, *Semiconductor International* **10**(1), 77 (1987).
262. M. S. Wang and J. B. Anthony, Oxygen Contamination Control for a $TiSi_2$ Atmospheric Thermal Anneal Process, in *Multilevel Metallization, Interconnection and Contact Technologies* (L. B. Rothman and T. Herndon, eds.), The Electrochem. Soc., Pennington, NJ, 114 (1987).
263. C. M. Osburn, H. Berger, R. P. Donovan, and G. W. Jones, The Effects of Contamination on Semiconductor Manufacturing Yield, *J. Environ. Sci.* **31**(2), 45 (1988).
264. H. K. Park, J. Sachitano, G. Eiden, E. Lane, and T. Yamaguchi, Mo/Ti Bilayer Metallization for A Self-Aligned $TiSi_2$ Process, *J. Vac. Sci. Technol. A* **2**, 259 (1984).
265. M.-Z. Lin, Y.-C. S. Yu, and C.-Y. Wu, An Environment-Insensitive Trilayer Structure for Titanium Silicide Formation, *J. Electrochem. Soc.* **133**(11), 2386 (1986).
266. R. S. Rastogi, V. D. Vankar, and K. L. Chopra, Silicide Formation By Solid State Reaction of Mo-Ni and Mo-Co Films with Si(100), *Thin Solid Films* **199**, 107 (1991).
267. P. Vandenabeele, K. Maex, and R. De Keersmaecker, Impact of Patterned Layers on Temperature Non-Uniformity During Rapid Thermal Processing for VLSI Applications, *MRS Symp. Proc.* **146**, 149 (1989).
268. J. T. Pan and I. Belch, Stress Measurements of Refractory Metal Silicides during Sintering, *J. Appl. Phys.* **55**(8), 2874 (1984).
269. V. L. Teal and S. P. Murarka, Stresses in $TaSi_x$ Films Sputter Deposited on Polycrystalline Silicon, *J. Appl. Phys.* **61**(11), 5038 (1987).
270. L. Van den hove, J. Vanhellemont, R. Wolters, W. Classen, R. De Keersmaecker, and G. Declerk, Correlation between Stress and Film-Edge Induced Defect Generation for $TiSi_2$ and $CoSi_2$ Formed by Metal–Si Reaction, *Proc. of First Int. Symp. on Advanced Materials for ULSI*, Electrochem. Soc. **88-19**, 165 (1988).
271. T. F. Retajczyk Jr. and A. K. Sinha, Elastic Stiffness and Thermal Expansion Coefficients of Various Refractory Silicides and Silicon Nitride Films, *Thin Solid Films* **70**, 241 (1980).
272. L. Krusin-Elbaum, J. Y.-C. Sun, and C.-Y. Ting, On the Resistivity of $TiSi_2$: The Implication for Low-Temperature Applications, *IEEE Electron Dev.* **ED-34**, 58 (1986).

273. C. A. Pico, Investigations of the Properties of TiSi(2), Ph.D. Thesis, University of Wisconsin (Nov. 1986).
274. C. Y. Wong, L. K. Wang, and P. A. McFarland, Transmission Electron Microscopy Investigation of the Thermal Stability of $TiSi_2$ Contacts in MOS Technology, *Electrochem. Soc., Extended Abstracts* **85**(2), 466 (1985).
275. C. Y. Wong, F. S. Lai, and P. A. McFarland, Cross-Sectional Transmission Electron Microscopy of Ti/Si Reaction of Phosphorus Doped Polycrystalline Silicon Gate, *Electrochem. Soc., Extended Abstracts* **85**(2), 468 (1985).
276. C. Y. Wong, L. K. Wang, P. A. McFarland, and C. Y. Ting, Thermal Stability of $TiSi_2$ on Mono- and Polycrystalline Silicon, *J. Appl. Phys.* **60**(1), 243 (1986).
277. C. Y. Ting, F. M. d'Heurle, S. S. Iyer, and P. M. Fryer, High Temperature Process Limitation on $TiSi_2$, *J. Electrochem. Soc.* **133**(12), 2621 (1986).
278. R. K. Shukla and J. S. Multani, Thermal Stability of Titanium Silicide Thin Films, *Proc. Fourth Intl. IEEE Multilevel Interconnection Conf.*, 470 (1987).
279. Z. G. Xiao, G. A. Rozgonyi, C. A. Canovai, and C. M. Osburn, Agglomeration of Cobalt Silicide Films, *MRS Symp. Proc.* **202**, 101 (1990).
280. T. Yoshida, S. Ogawa, S. Okuda, T. Kouzaki, and K. Tsukamoto, Thermally Stable, Low Leakage Self-Aligned Titanium Silicide Junctions, *J. Electrochem. Soc.* **137**, 1914 (1990).
281. H. Sumi, T. Nishihara, Y. Sugano, H. Masuya, and M. Takasu, New Silicidation Technology by SITOX (Silicidation Through Oxide) and Its Impact on Sub-half Micron MOS Devices, *IEDM Tech. Dig.*, p. 249 (1990).
282. S. Nygren and S. Johansson, Recrystallization and Grain Growth Phenomena in Polycrystalline Si/$CoSi_2$ Thin Film Couples, *J. Appl. Phys.* **68**(3), 1050 (1990).
283. H. Norstron, K. Maex, and P. Vandenabeele, Disintegration of $TiSi_2$ on Narrow Poly-Si Lines at High Temperatures, *J. Vac. Sci. Technol. B* **8**(1), 1223 (1990).
284. H. Norstom, K. Maex, and P. Vandenabeele, Thermal Stability and Interface Bowing of Submicron $TiSi_2$/Polycrystalline Silicon, *Thin Solid Films* **198**, 53 (1991).
285. J. Y. Tsai, B. Y. Tsui, and M. C. Chen, High Temperature Stability of Platinum Silicide Association with Fluorine Implantation, *J. Appl. Phys.* **67**, 3530 (1990).
286. B. Y. Tsui, T. S. Wu, J. Y. Tsai, and M. C. Chen, Effect of Fluorine Incorporation on the Thermal Stability of Pt/Si Structure, submitted to *IEEE Trans. Electron Dev.* (1991).
287. Z. G. Xiao, G. A. Rozgonyi, C. A. Canovai, and C. M. Osburn, Partial Agglomeration during Co Silicide Film Formation, *J. Mater. Res.* **7**(2) (1992).
288. J. P. Gambino, E. G. Colgan, and B. Cunninghan, *Electrochem. Soc. Extended Abstracts* **91-2** (1991).
289. W. W. Mullins, Theory of Thermal Grooving, *J. Appl. Phys.* **28**(3), 333 (1957).
290. W. W. Mullins, Grain Boundary Grooving by Volume Diffusion, *Transactions of the Metallurgical Society of AIME* **218**, 354 (1960).
291. J. R. Pfiester, T. C. Mele, Y. Limb, R. E. Jones, M. Woo, B. Boeck, and C. D. Gunderson, A TiN Strapped Polysilicon Gate Cobalt Silicide CMOS Process, *IEDM Tech. Dig.*, 241 (1990).
292. P. M. Smith, M. O. Thompson, The Effects of N_2 and Ar as the Ambient Gas During Rapid Thermal Annealing of Tungsten Silicide, *MRS Symp. Proc.* **106**, 707 (1988).
293. C. Canovai, C. M. Osburn, R. C. Chapman, Z.-G. Xiao, Q.-F. Wang, and G. E. McGuire, Dopant Diffusion and Interface Roughening Using $CoSi_2$ as a Diffusion Source, Presented at Spring Meeting of the Electrochemical Society, Washington, DC (May 1991).
294. M. Wittmer, C.-Y. Ting, I. Ohdonari, and K. N. Tu, Redistribution of As during Pd_2Si Formation: Ion Channeling Measurements, *J. Appl. Phys.* **53**(10), 6781 (1982).

295. M. Wittmer and K. N. Tu, Low-Temperature Diffusion of Dopant Atoms in Silicon During Interfacial Silicide Formation, *Phys. Rev. B* **29**(4), 2010 (1984).
296. J. Amano, P. Merchant, T. R. Cass, J. N. Miller, and T. Koch, Dopant Redistribution During Titanium Silicide Formation, *J. Appl. Phys.* **59**(8), 2689 (1986).
297. C. M. Osburn, T. Brat, D. Sharma, N. Parikh, W.-K. Chu, D. Griffis, S. Corcoran, and S. Lin. The Effects of Titanium Silicide Formation on Dopant Redistribution, *Proc. First Intl. Symp. on ULSI Sci. and Technol., Electrochem. Soc.* **87-11**, 402 (1987); *J. Electrochem. Soc.* **135**(6), 1900 (1988).
298. S. Batra, K. Park, S. Yoganathan, J. Lee, S. Banerjee, S. Sun, and G. Lux, Effects of Dopant Redistribution, Segregation, and Carrier Trapping in As-Implanted MOS Gates, *IEEE Trans. Electron Dev.* **ED-37**(11), 2322 (1990).
299. N. Yu, Z. Zhou, W. Zhou, S. Tsou, and D. Zhu, As Redistribution During Ti Silicide Formation, *Nucl. Instrum. Meth. Phys. Res. B* **19/20**, 427 (1987).
300. Y. Taur, J. Y. C. Sun, D. Moy, L. K. Wang, B. Davari, S. P. Klepner, and C. Y. Ting, Source-Drain Contact Resistance in CMOS With Self-Aligned $TiSi_2$, *IEEE Trans. Electron Dev.* **ED-34**, 575 (1987).
301. Y. Taur, B. Davari, D. Moy, J. Y.-C. Sun, and C. Y. Ting, Study of Contact and Shallow Junction Characteristics in Submicron CMOS with Self-Aligned Titanium Silicide, *IBM J. Res. Develop.* **31**(6), 627 (1987).
302. B. Davari, Y. Taur, D. Moy, F. M. d'Heurle, and C. Y. Ting, Very Shallow Junctions for Submicron CMOS Technology Using Implanted Ti for Silicidation, *Proc. of First Int. Symp. on Ultra Large Scale Integration Science and Technology, The Electrochem. Soc.* **87-11**, 368 (1987).
303. K. Maex, Interactions of Implanted Dopants and the Metal/Si System: Fundamental Aspects and Applications of Ion Beam Mixing and Dopant Redistribution, Ph.D. Thesis, Katholieke Universiteit Leuven (1987).
304. K. Maex, L. P. Hobbs, and W. Eichhammer, Silicided Shallow Junctions for ULSI, *Proc. 3rd Symp. on ULSI Science and Technol.* (J. M. Andrews and G. K. Cellar, eds.), The Electrochemical Soc. **91-11**, 254 (1991).
305. V. Probst, P. Lippens, L. Van den hove, H. Schaber, and R. De Keersmaecker, Limitations of $TiSi_2$ as a Source for Dopant Diffusion, *Appl. Phys. Lett.* **52**(21), 1803 (1988).
306. K. Maex, G. Ghosh, L. Dalaey, V. Probst, P. Lippens, L. Van den hove, and R. F. De Keersmaecker, Stability of As and B Doped Si with Respect to Overlaying $CoSi_2$, *J. Mater. Res.* **4**(5), 1209 (1989).
307. A Mitwalsky, V. Probst, and R. Burmester, Metal-Dopant Compound Formation in $TiSi_2$ Studied by Transmission and Scanning Electron Microscopy, *Proc. 6th Inter. Symp. on Silicon Materials Sci. and Technol., Semiconductor Silicon*, The Electrochemical Soc., Proc. Vol. **90-7**, 876 (1990).
308. P. Gas, V. Deline, F. M. d'Heurle, A. Michel, and G. Scilla, Boron, Phosphorus, and Arsenic Diffusion in $TiSi_2$, *J. Appl. Phys.* **60**(5), 1634 (1986).
309. O. Thomas, P. Gas, F. M. d'Heurle, F. K. LeGoues, A. Michel, and G. Scilla, Diffusion of Boron, Phosphorus, and Arsenic Implanted in Thin Films of Cobalt, *J. Vac. Sci. Technol. A* **6**(3), 1736 (1988).
310. O. Thomas, P. Gas, A. Charai, F. K. LeGoues, A. Michel, G. Scilla, and F. M. d'Heurle, The Diffusion of Elements Implanted in Films of Cobalt Silicide, *J. Appl. Phys.* **64**(6), 2973 (1988).
311. A. H. van Ommen, H. J. W. van Houtum, and A. M. L. Theunissen, Diffusion of Ion Implanted As in $TiSi_2$, *J. Appl. Phys.* **60**(2), 627 (1986).
312. C. Zaring, P. Gas, B. G. Svensson, M. Ostling, and H. J. Whitlow, Lattice Diffusion of Boron in Bulk Cobalt Disilicide, *Thin Solid Films* **193/194**, 244 (1990).

313. C. L. Chu, K. C. Saraswat, and S. S. Wong, Measurement of Lateral Dopant Diffusion in Thin Silicide Layers, *IEEE Trans. Electron. Dev.* **39**, 2333 (1992).
314. H. Okabayashi, E. Nagasawa, and M. Morimoto, A New Low Resistance Shallow Junction Formation Method Using Lateral Diffusion through Silicide, *IEDM Tech. Dig.*, 670 (1983).
315. L. C. Parillo et al., A Fine-Line CMOS Technology that Uses p^+ Polysilicon/Silicide Gates for NMOS and PMOS Devices, *IEDM Tech. Dig.*, 418 (1984).
316. D. T. Amm, D. Levy, M. Paoli, P. Delpech, T. Ternisen d'Ouville, H. Mingam, and G. Goltz, Avoiding Lateral Diffusion of Dopants in n^+/p^+ Polycide Gates, *ESSDERC*, Berlin, p. 561 (1989).
317. C. L. Chu, K. C. Sarswat, and S. S. Wong, Characterization of Lateral Dopant Diffusion in Silicides, *IEDM Tech. Dig.*, 245 (1990).
318. L. R. Zheng, L. S. Hung, and J. W. Mayer, Lateral Diffusion of Ni and Si through Ni_2Si Couples, *Appl. Phys. Lett.* **41**(7), 646 (1982).
319. Q. F. Wang, C. M. Osburn, and C. A. Canovai, Ultra-Shallow Junction Formation Using Silicide as a Diffusion Source and Low Thermal Budget, *IEEE Trans. Electron Dev.* **39**, 2486 (1992).
320. J. Lin, S. Banerjee, J. Lee, and C. Teng, Anomalous Current-Voltage Behavior in Titanium-Silicided Shallow Source/Drain Junctions, *J. Appl. Phys.* **68**(3), 1082 (1990).
321. I. Ohdomari, K. Konuma, M. Takano, T. Chikyow, H. Kawarada, J. Nakanishi, and T. Veno, *Materials Res. Soc. Symp. Proc.* **54**, 63 (1986).
322. D. S. Wen, P. L. Smith, C. M. Osburn, and G. A. Rozgonyi, Defect Annihilation in Shallow P^+ Junctions Using Titanium Silicide, *Appl. Phys. Lett.* **51**(15), 1182 (1987).
323. D. S. Wen, P. L. Smith, C. M. Osburn, and G. A. Rozgonyi, Elimination of End-of-Range Shallow Junction Implantation Damage during CMOS Titanium Silicidation, *J. Electrochem. Soc.* **136**(2), 466 (1989).
324. M. H. Wang, W. Lur, H. C. Cheng, and L. J. Chen. Interfacial Reactions of Titanium Thin Films on P^+-Implanted (001) Si, *MRS Symp. Proc.* **100**, 93 (1988).
325. W. Lur and L. J. Chen, Interfacial Reactions of Cobalt Thin Films on BF_2 Ion-Implanted (001) Silicon, *J. Appl. Phys.* **64**(7), 3505 (1988).
326. W. Lur, J. Y. Cheng, C. H. Chu, M. H. Wang, T. C. Lee, Y. J. Wann, W. Y. Chao, and L. J. Chen, Effects of Silicide Formation on the Removal of End-Of-Range Ion Implantation Damage in Silicon, *Nucl. Instr. Meth. Phys. Res. B* **39**, 297 (1989).
327. W. Lur, J. Y. Cheng, and L. J. Chen, Removal of End-Of-Range Ion Implantation Defects in Silicon by Near Noble and Refractory Silicide Formation, *Mat. Res, Soc. Symp. Proc.* **147**, 33 (1989).
328. S. M. Hu, Point Defect Generation and Enhanced Diffusion in Silicon Due to Tantalum Silicide Overlays, *Appl. Phys. Lett.* **51**(5), 308 (1987).
329. P. Fahey and R. W. Dutton, Investigation of Point-Defect Generation in Silicon During Oxidation of a Deposited WSi_2 Layer, *Appl. Phys. Lett.* **52**, 1092 (1988).
330. J. Honeycutt, Ph.D. Thesis, North Carolina State University (1991).
331. J. Lin, K. Park, S. Batra, S. Banerjee, and J. Lee, Enhancement of Boron Diffusion through Gate Oxides in MOS Devices under Rapid Thermal Silicidation, *Appl. Phys. Lett.* **58**, 2123 (1991).
332. R. B. Fair, PREDICT: Process Estimator for the Design of Integrated Circuits, Microelectronics Center of North Carolina (1986).
333. P. B. Moynagh, A. A. Brown, and P. J. Rosser, Modeling Diffusion in Silicides, *J. Physique* **C4**(9), 187 (1988).

334. J. F. Jongste, F. E. Prins, G. C. A. M. Janssen, and S. Radclaar, Modelling of the Formation of $TiSi_2$ in a Nitrogen Ambient, *European Workshop on Refractory Metals and Silicides*, p. B.3.2, Leuven Belgium, April (1989).
335. L. Borucki, R. Mann, G. Miles, J. Slinkman, and T. Sullivan, A Model for Titanium Silicide Film Growth, *IEDM Tech. Dig.*, 348 (1988).
336. P. D. Cole, G. M. Crean, J. Lorenz, and L. Dupas, Comparison of Models for the Calculation of Ion Implantation Moments of Implanted Boron, Phosphorus and Arsenic Dopants in Thin Film Silicides, *Nucl. Instrum. Meth. Phys. Res. B* **55,** 000 (1991).

8 Issues in Manufacturing Unique Silicon Devices Using Rapid Thermal Annealing

B. Lojek
Motorola Inc.
Advanced Technology Center
2200 W. Broadway Road
Mesa, Arizona

I. Impact of Patterned Layers on Temperature Nonuniformity during Rapid Thermal Annealing 314
   A. Stress, Warpage, and Slip 318
   B. Thermal Stress 319
II. Bipolar Transistor Processing 325
   A. Double Polysilicon Self-Aligned Structures 326
   B. Single Polysilicon Self-Aligned Structures 332
   C. Arsenic-Doped Polysilicon Emitter 334
III. MOS Transistor Processing 337
   A. Conventional MOS Structure 338
   B. MOS Transistor with Elevated Electrodes 340
   C. Ultrathin RTP Gate Dielectrics 342
IV. Conclusion 344
References 346

The increasing complexity and integration of both bipolar and complementary metal oxide semiconductor (CMOS) devices in modern processes lead to substantially longer thermal processing treatment. The necessity to minimize the thermal budget is driven by requirements for shallower junction formation as a result of geometry scaling. In order to solve this conflict a more advanced annealing and diffusion technology is necessary. Several approaches have been investigated. For instance, tremendous progress in vertical diffusion furnaces has led to a dramatic reduction of unwanted long heating times, necessary for convective heating in conventional horizontal systems. However, a conventionally heated system cannot prevent possible contamination from the hot wall of the processing chamber. State-of-the-art rapid thermal annealing (RTA) systems are cold-wall systems and they can easily meet strict requirements for low-level contamination.

Like any other equipment or technique, RTA has its own characteristics. RTA differs from furnace processing not merely in the manner in which wafers are heated, but also in the dynamics of heating and cooling. Some authors speculate that the heating dynamics, especially the fast cooling rate, which could be in the range of 1–125 °C/sec, is responsible for unusual annealing behavior. However, the experimental data do not support this hypothesis. Recent work [1, 2] shows that the so-called photonic contribution, which is present in RTA systems using an incoherent lamp source, creates a different mechanism of annealing. So far, annealing and diffusion processes in RTA are described in theory as a direct analogue of furnace processing. In order to match RTA results to the theory of thermal diffusion, several parameters without any physical justification were introduced. Due to a fundamental difference in the radiation spectrum (in the case of furnace processing, only photons in the infrared and longer wavelength region are available; in the case of RTA, the spectrum is from the ultraviolet to visible region) photochemical reactions are also present and differ from the thermal reaction.

On the basis of direct comparison of convectively and radiatively heated RTA systems [3], higher activation of implanted atoms in a radiative system is reported.

From the theory [4, 5] of basic mechanisms of radiation effects it can be concluded that electronic excitation in atomic processes plays a dominant role if the nonequilibrium carrier concentration is changed by illumination, injection, or irradiation. (In the case of RTA, we are consistently dealing with atoms in the excited state.) The dissociation of vacancy–interstitial pairs, association and dissociation of defect complexes, and migration of defects are essentially related to electron excitations. The general cause of the contribution of electronic excitation to the radiation effects under consideration is that the excitation changes the interatomic interactions.

On the basis of the extensive literature dealing with RTA one could suppose that RTA processing is a well-established technology and that it is used extensively during processing of very large-scale integrated circuits (VLSICs). In spite of many advantages, the semiconductor industry hesitates to use RTA systems on a wide-scale. There are several reasons for this situation. Unfortunately for RTA vendors, the concept of rapid thermal processing (RTP) employing incoherent light sources is so simple that almost any laboratory can build this system and generate experimental data. However, the industrial requirements are much stricter, and heating of silicon is only part of the problem. As a result, RTA system manufacturers often bring to market equipment that is not ready for industrial applications. This situation continues: for instance, of 20 or more RTA and RTP systems on the market in 1991, only two or three meet production needs. Fortunately for rapid thermal processing, there are at least two applications, silicide formation and polysilicon emitter anneal, that can not be duplicated with conventional processing. In addition, requirements for low impurity contamination will force industry to move from stand-alone process equipment to cluster tools in which RTA is one of the dominant processing steps. It is only a question of time and capital investment before this transition occurs, and then RTA will play a more important role in VLSIC processing. These trends underscore the importance of RTA and the necessity to understand all physical phenomena accompanying the process.

Since the introduction of the first RTA system, measuring and controlling temperature remains an unsolved problem. In most cases, the actual wafer temperature is not important, but temperature reproducibility is critical. Temperature reproducibility must be achieved for various circuit structures associated with different mask sets and for slight variations in the back side of wafers. This is a key to widespread acceptance of rapid thermal processing.

The RTA system employs "single wafer processing"; therefore, the process engineer has lost one of his most powerful tools—the monitor wafer. Temperature variations across a wafer can only be measured indirectly by observing process nonuniformity. Two approaches are commonly used for indirect temperature measurement:

1. Thermal oxidation and measurement of oxide thickness.
2. Annealing of high-dose implantation and measurement of sheet resistance.

Rapid thermal oxidation needs relatively long processing times (>40 sec) and processing temperatures above 1050 °C to achieve reasonable oxide thickness and to eliminate the impact of native oxide. Grown oxide thickness is dependent on oxygen concentration used and the thickness of native oxide before oxidation.

For annealing high-dose implantation (usually arsenic in the range of 1–2×10^{16} at 20–40 keV), much shorter processing times (~ 10 sec) and a wider range of temperatures can be used. The disadvantage of this method is that sheet resistance can not be reliably measured close to the wafer edge (<5 mm) where temperature nonuniformity is usually highest.

A common disadvantage to both indirect methods is that they integrate all variations from the defined processing recipe into a single parameter. Even if the final temperature distribution across the wafer is acceptable, there could be a part of the processing interval in which temperature nonuniformity is above the limit of acceptance. The tacit assumption in both methods is that the absorption, emissivity, and reflectivity are the same for a product wafer as for the wafer used for steady-state temperature calibration (usually a wafer with an attached thermocouple). However, this is not a realistic assumption for annealing structured wafers or for unpatterned wafers with layers stacked differently than on the calibration wafer. For example, the presence of lattice damage at the surface of a wafer has a significant effect on the optical properties of the material, particularly at photon energies well above band gap. The reflectivity of silicon disordered by high-dose implantation is about 15% higher than the reflectivity of a virgin polished wafer [6]. Therefore the heating rate for this wafer is different than the heating for a calibration wafer. A change in wafer reflectivity of about 10% will cause a variation in steady-state temperature of almost 50 °C.

The configuration of chamber, reflector and lamps in RTA systems largely dictates the uniformity of results and magnitude of temperature variations. A system that yields good results for one size wafer may not necessarily lead to the same results for a different size wafer. For these reasons RTA processing is equipment dependent, and one must be careful when judging and comparing published results from RTA processing. This is especially true for results achieved on the first generation of RTA systems with inaccurate temperature measurement capability. See Chapter 9 for a full discussion on equipment issues.

I. Impact of Patterned Layers on Temperature Nonuniformity during Rapid Thermal Annealing

One of the most important aspects of a rapid thermal annealing system is its generation and delivery of radiant energy to the wafer. Contrary to the conventional furnace processing, the wafer is not in thermal equilibrium with the cold walls of the RTA system.

The wafer is horizontally loaded into a chamber with a controlled ambient atmosphere. The wafer is then heated by lamps, either tungsten-halogen or arc lamp, that are mounted inside an optically polished reflector. The primary radiation (lamp radiation and multiply reflected wafer–reflector radiation) and secondary radiation (heat radiation from the hot wafer that is back reflected) are coupled to the wafer by the reflector cavity walls. A signal from an appropriate thermal sensor, thermocouple or pyrometer controls the feedback loop and determines the power to the lamps.

From this very simplified description, one could assume that the RTA concept is free of pitfalls. Unfortunately, the engineering realization is

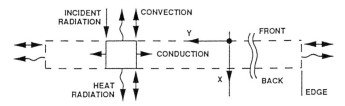

FIGURE 1. Schematic diagram illustrating energy balance across the control volume on a wafer.

rather difficult and knowledge (or prediction) of some key processing parameters is impossible. The major difficulties are:

- coupling primary and secondary radiation to provide a uniform heat flux across the wafer during the heating-up, the steady state and the cooling periods, and
- wafer temperature measurement and control of the heating system for arbitrary combinations of material layers on the front and back side of the wafers.

To deduce the basic heating dynamics, consider the energy flux balance in a volume element of unstructured silicon, irradiated only from one side without considering the secondary radiation. For the coordinate system as shown in Fig. 1 and from consideration of conservation of energy in a volume element of the wafer we can describe the temperature dynamics by the heat equation in the form [7]

$$\frac{\partial T(\bar{X}, t)}{\partial t} = \frac{\partial}{\partial \bar{X}} \left[k_{Si}(T) \frac{\partial T(\bar{X}, t)}{\partial \bar{X}} \right] + \frac{1}{\rho_{Si} c_p} [P_a(\bar{X}, T, t) - P_l(\bar{X}, T, t)] \quad (8.1)$$

where $\bar{X} = [x, y, z]$ or $\bar{X} = [r, \varphi, z]$ are the space coordinates.

ρ_{Si} is the density of silicon.
c_p is the specific heat capacity of silicon.
$k_{Si}(T)$ is the thermal diffusivity of silicon.
$P_a(\bar{X}, T, t)$ is the absorbed power distribution per unit of volume.
$P_l(\bar{X}, T, t)$ is the power lost by heat radiation or convection.

If multiple reflection in primary radiation is neglected, the absorbed power has the form

$$P_a(\bar{X}, T, t) = \int_0^\lambda I_\lambda(T)[1 - R_\lambda(\bar{X})][1 - \exp(-\alpha_\lambda(\bar{X})x)] \, d\lambda \quad (8.2)$$

where $I_\lambda(T)$ is the incident radiation intensity.
$R_\lambda(\bar{X})$ is the reflectivity.
$\alpha_\lambda(\bar{X})$ is the absorption coefficient of the silicon.

I_λ, R_λ, and α_λ are complicated functions of several physical parameters and they are not very well known, especially at elevated temperatures.

If the convective term can be neglected (this is true almost in all cases except during the cooling down period), the power lost by radiation is

$$P_1(\bar{X}, T, t) = \frac{\sigma}{t_w} [\varepsilon_F(\bar{X}, T) + \varepsilon_B(\bar{X}, T)] [T(\bar{X}, t)]^4 \quad (8.3)$$

where σ is the Stefan–Boltzman constant.

t_w is the thickness of silicon wafer.

$\varepsilon_F(\bar{X}, T)$ and $\varepsilon_B(\bar{X}, T)$ are the emissivity of the front and back side of the wafer, respectively.

Without any internal heat sources, the heating dynamics are given by boundary condition

$$\frac{\partial T(x, y = R, t)}{\partial y} = -\frac{\varepsilon_B(t_w, t)\,\sigma[T(x, y = R, t)]^4}{k_{Si}(T)\rho_{Si}c_p} + P_a(x, y = R, t) \quad (8.4)$$

The solution of Eqs. 8.1–8.4 has been presented by several authors; see for example [8, 9]. The result of these calculations is that because of the boundary condition at the edge of the wafer Eq. 8.4, uniform intensity of incident radiation across a wafer does not result in a uniform temperature distribution across the wafer. The edge of the wafer will be hotter than the center of the wafer during the heating-up period and cooler during the steady-state and cooling-down periods.

This result has very important consequences. Rapid thermal annealing is frequently classified [10] as an isothermal type of processing. This is true in the vertical (across thickness of wafer) direction. The vertical temperature profile is almost uniform, but this is not necessarily true in the horizontal direction.

So far, all RTA systems are intentionally designed to achieve a uniform steady-state distribution of temperature. Using reflector design or lamp configuration they compensate for edge losses during the steady-state period by creating a nonuniform distribution of incident radiation (see Chapter 9). This energy loss compensation leads to large transient nonuniformities (overshoot) in temperature, producing the difference between the temperature in the center and edge [11]. This result follows from the fact that heat conduction is proportional to temperature and that heat radiation is proportional to T^4. The overshoot in temperature is a function of the steady-state temperature (annealing temperature), and the heating rate. With increasing steady-state temperature, the overshoot is reduced because the relaxation from overshoot temperature to steady-state temperature occurs faster (due to the increase of heat conduction). Therefore, the

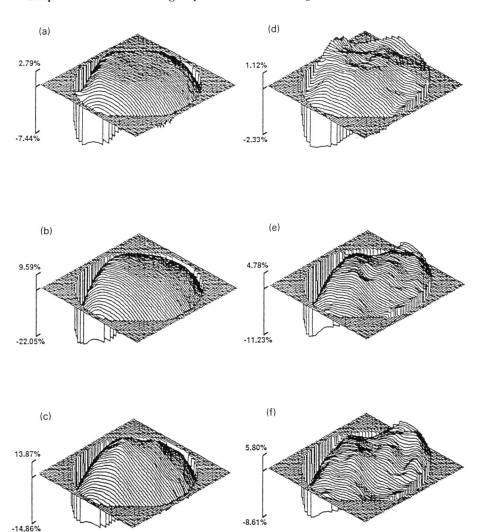

FIGURE 2. Mapping of the temperature transient nonuniformities for a 100 mm wafer: Fast heating rate and steady-state temperature (a) 800 °C, (b) 1000 °C, (c) 1200 °C; slow heating rate and steady-state temperature (d) 800 °C, (e) 1000 °C, (f) 1200 °C.

overshoot and lateral thermal gradient is worse for fast heating up and intermediate annealing (~1000–1100 °C) temperatures.

Experimental data show very good agreement with the above model. Figure 2 shows the mapping of temperature overshoot for 100 mm unstructured wafers at different steady-state temperatures for the fast (>80 °C/sec) and slow (50 °C/sec) heating rates. The sample processed with a fast heating rate and annealing temperature of 1000 °C had an unacceptable temperature distribution, with the lateral temperature gradient higher than 40 °C. Faster heating up will enhance the temperature transient nonuniformity, and extending the duration of the steady-state period will enhance the stationary nonuniformity. It is not clear at this time how this dilemma will be solved in upcoming generations of RTA equipments.

Further study of Eqs. 8.1–8.4 reveals another important conclusion. In RTP systems, the temperature is monitored almost solely by an optical pyrometer. The use of thermocouples is limited because of their degradation with time and their incompatibility with reactive gases. Pyrometers sense the radiation (the power lost by radiation, Eq. 8.3), which is critically dependent on the wafer emissivity. The emissivity of silicon varies not only with temperature and wavelength, but mainly with the back side roughness of the wafer. In the production environment, the back side of the wafer is subject to wide variations that are not easy to control. In order to define the optical properties of the back side of the wafer, quite often the back side etch is used. Unfortunately, this is not a very good practice because the wafer edges become thinner and the wafer is more vulnerable to slip. In addition, there is direct correlation between the degree of wafer warpage and the etching of the wafer back side.

A. Stress, Warpage, and Slip

Mechanical stress in a material is one factor that should be taken into consideration during the development of VLSI devices. A correlation between device structure and mechanical properties has been found, but a quantitative theory of stress mechanics has evolved very slowly.

The preliminary data show that mechanical stress is more pronounced for MOS devices. For instance, several authors [12–14] report several kinds of device degradation resulting from process-induced mechanical stress. This stress may cause device reliability problems, and during processing, diffusion behavior of impurities may be affected.

Mechanical stress can be divided into the following components:

- intrinsic stress resulting from film structural properties, which is a function of deposition conditions.
- extrinsic stress
  - thermal stress that is caused by a mismatch in film and substrate thermal expansion coefficients.
  - internal thermal stress due to the presence of a thermal gradient across the wafer.

For most physical models of extrinsic stress it is assumed that the films have no surface discontinuity. It is necessary to note, however, that the stress at the edges of structured films can be many times larger than the average stress in the film [15, 16]. High stress can cause device degradation by enhancing junction motion, by forming dislocations, or by causing adhesion failures.

It is obvious that all stress components are combined in the final stress distribution.

Usually the internal thermal stress is dominant if RTA processing is used.

B. Thermal Stress

In an RTA system, as the wafer temperature gradually increases from room temperature to annealing temperature any nonuniformity in temperature distribution induces a thermal stress in the wafer. Once the temperature distribution $T(x, y, z, t)$ is determined, the thermal stress may be determined from the laws of conservation of linear and angular momentum and constitutive relations between strain, stress, and temperature (generalized Hooke's law). Thus, the thermoelastic behavior of the wafer can be formulated as a plane stress problem based only on thermal loading.

As silicon crystallizes in the diamond cubic structure, the slip planes are {111} (where { } indicates the family of equivalent (111) planes) and the slip directions are <110> (where < > indicates the family of equivalent [110] directions). For the case of silicon with <100> orientation it is convenient to introduce a rectangular coordinate system as shown in Fig. 3a. Because four different orientations of (111) planes exist, 12 distinct slip directions should be considered. Because of symmetry in the crystal lattice, the stress in all directions can be deduced from the coordinate components of the force $\bar{F}$, which are exerted on the plane with the unit normal vector $\bar{n}$ due to plane stress components $\bar{\sigma}$ [8]:

$$\bar{F}_i = \sum_{j=x, y, z} \bar{\sigma}_{ij} \bar{n}_j \quad i \in (x, y, z) \tag{8.5}$$

where $\bar{n}_j$ is the normal vector to the plane where the component of force is solved. The shear stress resolved (projected) in a given plane in the slip direction defined by the unit vector $\bar{l} = [l_x, l_y, l_z]$ is given by

$$S_\alpha = \sum_{i=x, y, z} \bar{F}_i \bar{l}_{\alpha i} \tag{8.6}$$

where $\alpha = [BC_{111}, CE_{111}, BE_{111}, AE_{1\bar{1}1}, BE_{1\bar{1}1}]$.

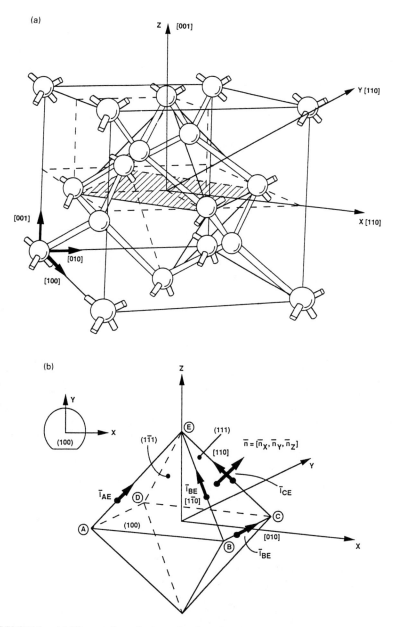

FIGURE 3. (a) Elemental octahedron showing slip planes in a silicon diamond lattice. (b) Elemental octahedron showing slip planes and slip direction in a silicon lattice. Using the unit normal vector $\bar{n}$ on the slip plane (111) and unit vector $\bar{l}$ in slip direction <110> the resolved shear stress in the xyz coordinate system can be calculated.

It is assumed that slip occurs instantaneously when shear stress exceeds the critical shear stress value (so-called yield stress) for dislocation generation at a given temperature. To compare actual stress with the critical shear stress, only the absolute value of $|S_\alpha|$ is required. It is also assumed that shear stress is independent of the tensile or compressive nature of the stress. From this assumption, for the plane (111), Eqs 8.5 and 8.6 yield:

$$|S_{BC}| = \sqrt{2/3}\,|\sigma_{xx}|$$
$$|S_{CE}| = \sqrt{1/6}\,|\sigma_{xx} + \sigma_{xy}|$$
$$|S_{BE}| = \sqrt{1/6}\,|\sigma_{xx} + \sigma_{xy}|$$

Because of lattice symmetry the remaining nine slip directions yield only two additional values of shear stress in plane $(1\bar{1}1)$:

$$|S_{AE}| = \sqrt{1/6}\,|\sigma_{yy} + \sigma_{xy}|$$
$$|S_{BE}| = \sqrt{1/6}\,|\sigma_{yy} + \sigma_{xy}|$$

From Fig. 3b we can see that stress $|S_{BC}|$ would produce glide along the edges of the base crystal orientation $<100>$ and that this kind of displacement would not be visible on the surface of the wafer. The stress components $|S_{CE}|$ and $|S_{BE}|$ would produce glide on the plane (111) and $(\bar{1}\bar{1}1)$ along the lateral edges giving the visible slip band along the y direction. The stress $|S_{AE}|$ and $|S_{BE}|$ produce glide on the plane $(\bar{1}11)$ and $(1\bar{1}1)$ with visible slip bands along the x direction.

For the case where the wafer is an isotropic and homogeneous elastic solid circular body having a one-dimensional temperature distribution (no circumferential variations), the components of thermal stress in polar coordinates, radial σ_r and tangential σ_t, can be calculated from analytical functions [17]. When more realistic assumptions are introduced only the numerical solutions are available. Transforming polar coordinates to cartesian coordinates, we obtain the stress components

$$\sigma_{xx} = \frac{\sigma_r + \sigma_t}{2} + \frac{\sigma_r - \sigma_t}{2}\cos(2\varphi)$$

$$\sigma_{yy} = \frac{\sigma_r + \sigma_t}{2} - \frac{\sigma_r - \sigma_t}{2}\cos(2\varphi) \qquad (8.7)$$

$$\sigma_{xy} = \frac{\sigma_r - \sigma_t}{2}\sin(2\varphi)$$

where φ is the angle in the xy plane (counted counterclockwise from the x [010] direction) axis. From Eq. (8.7) it is evident that the resolved shear stress distribution has a $\pi/2$ periodicity and is symmetrical around $\pi/4$.

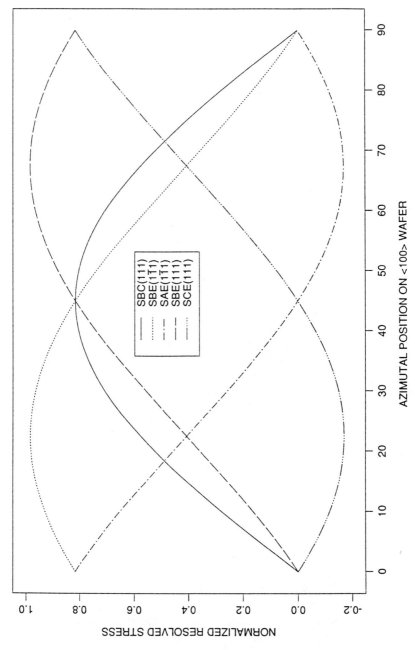

FIGURE 4. Azimutal distribution of the normalized stresses resolved on the five independent slip systems for the <100> oriented wafer.

Maximum stress occurs in the direction where stress components of all of the combined slip planes are additive. Thus there are eight primary equispaced topographic directions where stress in the wafer occurs and where the limit of elasticity is expected to be exceeded first ($S_{CE}(111)$, $S_{BE}(111)$, $S_{AE}(1\bar{1}1)$, $S_{BE}(1\bar{1}1)$. Figure 4 schematically illustrates azimuthal distributions of the normalized shear stress components. It is clear that shear stress $S_{BE}(111)$ and $S_{BE}(1\bar{1}1)$ intersect at $\varphi = 45°$ where the shear stress component $S_{BC}(111)$ also indicates the maximum value. Therefore, cross slip (simultaneous slip on two glide systems) can easily take place in this region.

From a practical point of view, it is very instructive to compare the magnitude of the resolved shear stress to the critical shear stress. When judging these results we must, however, keep in mind that the material behavior may be more complex than assumed and that the RTA system may generate thermal distributions that also contradict our assumptions. Nevertheless, the magnitude of critical shear stress can be used to predict the onset of slip conditions. Review of successful attempts to describe the plastic behavior of silicon can be found in Ref. 18. According to this work the measured critical shear stress is given as

$$\sigma_{\text{crit}}(T) = C\varepsilon^{1/n} \exp\left(\frac{U}{nkT}\right) \qquad (8.8)$$

where U is activation energy of glide movement.
ε is the strain rate.
n, C are material constants.

The result can be summarized as follows: conditions are favorable for slip to occur when the thermal stress becomes larger than the yield stress for a given temperature. The yield stress depends on (1) geometrical correspondence of the glide plane and distribution of the radial and the tangential components of thermal stress, (2) the temperature, (3) the initial density of dislocations, and (4) mechanical surface and edge damage.

There is controversy about which segment of processing induces slip. In most systems, slip is generated during the heating-up phase. It must be emphasized, however, that this is dependent on reflector and chamber geometry. A clear difference in slip formation between different systems has been confirmed experimentally. No unique processing temperature can be defined as a threshold temperature for slip formation, but slip has never been observed on 100 mm wafers for processing temperatures below 1000 °C.

Figure 5 presents the topography of a production wafer that was processed at 1050 °C. After sequential elimination of all possible sources of slip (epitaxial layer—no epi layer, trench—no trench, no epi—no trench,

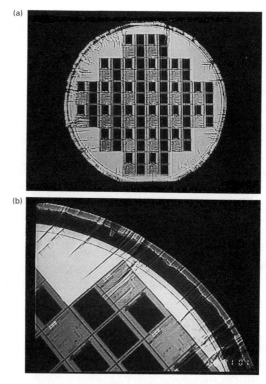

FIGURE 5. (a) Topography of a 100 mm product wafer after one RTA cycle at 1050 °C. Note that the position of the poles in the 22.5°, 45°, and 67.5° directions agree very well with the mapping of the temperature distribution shown in Fig. 1.2. (b) Simultaneous slip on two glide systems (conjugate slip).

etc.), transient nonuniformities in temperature distribution were identified as the unique cause of induced slip. For the identical processing of two batches where large-scale (>12 mm) "not printed" areas around wafer edges were prepared (see Fig. 5), no statistically meaningful difference in slip behavior was found in comparison with a batch of wafers with "print" close to wafer edge.

Slip remains as the major roadblock to industry acceptance of RTA. Currently only systems where the wafer is heated using a susceptor (usually silicon carbide) do not generate slip. However, such systems can not heat a wafer faster than 20 °C/sec, and convert a cold-wall system to a hot-wall system with all the disadvantages of hot-wall processing. Little data are available about higher warpage (i.e., deviations from planarity) generated in these systems due to possible poor thermal contact between the wafer and susceptor.

II. Bipolar Transistor Processing

We have already mentioned that RTA processing is equipment dependent. In the production environment, other factors alter rapid thermal processing: the presence of device patterns on the wafer, composition of layers deposited on the wafer, roughness of the wafer back side, doping levels, and so forth, can alter energy absorption and change reflectivity and emissivity of the wafer. For these reasons, RTA processing must be tailored to each specific application. Rather than discuss a specific process, we will show here the possible applications for RTA processing and the effect RTA processing will have on doping profiles and devices.

Generally, a process flow can be sectioned into three modules:

- substrate preparation and the isolation module.
- active device module.
- interconnect and metalization module.

In the substrate and isolation module, there are no serious limitations on thermal treatment. Usually all thermal steps can be performed by conventional processing. The application of RTA to the interconnect and metalization module is covered in Chapter 7 of this book. Thus in this chapter we will focus only on the active device module.

Scaling the bipolar transistor to submicron sizes requires a reduction in both lateral and vertical geometries. Several basic changes to the conventional transistor structure are necessary in order to overcome the disadvantages of traditional ion-implanted emitter and base junctions. The major problem with an implanted emitter is that the required implantation dose creates significant lattice disorder. Subsequent annealing treatment for lattice recrystalization will cause significant redistribution of impurities in the base. The secondary problem is that if we scale the emitter-base junction depth, the hole diffusion length in the heavily doped emitter (typically 0.35–0.45 μm) is comparable to the thickness of the emitter region. Thus the distribution of holes in the emitter region is linear. This type of scaling leads to an increase in base current and decrease in device gain, because the diffusion component of the base current is directly proportional to the gradient of the hole distribution.

A new device structure has been developed to solve this dilemma—the polysilicon contacted bipolar transistor. The arsenic-doped polysilicon emitter was introduced by Tagaki *et al.* [19] and Graul *et al.* [20] in 1972 and 1976, respectively. In 1978 Okada *et al.* [21] also utilized polysilicon as a base electrode.

Self-aligned processes with polysilicon emitters have become standard techniques for high-performance bipolar transistors. The key advantage for the self-aligned base–emitter structure is a reduction in the extrinsic base area leading to a decrease in collector–base capacitance and base resistance.

Depending on the number of polysilicon layers used to form the emitter and base regions, two distinct groups can be found:

- double polysilicon self-aligned structures.
- single polysilicon self-aligned structures.

We will discuss both groups in detail and we will emphasize the possible application for RTA processing.

A. Double Polysilicon Self-Aligned Structures

A typical generic double polysilicon transistor structure is illustrated in Fig. 6. This type of structure has been most successful and is widely used in most "bipolar only" processes for high-performance bipolar transistors. Although this type of transistor is named differently in the literature, the key structure and stack of layers are the same. The first version of this structure was reported in 1980 by the NTT group [22]. Since that time, several process improvements have been introduced that utilize improved processing equipment and techniques.

Figure 7 shows the major fabrication steps for a typical double polysilicon transistor structure (MOSAIC III [23]). After an isolation oxidation,

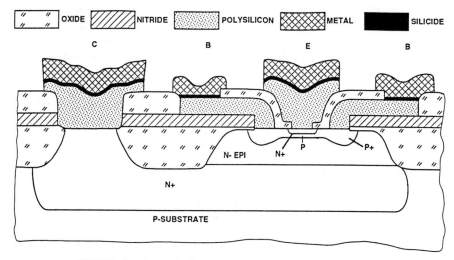

FIGURE 6. A generic double polysilicon self-aligned structure.

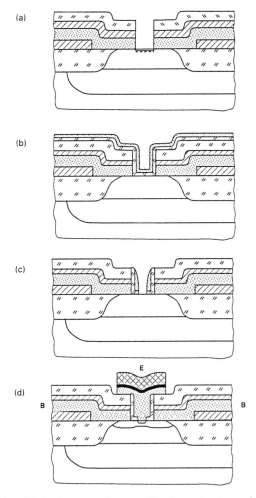

FIGURE 7. Major fabrication steps for a emitter/base structure of double polysilicon transistor (MOSAIC III).

nitride is deposited and regions where polysilicon will be in contact with monocrystalline silicon are patterned. The base polysilicon (this layer is commonly referred to as first poly) is deposited and capped by nitride and oxide layers. In the next step the emitter and collector areas are patterned. After oxide/nitride etch, the doped first polysilicon layer is removed by dry etch from emitter and collector regions. Subsequent annealing and oxidation remove RIE damage from the emitter monocrystalline silicon region and form a screen oxide for the intrinsic base implant (Fig. 7a). In the

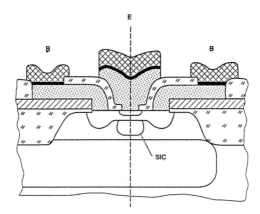

FIGURE 8. Self-aligned implanted collector (SIC).

following step, CVD oxide and polysilicon are deposited as a conformal layer in the emitter and collector openings (Fig. 7b). The spacer is formed by an anisotropic RIE etch of the polysilicon layer and the remaining oxide in the emitter window (Fig. 7c). The emitter polysilicon layer (second poly) is deposited and doped. The next masking step, after deposition and doping of the emitter polysilicon layer, defines the polysilicon caps that cover the emitter and collector areas. A final anneal redistributes the arsenic implanted into emitter polysilicon and finishes the intrinsic and extrinsic base anneal. In the final step, the metal system is prepared (Fig. 7d).

The recently introduced self-aligned implanted collector (SIC or pedestal collector) enables another degree of freedom in optimizing the transistor. A cross-sectional view of the base–emitter region, including SIC, is shown in Fig. 8. We will discuss SIC later in this chapter.

Significant reductions in emitter/base junctions depths have dramatically reduced the peripheral component of the base current. As a consequence, device performance is mainly governed by the intrinsic portion of the transistor part. However, critical trade-offs must be made on the degree of lateral diffusion redistribution between the intrinsic and extrinsic base regions, especially for devices in which the intrinsic base implant is performed after the spacer formation. The enormous requirements for controlling both lateral and vertical diffusion are beyond the capability of coventional processing. Therefore, well-controlled RTA is very promising for this part of transistor processing.

Careful optimization of the thermal treatment (dependent on equipment and type of structure) is necessary. Generally, the properties of concentration profile must comply with the following rules of thumb:

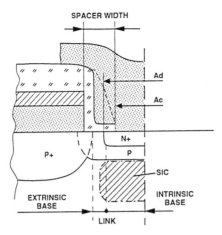

FIGURE 9. Schematic cross section of the emitter/base structure.

1. Concentration in the Emitter Region

The main advantage of the polysilicon emitter, unlike the conventional emitter, is that base current is only weakly dependent on x_{jE}. For the polysilicon emitter, base current is dominated by surface recombination at the polysilicon–silicon interface. In practical application, the monocrystalline portion of the polysilicon emitter can be very shallow, typically 0.04–0.08 μm with an active concentration at the interface of $3\text{--}8 \times 10^{19}$ cm^{-3}. For decreasing emitter depth (below approximately 30 nm), base current at low biases becomes nonideal because of recombination in the space charge region (typically the emission coefficient in SPICE nomenclature rises from 1 up to 1.5–1.8). Figure 9 shows a generic configuration for the sidewall base–emitter region. From Fig. 9 it is clear that there is another benefit from keeping the emitter junction as small as possible. The emitter region is characterized by two different areas: the emitter contact area A_c, which is defined by the physical contact between emitter polysilicon and silicon, and the diffusion emitter area A_d, which is always larger than A_c because of lateral diffusion. The area A_d controls collector current; base current is controlled by the surface recombination in the area A_c. Therefore the current gain is dependent on the ratio A_d/A_c. The ratio A_d/A_c approaches unity with vertical profile scaling.

The magnitude of the emitter resistance is the crucial limitation of polysilicon emitters. A thick interfacial oxide leads to high gain but large emitter resistance. On the other hand no interfacial oxide (polysilicon emitter with in situ clean) results in ideal base current characteristics but a significant reduction in gain. For the devices with an intentionally prepared thick interfacial oxide at higher forward biases (0.8–1.2 V), a kink and plateau in base current are observed. For devices with a broken-up oxide, the kink is less pronounced.

In Ref. 24 it is reported that speed and circuit performance degrade if the product of emitter resistance and collector current becomes comparable to a thermal voltage kT/q. To minimize performance degradation, the emitter resistance should be lower than kT/J, where J is the emitter current density. A high-speed bipolar transistor has a maximum f_T at current densities $\sim 5 \times 10^4$ A/cm^2. Therefore the specific emitter resistance must be lower than $60\,\Omega \cdot \mu\text{m}^2$. This emitter resistance puts a limit on scaling because it implies a minimum emitter area of about $1\,\mu\text{m}^2$.

Very tight control of heat treatment must be employed when the emitter is processed. Application of RTA for this part of thermal processing is a common practice now. Use of well-controlled RTA processing can solve conflicting requirements between current gain and emitter series resistance.

2. Concentration of the Intrinsic Base Region

The concentration of the intrinsic base region is determined by two requirements:

- to provide the necessary current gain.
- to prevent emitter–collector punchthrough.

Since the polysilicon emitter results in a higher current gain, this increase in gain can be traded off with a higher base concentration that will reduce the intrinsic base resistance. Therefore, a thinner base width and a higher cutoff frequency, f_T, can be realized without increasing the intrinsic base resistance.

The minimum base width is determined by the condition that prevents emitter–collector punchthrough. To avoid punchthrough at collector–base voltages of $V_{BC} = 3$ V, the base thickness should be at least 100 nm for a peak base concentration of 1×10^{18} cm^{-3} and 50 nm for a peak base concentration of 5×10^{18} cm^{-3}, respectively. The maximum base concentration is set by forward tunneling current, which leads to strongly nonideal base current characteristics. This occurs for base concentrations larger than 8×10^{18} cm^{-3}. To avoid a fall-off in current gain and f_T with high collector current density, the collector concentration should be increased proportionally with the base concentration. As collector concentration increases, the emitter–collector breakdown voltage (BV$_{CE0}$) is reduced and also the base–collector capacitance will increase. A compromise between speed and BV$_{CE0}$ is necessary, and a typical value for intrinsic base pinch resistance is 10 kΩ.

If the emitter–base junction lies deeper in the silicon than the peak concentration for the base boron concentration profile, the base Gummel number will be very sensitive to the emitter depth. A small variation in emitter junction depth will lead to a larger variation in base Gummel number. A large process sensitivity can be minimized only with very tight temperature control.

3. Concentration of the Extrinsic Base Region

The concentration profile of the external base is much more complicated than the profile of the intrinsic base region. From Fig. 9, it is clear that the base link region is controlled by lateral diffusion from the extrinsic and intrinsic bases. A large lateral diffusion from the extrinsic base would cause the extrinsic base to overlap the intrinsic base and emitter creating an unwanted p^+n^+ junction around the periphery of the emitter. This will result in a lowering of the emitter–base breakdown voltage (BV_{EB0}), sidewall injection leading to reduced current gain, and also to a higher electrical field at the emitter–base junction. A higher electric field will increase the peripheral tunneling current and increase the probability of field-induced breakdown. On the other hand, a small lateral diffusion would reduce the parasitic collector–base capacitance; but it would lead to a higher base resistance and increased gate delay. In addition small lateral diffusion would increase the undesirable component of the peripheral base current, eventually leading to punchthrough between the emitter and collector. For a given thickness of the oxide spacer the lateral diffusion is again defined only by thermal processing.

4. Concentration of the Epitaxial Layer and Self-Aligned Implanted Collector

Because the forward transit delay increases if the collector current reaches a critical current density [25], which is proportional to the product of collector doping concentration and carrier velocity, any reduction in collector concentration is undesirable. An increase in collector concentration (N_C) can suppress the high current effect, but collector–base and collector–emitter capacitances will increase in proportion to $\sqrt{N_C}$. The maximum collector concentration is determined in part by this limitation, and typical concentration is in the range of $1\text{--}2 \times 10^{16}$ cm^{-3}. Because most of the collector–base junction is located in the extrinsic area, which conducts only a small part of the collector current, the SIC implant can be used to increase intrinsic collector concentration (and therefore collector current density)

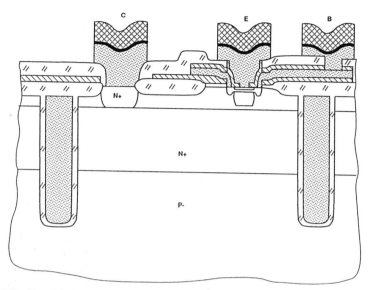

FIGURE 10. Advanced double polysilicon self-aligned bipolar transistor structure (MOSAIC V).

without an excessive increase in the extrinsic components of collector–base capacitance. Another benefit of the SIC implant is a base width reduction that occurs by counter-doping the tail of the base implant.

Devices based on the emitter–base structure described above have a common disadvantage in that the extrinsic collector–base areas are not effectively scaled either laterally or vertically. To overcome this problem, new double polysilicon structures have been proposed. An example of this type of advanced structure is the MOSAIC V [26] transistor shown in Fig. 10. The area of the extrinsic base is limited by the width of the polysilicon plug (100 mm) connecting the base polysilicon with the diffused region in the monocrystalline silicon. An additional advantage for this structure is that RIE damage is not introduced into the silicon when the emitter window is opened, because the emitter window can be opened by wet etch. All scaling rules described above are valid for this type of structure, and again, RTA is an important part of the processing.

B. Single Polysilicon Self-Aligned Structures

The common disadvantage of double polysilicon self-aligned structures is a very severe emitter topography with a very high aspect ratio (height to opening). This situation makes it difficult to control emitter polysilicon layer thickness, which is important for the subsequent diffusion into the monocrystalline silicon. This is especially true if more than one emitter width is used. Devices for output drivers frequently use a wider emitter. Polysilicon will fill the emitter openings of narrow devices before it fills the emitter openings of wider devices. Thus, the thickness of polysilicon could be much greater in narrow emitters than in wide emitters. After implantation, diffusion will result in different emitter depths for different emitter windows, but the base profile will be the same. Another drawback is that double polysilicon structures are not well suited for integration with advanced CMOS devices.

Advanced single polysilicon self-aligned processes originate from the following scenario: After isolation, which defines the active region, a pad oxide is grown and a nitride layer is deposited. After RIE patterning of

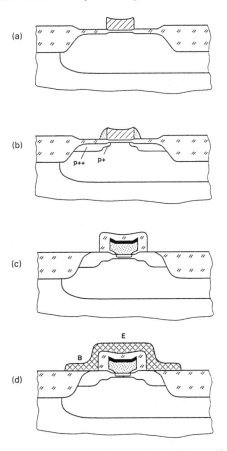

FIGURE 11. Fabrication steps of a typical single polysilicon self-aligned transistor.

emitter and collector regions on the nitride, a thicker oxide is grown over the extrinsic base region using the nitride as an oxidation mask (Fig. 11a). A shallow P^+ implant that serves as a base link is performed in the next step. A polysilicon spacer is then formed to serve as a mask for the extrinsic and intrinsic base. An RTA anneal is used to anneal both implants (Fig. 11b). Achieving a shallow base without perimeter punchthrough is a particularly important function of this anneal. After nitride/oxide etch, the intrinsic base region and collector window is opened. The intrinsic base implant is followed by emitter polysilicon deposition, doping of emitter polysilicon, and diffusion of impurities from polysilicon into monocrystalline silicon in order to form the monocrystalline emitter region. The silicide layer and capping CVD oxide is deposited and patterned as shown in Fig. 11c. The extrinsic base contact is opened at the same time the capping oxide is etched. Then the metal wraps the emitter stack (Fig. 11d).

The process requires different masks for the definition of the intrinsic base region and emitter polysilicon contact; therefore, the emitter width is lithography limited. Furthermore, because the emitter current is conducted along the length of the emitter, a high emitter resistance that increases with the emitter length can limit this type of transistor. Because of these disadvantages, general development efforts are focused on improving the double polysilicon self-aligned structures.

C. Arsenic-Doped Polysilicon Emitter

Emitter–base junctions formed by diffusion from arsenic-doped polysilicon led to important improvements in bipolar technology. A number of theoretical works have been published on this subject (for summary see Ref. 27). However, because of the poor characterization of key parameters, namely the polysilicon/silicon interface and the polysilicon grain boundary, no final conclusion can be drawn. On the other hand, experimental data show the unambiguous result that base current and emitter resistance of transistors with polysilicon emitter contacts depend strongly on the morphology of the interfacial layer present between the silicon and the emitter polysilicon. This fact is responsible for the major drawback of the polysilicon emitter contact: process sensitivity. Within a batch of wafers processed under the same process conditions, wide fluctuations in base current and emitter resistance may occur.

A larger current gain occurs if an oxide film is present at the polysilicon/silicon interface. This can be explained in terms of tunneling through the oxide film that forms a barrier to both electrons and holes. Devices intentionally prepared with a continuous oxide layer (10–20 Å) between polysilicon and monocrystalline silicon show very high current gain; however, they also have high emitter resistance. This type of polysilicon emitter is reasonably well understood and quite good agreement between theoretical models and experimental data can be obtained.

For devices where a dip etch or other methods are used to remove any interfacial oxide, the gain enhancement is attributed to the transport properties of the bulk polysilicon or the segregation of impurities at the interface. A lower emitter resistance for this type of processing makes this device more acceptable for submicron scaling. Unfortunately, the dependence of the emitter resistance and current gain on the structure of the interface is not fully understood.

If the interfacial oxide is broken up over the whole area of the emitter contact, the processing sensitivity can be alleviated. Experimental data show that an emitter anneal at temperatures of 900 °C and higher will break up the thin (< 6–8 Å) interfacial oxide layer. During thermal processing the process of interfacial oxide breakup can be divided into three stages:

1. The initial uniform oxide thickness degrades, with some parts of the layer becoming thicker and others thinner.
2. "Holes" open in the oxide layer.
3. "Holes" widen in the oxide layer accompanied by the epitaxial alignment of the polysilicon layer and an increase in the oxide thickness in the remaining areas.

Wolstenholme [28] shows the influence of conventional furnace emitter annealing on the polysilicon/silicon interface. Although his data may change for different polysilicon structures, a unique conclusion can be made: thermal treatment at temperatures above 900 °C leads to the breakup of thin interfacial oxides (< 8 Å). The rate of the oxide breakup depends upon the initial oxide thickness, the polysilicon structure, and the arsenic concentration in the polysilicon. This suggests that different thermal processing of the polysilicon interface may be used to optimize current gain and emitter resistance. If higher gain is required, a thick interfacial oxide can be used along with an emitter anneal performed at lower temperatures. Alternatively, for low emitter resistance, the dip etch before polysilicon deposition along with a high-temperature anneal could be performed. The only problem with this approach is that the intrinsic base implant and extrinsic base region are already in place. Because base transit time τ_B is proportional to the square of base width, and because parasitic components of base–collector capacitance C_{BC} (for extrinsic base) will increase with any thermal processing, device speed optimization does not allow excessive thermal processing.

One alternative strategy is based on the following assumption: Because of the low diffusivity of oxygen, once the interfacial oxide is systematically broken up over the whole emitter area it will not change with subsequent thermal processing. Therefore a suitable fabrication sequence is to perform the base implant anneal after dip etch and emitter polysilicon deposition in order to (1) activate the boron implant and (2) break up residual interfacial oxide. In the next step, arsenic is implanted and diffusion from the emitter polysilicon is performed. All thermal processing can be accomplished by a short high-temperature (~ 1050 °C) RTA process. An additional free benefit is that the emitter polysilicon is partially recrystallized during the first thermal cycle; thus arsenic diffusion is slowed down leading to shallow emitter junction x_{jE}. In this way, the interfacial oxide layer could be subjected to temperature treatment without affecting the emitter junction depth and the base current, which is essentially determined by the emitter anneal.

Another method for achieving a scaled intrinsic base and emitter region is based on subsequent diffusion of boron and arsenic from the emitter polysilicon (so-called double-diffused poly). The advantage is that the reference plane for both base and emitter diffusion is exactly the same. Because of the very low diffusivity of boron in the presence of high arsenic concentrations it is not possible to process this with one heat cycle; two-step diffusion must be used consisting of boron diffusion from boron-doped polysilicon followed by arsenic implantation and diffusion. Therefore, very tight control of the processing employed is required. More specifically, we have already mentioned that the processing temperature must be above 1000 °C to break up the interfacial oxide. Because of the diffusivity of arsenic and boron in silicon at this temperature, a 5 °C deviation in processing temperature will correspond to a 5% deviation in emitter junction depth and a 20% deviation in base junction depth. In the case where more than one emitter size is used, the processing "window" is even tighter. The disadvantage of double-diffused structures is the need for a separate base link in order to keep base resistance low. A more important drawback is that the current gain and f_T decrease because of the high compensating concentration created in the emitter region.

This summary of the application of RTA processing for homojunction bipolar transistors is necessarily incomplete. In practice, the fabrication steps used to produce the emitter and base regions are considerably more complicated than suggested here, and scaling requirements for these structures force precise processing control where reproducibility is of paramount importance. To this point, RTA processing could be a superior alternative to furnace processing. However, the full potential of RTA has yet to be explored because RTA equipment is in its infancy.

III. MOS Transistor Processing

RTA is not applied to MOS structures as frequently as it is used in bipolar technology. Despite a long development time, subhalf-micron scaling of MOS devices, which has already been achieved in bipolar self-aligned devices (now in production), has not yet been achieved for the CMOS counterpart, where submicron processes are moving into production only now.

Until recently, requirements toward shorter thermal processing have not been an absolute necessity in CMOS processing. Therefore, full process flows could be realized with conventional processing.

Concern about the effect of rapid thermal processing on gate oxide integrity is one of the key issues that contributes to low acceptance of RTA in MOS fabrication. Unfortunately, this aspect of rapid thermal processing is still in the exploratory stage of development, and processing interactions are not fully known or understood. For example, in Ref. 29 it is reported that even a thick gate oxide fails during RTA processing. In Ref. 30 it is concluded that under accelerated stress conditions CMOS devices fabricated with RTA show a greater degradation of threshold voltage than devices fabricated with conventional furnace anneals. In all fairness, we must point out that all of these results were achieved on obsolete RTA equipment, which definitely contributed to the failures. More work using state-of-the-art RTA tools must be done to clarify or confirm any previous conclusions.

Perhaps for these reasons, rapid thermal processing in CMOS fabrication is primarily used for silicide formation and less often used for reflow of interlevel glass dielectrics (BPSG). These two applications account for 80% of current industry installations [31]. This may seem surprising because annealing of implanted layers is traditionally the driving force behind RTA, and CMOS fabrication is based almost exclusively on doping by ion implantation.

The success of $TiSi_2$ formation in RTA systems needs to be interpreted correctly. Outstanding results are not due to the RTA but to another reason. The reaction to form $TiSi_2$ from a layer of titanium over silicon must be performed in an oxygen-free ambient (below 10 ppm), because oxidation of titanium is a much faster reaction than silicide formation. It is very difficult to process this step in a big-volume furnace tube without introducing oxygen. The temperature sensitivity and sensitivity to the heating rate are rather low for this operation. Therefore, this processing works very well even in RTA systems that may be less adequate for annealing applications.

The equipment aspects that have been discussed need to be solved before the application of RTA is extended further.

Analogously to its bipolar counterpart, in the next section we will briefly discuss generic CMOS device fabrication with a emphasis on thermal processing and the potential application of RTA processing.

A. CONVENTIONAL MOS STRUCTURE

Fabrication and processing issues of conventional CMOS devices are discussed in numerous works (see for example Ref. 32 and references therein).

For a given gate oxide thickness, the threshold voltage is determined by the surface concentration of the shallow implants in the channel region. The channel implants are designed to achieve a threshold voltage low enough to provide high current drive, but they must be high enough to avoid surface drain source current (subthreshold current) at a gate voltage lower than the threshold voltage. The subthreshold channel current is proportional to the concentration of minority carriers in the channel region, which depends exponentially on gate voltage. If the electric field of the drain penetrates into the source region, an undesirable current (punchthrough current) can flow in the deep channel region. To prevent punchthrough current in short-channel NMOS transistors, an additional boron implant (punchthrough implant) is designed to reduce the lateral extension of the drain depletion region (drain-induced barrier lowering). Punchthrough current in PMOS devices is much more pronounced because of the buried channel that typically forms (p-type region near the n well surface). The n-type implant located below the channel pn junction is employed to reduce the drain potential in the channel region. In both types of devices, the source/drain junction depth should be as shallow as possible to minimize short channel device degradation.

In order to reduce source/drain junction charging delay, it is necessary to keep the source/drain-to-well capacitance at a minimum. Any lateral distribution of the punchthrough implant will increase this capacitance. Therefore the thermal treatment for both channel implants should be kept to a minimum. A common practice to eliminate the relatively long thermal treatment needed for gate oxide growth (and redistribution of impurities into the gate oxide) is to perform both the channel and punchthrough implants through the gate oxide. To avoid gate oxide contamination by the photoresist or implant processing, a thin layer of gate polysilicon (< 500 Å) is deposited immediately after gate oxide growth. After channel implantations the remaining gate polysilicon is deposited and doped. All subsequent processing must maintain low-temperature thermal processing.

As a result of the short-channel length, carriers are accelerated by the electric field at the drain–channel boundary region and generate free hot electron and hole pairs through the impact ionization in the silicon lattice. The density of hot carriers is determined by several factors—current density and impact ionization probability. Impact ionization probability is proportional to $\exp(-1/E)$, where E is the intensity of the electric field. The impact ionization efficiency is three to four times higher for electrons than it is for holes. Therefore, hot carrier generation is much more important for NMOS transistors. The device degradation as a result of the hot carriers is due to (1) injection of the hot carriers into the gate oxide where they cause instabilities such as a shift in threshold voltage and drain current reduction and (2) accumulation of hot holes in the substrate leading to increased substrate current, which may act as a substrate bias generator or induce snap-back and latch-up in CMOS structures.

Problems associated with hot carriers have been recognized as one of the major device limitations. Several structural modifications, such as double-diffused drain or lightly doped drains (LDD), have been proposed to improve hot carrier reliability and short-channel effects. The common goal of these modifications is to spread the voltage drop over a long distance, reducing the peak electric field. Some of the buried LDD structures emphasize separating the maximum current path from the maximum electric field. LDD structures are the most frequently used, forming a lateral grading of the doping level between the drain contact and the channel, using a self-aligned sidewall spacer. However, a disadvantage of using LDD structures is an increased source/drain parasitic resistance which degrades device performance (g_m).

The threshold voltage (V_T) and its sensitivity to effective channel length (dV_T/dL_{eff}) strongly depends on the L_{eff} and the channel doping level. Because the separation between the lightly doped regions under the gate defines L_{eff}, maximum control of L_{eff} is achieved when the lightly doped regions are self-aligned to the gate. There are two possibilities: nonremovable and removable spacer LDD processes. In a nonremovable spacer process after gate definition, the n^- and p^- regions are implanted before spacer formation. An n^+ implant follows the spacer formation. A thermal anneal is performed to activate the n^+ source/drain region before the p^+ implant and anneal in order to prevent additional dopant redistribution in the p^+/p^- regions. However, during the p^+ implant anneal, p^- regions can diffuse undesirably far under the gate. Because of the need for accurately defined thermal cycles it is expected that rapid thermal processing will be used extensively for this part of processing.

For a thin gate oxide a disadvantage of a nonremovable spacer is possible overetching of the source/drain region during the spacer formation. Any overetch would result in radiation damage to the crystal lattice. A second option is to use an LDD process with a removable spacer. In this option

n^+/p^+ source/drain regions are implanted first. In this case the thermal treatment of n^+ regions, which is performed before the p^+ implant, is independent of processing of the p^+ and p^- regions.

It is clear from this discussion that conventional thermal processing of the LDD structure, especially the p-type source/drain, is a very serious limitation to scaling of subhalf-micron devices.

B. MOS Transistor with Elevated Electrodes

The combination of RTA and RTP enables an alternative concept of manufacturing of MOS devices.

It is well known that the source/drain junction depth plays an important role in scaling of MOS devices because of its influence on the channel potential. For example, an increase of x_j from 0.1 to 0.2 μm causes an increase in the subthreshold current of almost two orders of magnitude and a decrease in threshold voltage. The LDD approach is not a fully satisfying response to this problem because the highly doped source/drain regions are quite deep. This leads to an alternative approach to improving the short-channel effects and performance of submicron CMOS devices—elevated electrodes. The source/drain structure could be self-aligned with elevated contacts over the field oxide, as in bipolar devices.

The approaches proposed in Refs. 33-35 employ a selective epitaxial deposition after formation of the gate sidewall spacer. Practical realization faces several problems. Faceting (i.e., thinning of the epitaxial layer at the edges [36]) during the selective silicon deposition can occur near the sidewall spacer oxide and near the field oxide edge. The angle of faceting is about 72° for a <100> substrate orientation. Implantation through the thinner faceted regions near the sidewall spacer at the gate edge will result in a deeper source/drain junction depth than desired. It has also been found that selective epitaxial deposition prefers large islands of exposed silicon to small islands, resulting in different thicknesses of deposited silicon.

Elevated source/drain contacts based on selective epitaxial layers employ conventional implantation techniques for the p^-/n^- LDD regions, and thus do not allow any possibility for reducing junction depth (especially for the p^- region).

Numerical simulations [37] have shown that no significant benefits are obtained from LDD structures for devices with channel lengths below 0.3 μm. The result of a study [38] shows that the conventional drain is superior to an LDD under lower operating voltage conditions. On the basis of these facts and on analysis of the disadvantages of structures described above, a transistor structure has been proposed [39] that uses polysilicon

Unique Silicon Devices Using Rapid Thermal Annealing

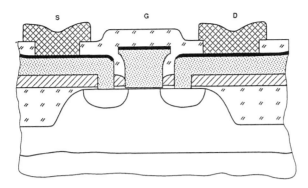

FIGURE 12. MOS transistor with elevated electrodes (FAST I).

as a diffusion source for scaled source/drain regions and also uses this polysilicon to serve as a connection to the source/drain contact. A cross section of this structure is shown in Fig. 12.

Further elaboration of the structure [40], which is fully compatible with a flow for a double polysilicon self-aligned bipolar transistor, is shown in Fig. 13. In this flow the RTP reoxidized nitrided oxide is used as the gate dielectric. All the channel and punchthrough implants are performed before the rapid thermal oxidation and nitridation.

In most cases, it has been demonstrated that the short-channel effects were found to be significantly smaller than for conventional transistors, but the parasitic capacitances C_{gs} and C_{gd} are slightly larger. Although successful fabrication of many new transistor structures with elevated electrodes has been demonstrated, it is recognized that further experimentation remains to be performed.

Processing used during fabrication of MOS devices with elevated electrodes is on the edge of current possibilities for not only RTA but for all

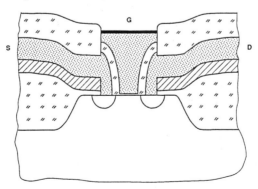

FIGURE 13. MOS transistor with elevated electrodes and RTA ONO gate dielectric.

of the currently used processing equipment. Even so, because the physical limits of the discussed devices have not yet been reached, the complexity and enormous process sensitivities are major obstacles in faster progress in this direction. However, preliminary results indicate that rapid thermal processing has a paramount importance for this type of fabrication.

C. Ultrathin RTP Gate Dielectrics

Thermally grown thin oxide films are now used solely as a gate dielectric in MOS devices. As the MOS devices are scaled down, the growth of high-quality oxide layers is rather difficult because of defect density and integrity problems. The properties of thin oxides are affected by the following phenomena:

1. Quality of the silicon surface determines the structural homogeneity and breakdown uniformity of thermal oxide.
2. Redistribution of impurities and penetration of gate material into the oxide degrade structural homogeneity, especially for silicided gates.
3. Ion implantation and reactive ion etching lead to charge accumulation in oxide film.

The structural homogeneity of thin oxide layers is strongly influenced by the cleaning procedure. For ultrathin (<75 Å) high-quality gate oxide, the thickness of the native oxide (typically ~ 10 Å measured by elipsometry) is not a negligible source of variation in the final oxide thickness. A good chemical clean with a time limit between the clean and the gate oxidation is required. After loading wafers into an RTA chamber and purging with a high flow of purging gas in order to minimize uncertainties in the oxidation time, the wafer is heated up to 800 °C in a nonoxidizing ambient. After switching to the oxidation gas flow, the wafer should be quickly heated to oxidation temperature. With this approach the uniform gate oxide can be prepared if the RTP system does not suffer from the temperature uniformity drawback.

It is interesting to point out that, like rapid thermal annealing, rapid thermal oxidation is equipment dependent. For arc lamp systems an activation energy for initial oxide growth rate of 1.31 eV has been found [41]; the same parameter for the tungsten halogen lamp systems is 1.44 eV [42]. The explanation can be found in the analysis of physical parameters determining the heating dynamics. Because of the spectral distribution of intensity and absorption, for equal power levels, an arc lamp will create more electron–hole pairs per unit area of silicon surface than the tungsten halogen lamp. Several proposed models of oxidation [43] have emphasized the importance of:

1. the number of silicon bonds available for reaction with oxygen, and
2. the number of electrons in the conduction band of silicon, which may be further excited into the conduction band.

Both of these parameters are altered by changes in the concentration of electrons in the conduction band (which is proportional of the number of electron–hole pairs generated).

There are, however, several additional technological and reliability problems with oxides in the very thin regime. Among the major limitations of reduced oxide thickness are (1) its tendency to react with the gate electrode material, and (2) the fact that very thin oxide is not an efficient barrier against impurity diffusion into the substrate from the gate electrode (especially if a silicided gate is used).

Thermal nitridation of oxides, first introduced by Ito *et al.* [44], has been studied as an alternative gate dielectric. The nitridation of thermally grown oxide films and reoxidized nitrided oxide (ONO) are currently of high technological interest as an alternative to thermal oxide because of their unique low thermal budget properties and processing flexibility. It has been reported that nitridation of thermally grown oxide increases the dielectric strength, provides a better barrier against dopant diffusion, and is less sensitive to ionizing radiation.

Previously, nitridation was performed using conventional furnace processing at high temperatures (typically 1100–1200 °C) in pure ammonia or using a plasma enhanced process. Both techniques suffer from the inability to control the spatial nitrogen distribution in the oxide and a need for long processing time at elevated temperatures. The disadvantage of nitridation in a conventional furnace is segregation of nitrogen at the oxide–silicon interface, which may create electronic traps. Rapid thermal processing of nitrided thermal oxide allows better control of nitrogen incorporation and a very short nitridation time. The evidence of an increased trap density has been frequently noted as a drawback of thermal nitridation. However, there is a very strong indication that the higher concentration of traps is a result of furnace processing where only heavily nitrided films (nitrided on both surface and interface) and oxynitrides can be prepared. For example, Terry [45] noted that furnace contamination is a factor that causes properties of nitridated oxides to deteriorate. On the other hand, recent studies [46–48] using rapid thermal processing report superior characteristics with comparable charge trapping in nitrided oxides and pure oxides.

The processing issues are clearly favorable for rapid thermal processing. However, again the equipment issues need to be solved first. Problems with stress and slip have been already discussed. During oxidation and nitridation, the wafer emissivity can be affected by film growth on the back side of the wafer. Finally, one must realize that rapid thermal oxidation and nitridation are not isothermal processes in systems that use feedback to control, monitor, and maintain a constant temperature. The normal incidence reflectivity of silicon covered by a thin optically transparent film of thickness x is defined by the standard thin film interference formula [49];

$$R_\lambda(x) = \frac{n_f^2(1 - n_s)^2 \cos^2\varphi + (n_s - n_f^2)^2 \sin^2\varphi}{n_f^2(1 - n_s)^2 \cos^2\varphi + (n_s + n_f)^2 \sin^2\varphi} \tag{8.9}$$

where n_f and n_s are the film and substrate refractive indices, respectively, and $\varphi = 2\pi n_f x/\lambda$. Using the known optical properties of silicon and oxide it has been found that R_λ changes from 0.38 for native oxide to 0.1 for 1000 Å of grown oxide. In Section IA it was shown that absorbed power is proportional to the $(1 - R_\lambda)$ term. As a consequence of the decrease of reflectivity on the front of the wafer during film growth, the power (or photon rate) has to be gradually reduced (at least during initial phase of growth) in order to keep the processing temperature constant. Therefore any type of temperature measurement based on emissivity can introduce some error. The processing recipe is developed on an experimental basis; consequently the "reasonable" emissivity variation may not be a serious problem unless the optical properties of the chamber window are also changed by deposition of material on the chamber wall. In this situation, the batch-to-batch as well as the wafer-to-wafer reproducibility can be limiting factors.

The reoxidized nitrided oxide prepared by rapid thermal processing is the most promising candidate as a replacement of thermally grown gate oxide. This application can not be done now any way other than with RTP. It is expected that this application could accelerate faster progress in equipment improvement and solve problems that are, in fact, the consequences of the transient temperature effects and poor processing reproducibility.

The most apparent opportunity that arises from the application of rapid thermal processing on fabrication of scaled MOS devices is greater processing control and flexibility. Rapid thermal processing has demonstrated its significant capability in an experimental environment. However, the production environment dictates additional requirements that unfortunately are not yet solved.

IV. Conclusion

Rapid thermal processing can replace, in principle, most of the annealing and diffusion steps used in VLSI device fabrication, except for the long processing steps in the front end (well anneal, field oxidation, etc.). Actually, the use of RTA can be superior because of

1. shorter processing times and
2. higher solubility and electrical activation of dopants.

To utilize the possible advantage of rapid thermal processing, the RTA techniques now being developed in the experimental environment need to be transformed into manufacturing processing.

It was shown that in contrast to conventionally heated processing, in which thermal equilibrium exists between the processing chamber and the wafers, the wafer is not in thermal equilibrium with the cold walls of the processing chamber during the RTA cycle. In addition, the presence of a pattern layer can locally change physical parameters such as the absorption, reflectivity, and emissivity, resulting in lateral temperature gradients that could contribute to final temperature nonuniformity.

When rapid thermal processing is used several times during the fabrication flow, requirements on temperature uniformity across a wafer are beyond the capability of the current generation of RTA systems. Recent tests [31] reveal that 100 mm product wafers after two to three thermal cycles above 1000 °C show slip in 18 out of 19 different systems tested.

We have discussed here only basic aspects connected with doping profile engineering. The electrical performance of devices is defined not only by the doping profile and structural construction of the device but also by its material properties. For example, the defects in the crystal lattice, and more precisely, the energy band structure of the intrinsic device, need to be optimized. Very little is known about recombination centers that are observed in rapid thermal processed implanted silicon. The majority of work dealing with annealing of defects introduced by radiation damage and annealed in RTA are based on experiments with blanket wafers. In this case, the conditions can be chosen such that most defects are annealed out. However, when a patterned wafer is used and implantation of dopants is restricted to local areas, the pattern edges are a source of mechanical stress that has an impact on dislocation glide motion. Defects such as interstitials or interstitial loops can move away from the implanted areas into regions where they stop when the driving force for moving (which exists in implanted areas) disappears. Because the stress load is significantly higher in RTA processing, one must consider this issue when judging rapid thermal processing.

More research is needed, both in terms of reliable and reproducible processing as well as theoretical studies of fundamental physical processes. Regardless of many temporary problems it is highly likely that most (if not all) high-temperature processing steps during the active device fabrication will eventually be performed in radiatively heated rapid thermal systems.

References

1. I. A. Boyd and F. Micheli, "Confirmation of the wavelength dependence of silicon oxidation induced by visible radiation," in D. J. Ehrlich and V. T. Nguyen (eds.), *Emerging Technologies for in Situ Processing*, Martinus Nijhoff, (1988) p. 171-178.
2. R. Singh, S. Sinha, R. P. S. Thakur and P. Chou, "Some photoeffect roles in rapid isothermal processing," *Appl. Phys. Lett.* **18** (1991), p. 1217-1219.
3. B. Lojek, "The behavior of free carriers during rapid thermal annealing of doped silicon," MRS spring meeting, Anaheim 1991, MRS Symposium Proceedings Vol. 224.
4. N. A. Vitovskij, M. I. Klinger, T. V. Mashovets, D. Mustafakulov and S. M. Ryvkin, "Possibility of the ionization mechanism of defect formation in extrinsic semiconductors under subthreshold irradiation conditions," (in Russian); Physica i Technica Polupravodnikov, **13** (1979), c.925-930.
5. V. V. Emtsev, M. I. Klinger, T. V. Mashovets, E. K. Nazaryan, S. M. Ryvkin, "Impurity Ionization Mechanism of Defect Formation in Above-Threshold Irradiation of Germanium" (in Russian), Physica i Technica Polupravodnikov, **13** c.933-925 (1979).
6. B. Lojek, "The impact of the Wafer Backside on RTA Processing," 2nd International Symposium on Reduced-Thermal-Budget Processing for Fabrication of Microelectronic Devices, St. Louis, MD (May 1992).
7. E. R. G. Eckert and R. M. Drake, Jr., *Analysis of Heat and Mass Transfer*, McGraw-Hill, NY (1956).
8. H. A. Lord, "Thermal and Stress Analysis of Semiconductor Wafers in a Rapid Thermal Processing," *IEEE Trans. Semicon. Manuf.* **1**, 105-114 (1988).
9. T. J. Shieh and R. L. Carter, "RAPS—A Rapid Thermal Processor Simulation Program," *IEEE Trans. Electron Dev.* **ED-36**, 19-24 (1989).
10. T. O. Sedgwick, "Short Time Annealing," *J. Electrochem. Soc.* **130**, 484-493 (1983).
11. R. Kakoschke, "Temperature Problems with Rapid Thermal Processing for VLSI Applications" *Nucl. Instr. Meth. Phys. Res.* B **37/38**, 753-759 (1989).
12. Y. Ohno, A. Ohsaki, T. Kaneoka, J. Mitsubashi, M. Hirayama and T. Kato, "Effect of Mechanical Stress for Thin SiO_2 Film in TDDB CCST Characteristics," in *Proc. Int. Reliability Phys. Symp.* (1989), pp. 34-38.
13. D. A. Baglee, "Characteristics and Reliability of 100 Å Oxides," *Proc. Int. Reliability Phys. Symp.* (1984), pp. 152-155.
14. J. Mitsuhashi, K. Sugimoto, M. Hirayama, S. Sadahiro and T. Matsukawa, *17th Conf. Solid State Dev. and Mater.* (1985), pp. 267-270.
15. H. Booyens, G. R. Proto and J. H. Basson, "Strain Effects Associated with SiO Layers Evaporated onto GaAs," *J. Appl. Phys.* **54**, 5779-5784 (1983).
16. A. Blech and A. A. Levi, "Comments on Aleck's Stress Distribution in Clamped Plates," *Trans. ASME* **48**, 442-445 (1981).
17. B. A. Boley and J. H. Weiner, *Theory of Thermal Stresses*, Wiley, New York (1960).
18. W. Schroter, H. G. Brion and H. Siethoff, "Yield point Dislocation Mobility in Silicon and Germanium," *J. Appl. Phys.* **54**, 1816-1820 (1983).
19. M. Tagaki, K. Nakayma, C. Tevada and H. Kamioko, "Improvement of Shallow Base Transistor Technology by Using a Doped Polysilicon Diffusion Source," *J. Jap. Soc. Appl. Phys.* (suppl.) **42**, 450-454 (1973).
20. J. Graul, A. Glasl and H. Murman, "High Performance Transistor with Arsenic Implanted Polysilicon Emitters," *IEEE J. Solid-State Circ.* **SC-11**, 491-495 (1976).
21. K. Okada, K. Aomura, M. Suzuki and H. Shiba, "PSA—A New Approach for Bipolar LSI," *IEEE J. Solid-State Circ.* **SC-13**, 693-698 (1978).

22. T. Sakai, Y. Kobayashi, H. Yamaguchi, M. Sato and T. Makino, "High Speed Bipolar ICs Using Super Self-Aligned Process Technology," *12th Conf. Solid State Devices*, Tokyo (1980) pp. 155-159.
23. P. J. Zdebel, R. J. Balda, B. Y. Hwang, V. DelaTorre and A. Wagner, "MOSAIC III—A High Performance Bipolar Technology with Advanced Self-Aligned Devices," in *Proc. of the 1987 Bipolar Circuits and Technology Meeting*, pp. 172-175.
24. J. M. Storck and J. D. Cressler, "Performance Degradation Due to Emitter Resistance in Polysilicon Emitter Bipolar Transistors," in *Symp. VLSI Technology Dig. Tech. Papers 1985*, pp. 47-48.
25. C. T. Kirk, "A Theory of Transistor Cutoff Frequency (f_T) Falloff at High Current Densities," *IRE Trans. Electron Dev.* **ED-9**, 164-174 (1962).
26. V. DelaTorre, J. Foerstner, B. Lojek, K. Sakamoto, S. L. Sanduram, N. Tracht, B. Vasquez and P. J. Zdebel, "MOSAIC V—A Very High Performance Bipolar Technology," in *Proceedings of the 1991 Bipolar Circuits and Technology Meeting*.
27. A. K. Kappor and D. J. Roulston, "Polysilicon Emitter Bipolar Transistors," IEEE Press, New York (1990).
28. G. R. Wolstenholme, N. Jorgensen, P. Ashburn and G. R. Booker, "An Investigation of the Thermal Stability of the Interfacial Oxide in Polycrystalline Silicon Emitter Bipolar Transistors by Comparing Results with High-Resolution Electron Microscopy Observations," *J. Appl. Phys.* **61**, 225-233 (1987).
29. N. E. McGruer and R. A. Oikari, "Polysilicon Capacitor Failure During Rapid Thermal Processing," *IEEE Trans. on Electron Dev.* **ED-33**, 929-932 (1986).
30. Z. A. Weinberg, D. R. Young, J. A. Calise, S. A. Cohen, J. C. DeLuca and V. R. Deline, "Reduction of Electron and Hole Trapping in SiO_2 by Rapid Thermal Annealing," *Appl. Phys. Lett.* **45**, 1204-1206 (1984).
31. B. Lojek, *Evaluation of RTA Annealing Systems*, internal report, Motorola Advanced Technology Center (May 1991).
32. L. C. Parrillo, "VLSI Process Integration," in "VLSI Technology," (S. M. Sze, ed.), McGraw-Hill, New York (1983).
33. S. S. Wong, D. R. Bradbury, D. C. Chen and K. Y. Chin, "Elevated Source/Drain MOSFET," *IEDM Tech. Dig.* (1984), pp. 634-637.
34. H. Shibata, Y. Suizu, S. Samata, T. Matsuno and K. Hashimoto, "High Performance Half-Micron PMOSFET's with $0.1\,\mu m$ Shallow p^+n Junction Utilizing Selective Silicon Growth and Rapid Thermal Annealing," *IEDM Tech. Dig.* (1987). p. 590.
35. J. R. Pfiester, R. D. Sivan, H. M. Liaw, C. A. Seelbach and C. D. Gunderson, "A Self-Aligned Elevated Source/Drain MOSFET," *IEE Electron Dev. Lett.* **11**, 365-367 (1990).
36. A. Ishitani, H. Kitajima, N. Endo and N. Kasai, "Facet Formation in Selective Silicon Epitaxial Growth," *Jap. J. Appl. Phys.* **24**, 1267-1269 (1985).
37. W. Engl, "Device Simulation Programs MEDUSA, GALENE and TOSHIE," unpublished, presentation at Motorola Advanced Technology Center, (1990).
38. B. Lojek, "Drain Structure Study with SILVACO MINIMOS IV," Motorola Advanced Technology Center, internal report (January 1990).
39. P. J. Zdebel, "FAST I—BiCMOS Technology Process Sequence," Motorola Advanced Technology Center, internal report (April 1990).
40. B. Lojek, N. Tracht and P. J. Zdebel, "FAST II—$0.6\,\mu m$ BiCMOS process definition," Motorola Advanced Technology Center, internal report (June 1990).
41. S. E. Lassing, T. J. Debolske and J. L. Crowley, "Kinetics of Rapid Thermal Oxidation of Silicon," in *Rapid Thermal Processing of Electronics Materials*, MRS Symposia Proc., Vol. 92 (1987), pp. 103-108.

42. C. A. Pax de Araujo, J. C. Gelpey and Y. P. Huang, "Comparison of the Growth Kinetics of Oxides Grown in Tungsten-Halogen and Water Cooled Arc Lamp Systems," in *Rapid Thermal Processing of Electronics Materials*, MRS Symposia Proc. Vol. 92 (1987), pp. 133-140.
43. S. T. Pantelides, Ed., *Physics of SiO_2 and its Interfaces*, Pergamon Press, New York (1978).
44. T. Ito, T. Nakamura and H. Ishikawa, "Advantage of Thermal Nitride and Nitrioxide Gate Films in VLSI Process," *IEEE Trans. on Electron Dev.* **ED-29**, 498-502 (1982).
45. F. L. Terry, R. J. Aucoin, M. L. Naiman and S. D. Senturia, "Radiation Effects in Nitrided Oxides," *IEEE Electron Dev. Lett.* **EDL-4**, 191-193 (1983).
46. T. Hori and H. Iwasaki, "Ultra Thin Re-Oxidized Nitrided-Oxides Prepared by Rapid Thermal Processing," *IEDM Tech. Dig.* (1987), pp. 570-573.
47. D. K. Shih, W. T. Chang, S. K. Lee, Y. H. Ku, D. L. Kwong and S. Lee, "Metal-Oxide-Semiconductor Characteristics of Rapid Thermal Nitrided Thin Oxides," *Appl. Phys. Lett.* **55**, 1698-1700 (1988).
48. D. K. Shih, D. L. Kwong and S. Lee, "Study of the SiO_2/Si Interface Endurance Property during Rapid Thermal Nitridation and Reoxidation Processing," *Appl. Phys. Lett.* **54**, 822-824 (1989).
49. M. Born and E. Wolf, *Principles of Optics*, Pergamon Press, New York (1970).

9 Manufacturing Equipment Issues in Rapid Thermal Processing[1]

Fred Roozeboom
Philips Research Laboratories
PO Box 80,000
NL-5600 JA Eindhoven
The Netherlands

I. Historical Survey of Rapid Thermal Processing	351
II. Fundamental Thermophysics in Rapid Thermal Processing	352
A. Emissivity	353
B. Light Pipe and Cavity Concepts	358
III. General Rapid Thermal Processing System Components	359
A. Heat Source and Reflector Designs	360
B. Chamber Designs	365
C. Temperature Sensing, Calibration, and Control	368
IV. Survey of Commercial Rapid Thermal Processing Equipment	381
A. The Middle Infrared (3–6 μm)	387
B. The Far Infrared ($\geq 6\,\mu$m)	388
C. The Near Infrared (0.8–3 μm)	390
V. Temperature Nonuniformity, System Modeling, and Effective Emissivity	390
A. Nonpatterned Wafers	391
B. Patterned Wafers	393
C. Effective Emissivity and Cavity Design	397
D. Miscellaneous Design Aspects to Improve Process Uniformity	400
VI. Noncontact *In Situ* Real-Time Process Control Options	401
A. Novel Temperature Measurement Techniques	401
B. End-Point Detection Techniques	404
VII. Recent Developments and Future Trends in Rapid Thermal Processing	407
A. Equipment Architecture and Design	408
B. Scalar vs. Multivariable Process Control	410
C. Thermal Switching vs. Rapid Gas Switching	413
VIII. Technology Roadmap and Concluding Remarks	414
Acknowledgments	417
References	417

---

[1] Based in part on an invited presentation at the SPIE Workshop on Rapid Thermal and Integrated Processing, San Jose, California, September 9, 1991.

Rapid thermal processing (RTP) started in the late 1960s with IBM's pioneering work on making silicon with submicrometer details. They used pulsed laser beams on boron-doped silicon with paint-on phosphorus [1]. Today RTP, with its inherently smaller thermal mass and more stringent ambient control, has the potential to become a core technology step in the development and manufacturing of ULSI devices [2-4]. Process steps that are under study are source/drain implant annealing [5], contact alloying, formation of refractory nitrides and silicides [6-8], thin gate dielectrics formation [9, 10], and glass reflow [2]. Relatively immature are depositions in reactive gases such as chemical vapor deposition of amorphous and polysilicon [11, 12], epitaxial silicon [13, 14] and Ge_xSi_{1-x} [15-21], and tungsten [22, 23]. Selective depositions, recently published, are those of titanium disilicide [24], titanium nitride [25], epitaxial silicon [26], epitaxial Ge_xSi_{1-x} [27], etc. Other application fields just opening up are in the back-end processing, such as packaging or solder reflow, and in the processing of thin films of III-V and other semiconductors [28], ceramic [29] and magnetic films [30], and in flat panel display processing [31].

By the beginning of 1992 the worldwide number of single-wafer RTP systems was just above 1500, made by some 22 manufacturers and with incoherent lamps as the predominant type of heat source. This number shows a large annual growth [32], yet one should realize that the vast majority of these systems is not advanced enough for production and used in research only. Although complete single-wafer process flows, including many RTP steps, have been demonstrated, as in half- and subhalf-micrometer CMOS [2-4], a surprisingly small minority is used in a full-production environment. Even then this is mainly limited to titanium disilicide formation, in which any annealing system is successful as long as the oxygen content is kept below 5-10 ppm in a 1 atm ambient. Illustrative is the fact that Japan has not even 200 systems—mostly manual research systems.

The major obstacles for full acceptance remain *temperature reproducibility and uniformity* during all processing, namely the dynamic and stationary parts of the thermal cycle, in which layers are being (re)formed. It is the challenge of this decade to overcome these crucial obstacles such that, by the turn of this century, the semiconductor industry can successfully start the anticipated processing of 1-Gbit DRAM devices on 300 mm diameter wafers with 0.15 μm design rules [33]. It is obvious that many steps that can or must be performed in batch furnaces at the present time may eventually have such narrow process windows that single-wafer processing provides the best, if not the only, processing route. Additionally, cycle time and yield considerations make RTP the technology of choice.

This chapter aims at giving a review of the status, problems, and options in system design and the aforementioned obstacles of temperature reproducibility and process uniformity, with focus on, but not limiting to, silicon technology. For earlier reviews and selection guides one is referred to Refs. 34–46.

After a historical review in Section I, we will treat the fundamental thermophysics of RTP in Section II to provide a basis of understanding the aspects in RTP system and process design. Next, the general characteristics for the different designs of heat sources, reactor chambers and temperature sensors are treated in Section III. A survey of commercially available systems follows in Section IV.

The thermal performance of an RTP system is complicated by the total system configuration and the optical wafer properties. This is treated in Section V under temperature nonuniformity and effective emissivity; it includes the highlights of recent, important work on system modeling.

Section VI gives a compilation of noncontact, *in situ*, real-time process control options. Next, Section VII describes some recent developments and trends that aim at increasing the throughput and yield in RTP. Here, trends in system architecture and design, scalar vs. multivariable control—that is fixed, dependent vs. independent multizone control—and thermal switching vs. rapid gas switching are included.

A technology roadmap and some final remarks in Section VIII conclude this chapter.

Finally, it should be noted that the use of product names in this chapter does not in any way constitute an endorsement of the products by the author, the editor or the publisher.

I. Historical Survey of Rapid Thermal Processing

As mentioned above, the origins of RTP date back to the late 1960s, when IBM used pulsed laser annealing [1]. During the 1970s rapid thermal annealing was dominated by laser annealing, using pulses of milliseconds to nanoseconds in inert ambients [36–39]. In the late 1970s other heating sources were studied, such as electron beams and flash lamps. This involved the milliseconds annealing range, still using inert gases and aiming at controlled diffusion and activation. In the 1980s graphite strip heaters and flood heating sources, such as tungsten–halogen lamps and long-arc lamps, have taken the pulse annealing into the 1–100 sec processing regime [36–39].

Recent review articles [36–39] have treated the advantages of *isothermal* heating (1–100 sec pulse using lamps, resistances, or e-beams) over *thermal flux* (0.1–10 ms pulse using scanned cw lasers or e-beams) and *adiabatic* (1–1000 ns laser or e-beam pulse) heating. The three characteristic heating

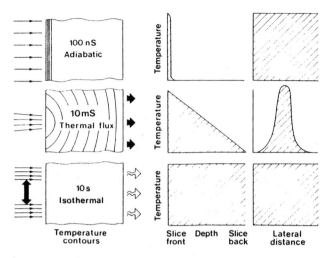

FIGURE 1. Temperature contours and profiles for three typical modes of beam heating. (From Hill [38]).

modes are illustrated in Fig. 1. It is obvious that only the isothermal mode gives rise to uniform lateral and transverse temperature profiles throughout the wafer, thus minimizing the chance of disruption of layers and circuits. Details on the process of absorption of energy-carrying beams by grey bodies are found elsewhere [38, 39].

The addition of reactive ambients in the reactors opened up many other applications, continuously replacing, or better, merging with batch annealing and chemical vapor deposition (CVD) processes, using susceptors. This merger of single-wafer RTP and single-wafer CVD (especially low pressure, but also atmospheric and ultrahigh vacuum) has progressed sufficiently that now the only difference might be the presence of a susceptor in case of CVD. For simplicity we will consider both systems as RTCVD or RTMOCVD (rapid thermal metalorganic chemical vapor deposition) systems. In that case the basic common characteristics reduce to a single-wafer processing system with incoherent flood heating with processing times of the order of 1 min.

II. Fundamental Thermophysics in Rapid Thermal Processing

RTP is based on the energy transfer between a radiant heat source and an object. At high temperatures the radiation term dominates the convection and conduction terms. For an RTP reactor, because of the optical character of the energy transfer, the reactor wall is typically *not in thermal equilibrium* with the object. This is in contrast to a common batch reactor—or a chemical reactor—where convective and conductive heat flow play the dominant role. This explains the characteristic short processing times (seconds to minutes) that are possible with RTP, as compared to those (minutes to hours) for batch systems.

This section deals with the thermophysical aspects involved in RTP, emphasizing the *emissivity*, and *light pipe* and *cavity* concepts used in heat source and chamber design.

A. Emissivity

Infrared pyrometry is the most widely used form of noncontact temperature monitoring in RTP for many reasons. Like the human eye it is precise, fast, and noninvasive. Unfortunately, pyrometry is very dependent on the changes in wafer emissivity that occur upon processing. Thus emissivity is an unknown parameter. The emissivity of a material is a function of the optical properties of

- the starting wafer material (*intrinsic* emissivity),
- the layers on top of or buried in the wafer (*extrinsic* emissivity), and
- the specific optical properties of the reflective chamber with all components inside (*effective* emissivity).

Both intrinsic and extrinsic emissivity values should be measured in a cold, black environment, excluding the extra effects of the reflective chamber and stray light from lamps and other hot chamber parts, which together comprise the effective emissivity.

1. Intrinsic Emissivity of Silicon

The fundamental physical laws in radiative heat flow are those of Stefan-Boltzmann (Eq. 9.1), Planck (Eq. 9.2), Wien (Eq. 9.3) and Kirchhoff (Eqs. 9.5 and 9.6), which we will treat now. For a more detailed introduction one is referred to basic textbooks [47, 48].

The total radiant exitance M emitted per unit surface by an object with absolute surface temperature T is given by the Stefan-Boltzmann law

$$M(T) = \varepsilon \sigma T^4 \quad \text{W/m}^2. \tag{9.1}$$

where ε is the emissivity (or emission coefficient) and σ is the Stefan–Boltzmann radiation constant of 5.6697×10^{-8} W/m^{2}K^4. For a perfect blackbody ε equals 1.

The spectral distribution of the radiant exitance is given by Planck's radiation law

$$M_\lambda(\lambda, T) = \frac{\varepsilon c_1}{\lambda^5 (e^{c_2/\lambda T} - 1)} \quad \text{W/m}^3, \tag{9.2}$$

where λ is the wavelength in m, and the first and second radiation constants c_1 and c_2 are equal to 3.7415×10^{-16} W · m^2, and 1.43879×10^{-2} m · K, respectively.

The wavelength of the maximum intensity in the Planckian distribution is inversely proportional to T, according to Wien's displacement law

$$\lambda_{\text{peak}} T = 2.89783 \times 10^{-3} \quad \text{m} \cdot \text{K}. \tag{9.3}$$

From Eq. 9.3 we can easily derive a theoretical blackbody color temperature of 6000 K for an arc lamp with peak emission at $\lambda_{\text{peak}} = 0.5\,\mu$m, of 2900 K for a tungsten filament with $\lambda_{\text{peak}} = 1.0\,\mu$m, and of 1000 K for a wafer with $\lambda_{\text{peak}} = 2.9\,\mu$m.

Only perfect blackbody surfaces (emissivity $\varepsilon = 1$) emit the maximum possible amount of radiation, according to Eq. 9.1. It may be obvious that the emissivity of any Lambertian (diffuse) radiator in thermal equilibrium with an isothermal enclosure is numerically equal to its absorptivity: $\varepsilon = \alpha$.

Practical surfaces such as wafers, and the components of an RTP chamber, deviate from ideal blackbody behavior and have emissivities smaller than 1. In that case one has at best grey surfaces that have constant emissivity values between $\varepsilon = 1$ (perfectly black) and $\varepsilon = 0$ (perfectly white), or nongrey surfaces with varying $\varepsilon(\lambda, T)$ values.

Above 600 °C a bare silicon wafer will behave as an opaque grey body with $\varepsilon = 0.7$ over the entire wavelength range [49]. Below this temperature silicon behaves as a transparent nongrey body.

The *intrinsic* emissivity value depends not only on the roughness of the wafer surface and on the dopant concentration used, but also on the surface temperature and on the wavelength of the absorbed and emitted radiation, as shown by Sato [49]. Figure 2b shows the deviation of silicon from true greybody behavior. Electronic valence-to-conduction-band absorption is efficient for wavelengths below the silicon band gap of 1.2 μm. Above 6 μm the absorption becomes efficient due to lattice vibrations [49]. However, in the 1.2–6 μm wavelength range and below 600 °C the main heat absorption mechanism is by intrinsic free-carrier absorption. Thus, in this range the emissivity is a strong function of the free-carrier density, and therefore

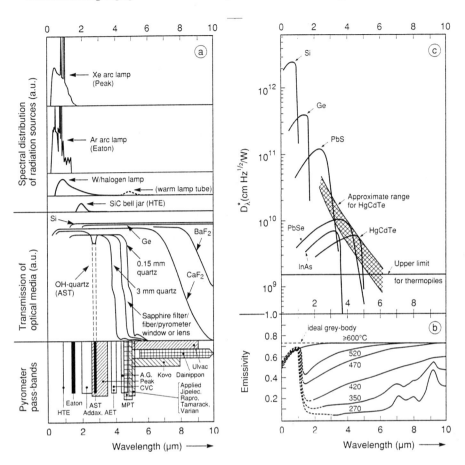

FIGURE 2. Spectral features in RTP systems. (After Roozeboom [42].) (a) Heat sources, optical media, and pyrometers; (b) emissivity of 1.8 mm thick, 15 Ω cm, phosphorus doped silicon, cf. Sato [49]; and (c) detectivity/sensitivity for temperature sensors at room temperature, cf. Dimmock [87].

on the thickness, dopant concentration, and temperature of the wafer. This was one of the early problems of pyrometry in RTP; from Fig. 2 one can see that a fraction of the wavelengths radiated by the lamps is transmitted in this region of transparency during the initial heating. This fraction is the smallest for arc lamps, since they emit most of their spectra ($\lambda_{peak} = 0.5\,\mu m$) at photon energies above the Si band gap. For tungsten-halogen lamps ($\lambda_{peak} = 1.0\,\mu m$) the fraction increases; even more so for the silicon carbide bell jar ($\lambda_{peak} = 2.0\,\mu m$). Thus, the initial heating

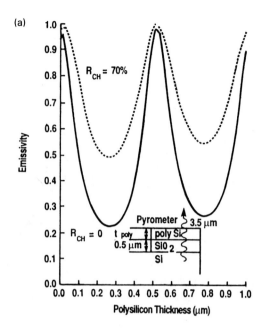
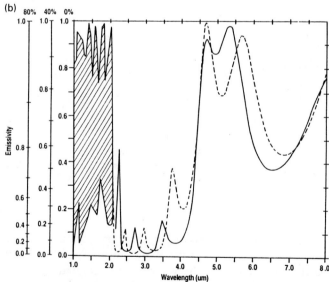

FIGURE 3. Extrinsic (0% chamber reflectivity) and effective ($R_{CH} \geq 40\%$) emissivity changes for silicon with planar layer structures. (After Hill et al. [39].) (a) Varying polysilicon layer thickness on 0.5 μm SiO_2, with pyrometer at 3.5 μm, in a black (—) and in a 70% reflecting (----) chamber. (b) Complex 10-layer structure (poly-Si, oxide and nitride layers) with two layers changing in thickness in a black, a 40 and an 80% reflecting chamber.

rate depends heavier on the dopant level, but *only* up to 600 °C. Consequently, only below this temperature the lamp choice affects the initial heating rate, provided that heat source geometry and installed power are identical. Typical total heating rates range from 300 K/s for arc lamps to 100 K/s for the resistively heated bell jar [44].

2. Extrinsic Emissivity of a Silicon Wafer

On top of the above parameters the emissivity of a wafer is further complicated and determined by *extrinsic* parameters [50, 51]. Front side layers such as dielectric, metallic or patterned films, and buried areas will have different refractive indices and buried areas have different dopant concentrations.

The presence of overlayers will influence the optical properties of the silicon surface. Optical interference will modulate the emissivity and thus the absorptivity. Many authors have put forward emissivity models [11, 39, 50–54]. Hill *et al.* [39] showed by theory and experiment that the spectral emissivity oscillates as a function of the thickness of deposited layers. From Fig. 3a one can see that for a simple two-layer structure with one layer varying in thickness the extrinsic emissivity oscillates in a regular pattern between 0.25 and 0.95. From Fig. 3b, with two out of ten layers varying in thickness, we can conclude that one certainly needs *in situ*, real-time emissivity compensation here.

Similar oscillating behavior was modeled and confirmed experimentally by Nulman *et al.* [50–52]. Vandenabeele *et al.* [53, 54] showed that the temperature difference across one patterned wafer can vary by 80 °C over small distances. Oscillations of the substrate temperature were also found by Liao and Kamins [11] during the deposition of polycrystalline silicon on oxidized silicon. The initial drop in temperature exceeded 100 °C, below the polycrystalline-to-amorphous transition value. This led to the formation of an amorphous layer, sandwiched in the polycrystalline film.

3. Effective Emissivity of a Wafer in a Chamber

From reflectance measurements in a black environment Pettibone *et al.* [55] observed an emissivity increase from 0.71 to 0.87 during the growth of a thermal oxide layer of 0.4 μm on a wafer. Theoretically, in a black chamber, such an emissivity change should lead to a temperature difference of more than 100 °C, as calculated by the basic correction equation

$$1/T_{thermo} = 1/T_s + (\lambda/c_2)\ln \varepsilon \tag{9.4}$$

where T_{thermo} is the correct thermodynamic temperature and T_s the spectral temperature, as detected by the pyrometer. Experimentally they found smaller errors, namely 10–50 °C. The discrepancy is due to the highly reflective chamber wall as an extra complicating factor: the extrinsic wafer parameters, together with the specific RTP chamber and the components inside, compose the overall, *effective* emissivity, which can be very system dependent [56–58] (see Fig. 3 and Section V). Recent developments to compensate for emissivity variations will be discussed later in Section III C2. Many relate to Kirchhoff's law [44, 45]:

$$\varepsilon(\lambda, T) = 1 - \rho(\lambda, T) - \tau(\lambda, T) \tag{9.5}$$

where ε is the emissivity, ρ reflectivity, and τ transmissivity. For opaque material ($\tau = 0$) Eq. 9.5 reduces to

$$\varepsilon(\lambda, T) = 1 - \rho(\lambda, T) \tag{9.6}$$

B. Light Pipe and Cavity Concepts

One of the problems in obtaining uniform temperature profiles across a wafer surface originates from the fact that one starts with a nonuniform primary radiant flux. The flux is emitted by light sources with line, point, or pseudo-ring symmetry. Consequently, the incident flux is not ideally uniform. Even in the ideal case the radiation from the wafer is not uniform, because of center-to-edge radiative differences [59] and to local differences in absorption and emissivity, induced by patterns on the wafers [39, 56, 57]. Reradiation back onto the wafer enhances these differences.

One way to obtain uniform radiation or reradiation is to redistribute the energy by properly shaping both the heat source and the reactor chamber. One such shape is that of the *light pipe*, illustrated in Fig. 4. A mirrored tunnel or *kaleidoscope* is placed with its entrance plane at a point (or line or ring) source of light and its exit plane at the sample. This light pipe collects the energy from the light source and reflects it many times. Because of the multiple reflections the energy is redistributed over the cross-sectional area of the light pipe. For a *square* light pipe with length L, side R, and wall reflectivity unity, it can be derived by geometric optics that the flux density at the exit plane converges to unity as $L/R \to \infty$ [60]. In other words, the larger the pipe's aspect ratio and reflectivity and the more a light ray is reflected before reaching the exit, the more uniform flux distribution is obtained. It can be derived that other useful cross sections to obtain

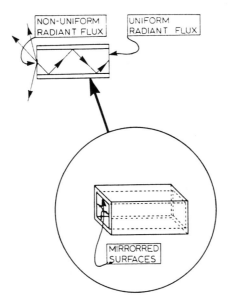

FIGURE 4. Principle of a light pipe. (After M. M. Chen, J. B. Berkowitz-Mattuck, and P. E. Glaser. *Appl. Opt.* **2**, 265-271, 1963.)

uniform flux with specularly reflective surfaces are the *equilateral triangle*, the *rectangle* and the regular *hexagon*. Chen *et al.* [60] have shown theoretically and experimentally that a specularly reflecting, *cylindrical* light pipe is not effective in achieving uniform illumination: a diffuse light source at the pipe entrance will result in a hot spot of the same size at the exit plane. In this case the uniformity can only be increased by using diffusely reflecting walls, for example, by sandblasting the inner walls.

The above concept of multiple reflections by light pipes with high reflectivity and high aspect ratio relates to the *reflective cavity* concept, described by Sheets [61-63] and to recent Monte Carlo modeling on RTP reactor design by Kakoschke [57]. They also concluded that a reactor chamber with high aspect ratio and high reflectivity walls leads to increased effective emissivity and thus to better uniformity. More details on this appear in Section V.

III. General Rapid Thermal Processing System Components

Temperature repeatability and temperature uniformity are mainly influenced by

- the heat source characteristics (type, size, shape, location, aging, reflectors used)

- the process chamber characteristics (optical wall/window material parameters, size, shape, gas flow patterns, etc.) and
- the temperature sensor and control system (sensor type and wavelength pass-band, scalar vs. multivariable control, etc.)

In this section we will discuss each component with focus on spectral and optical features.

A. HEAT SOURCE AND REFLECTOR DESIGNS

In commercial RTP systems with isothermal heating one uses three heat source types: the *tungsten–halogen lamp*, the *long-arc noble gas discharge lamp*, and the *continuous resistively heated bell jar*. In Section II we saw that their spectral temperatures are 3300, 6200, and 1700 K, respectively. The three heat source types have several differences, the first being the *spectral temperatures*, just mentioned. Consequently, the initial heating rates differ, albeit only up to 600 °C (see Section II). Other differences are their price, lifetime, thermal mass and most importantly, their *symmetry*. This is described below, after a detailed description of each heat source.

1. The Tungsten–Halogen Lamp

The mainstay in RTP heaters is the tungsten (W)–halogen lamp. A good introduction on this type of lamp is Ref. 64. Usually this lamp consists of a linear double-ended quartz tube around a tungsten filament that is heated resistively. The quartz envelope may also be a single-ended quartz bulb, in a parabolic or light pipe reflector. The quartz transmits the entire spectrum emitted by the filament, up to an absorption wavelength of 4–5 μm (see Fig. 2).

The quartz heats up to 400 °C [34] or more, the maximum allowable temperature being 900 °C. This can lead to perturbations in the temperature detection at wavelengths longer than 5 μm, because the warm quartz envelope becomes a secondary radiator with $\lambda_{peak} \geq 5\ \mu$m. This is indicated by the upper dotted line in Fig. 2a.

The quartz envelope is filled with halogen gas, often $PNBr_2$ [64], to increase the filament's temperature and lifetime. At quartz wall temperatures above 250 °C the halogen prevents the deposition on the envelope, thanks to a regenerative tungsten halide chemical transport cycle. This cycle suppresses the nonuniformity of the emitted light due to envelope blackening.

The main advantages of the W–halogen lamp are its high life/price and power/filament length ratios, and its compact powering unit.

2. The Noble Gas Long-Arc Lamp

A good review on noble gas arc lamps is given by Rehmet [65]. Commercial RTP systems have used only two long-arc lamp types. One is the Vortek *dc water-wall argon lamp* [66–68], which consists of a single quartz tube. Cooling water and the argon discharge gas are both injected into the same inner space, such that the water forms a thin vortex wall on the inner side of the tube. This water wall is very effective in internal cooling and in constricting an inner vortex of the Ar gas that is injected into the central space.

The second type is the *ac xenon lamp* [69, 70]. It consists of two concentric tubes. The inner fused silica tube contains the xenon fill gas, while the outer Pyrex tube comprises an external cooling water jacket around the inner tube. The water column in both lamps absorbs all radiation above 1.4 μm wavelength, hence the arc lamps emit practically no radiation beyond this wavelength (*cf.* Fig. 2a).

Both arc lamps need a high initial voltage pulse to ignite the gas discharge (arc) between the electrodes, which calls for a less compact powering unit. The advantages of the arc lamp are its efficient output coupling to silicon below 600 °C (see Section II), and its low thermal mass.

3. The Resistively Heated Bell Jar

An innovative option in resistive heating is offered by the continuously heated bell jar, made from silicon carbide or quartz [71, 72]. It serves both as a radiator and as a chamber (see later in paragraph III B 3). A heater module surrounds only the upper part of the bell jar and thus generates a constant temperature, typically 300 K above the desired process temperature. The lower part of the bell jar is connected with the transfer chamber, which is water cooled. This creates a well-defined temperature gradient along the central axis and a fully centrosymmetric profile along the wafer diameter. An elevator moves the wafer up and down to a position in the reactor chamber that corresponds to the desired temperature. Heating and cooling rates are controlled, and limited, by the speed of the elevator. Most of the initial heating is through intrinsic free-carrier absorption, but still, a heating rate of 100 K/s is possible [44].

The outstanding advantage of this heat source is its superior centrosymmetric temperature profile [71]. This facilitates upscaling to 300-mm wafer diameter. Upscaling can be done without complicated reflector redesigning or multizone lamp heating. Another advantage is that the hot bell jar serves as a *greybody cavity*, with wall emissivity equal to the intrinsic wafer emissivity: $\varepsilon_{SiC} = \varepsilon_{Si} = 0.7$. A higher aspect ratio for this type of cavity increases the wafer's effective emissivity and thus reduces its sensitivity

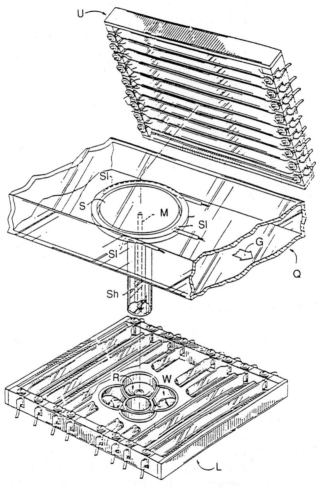

FIGURE 5. Example of line source and point source W-halogen lamps in one RTP system. Shown is the diagram of ASM's Paragon One system, after [75]. G = gas flow direction, L = lower lamp bank, M = master thermocouple, Q = quartz reactor, R = parabolic reflector, S = rotating SiC coated graphite susceptor, Sh = rotary drive shaft, SI = slave thermocouples in stationary graphite guard ring, U = upper lamp bank with 6 kW lamps, W = W-halogen bulb lamp.

to emissivity variations [73] (see Section V). A disadvantage is the sensitivity of the silicon carbide to gases that are corrosive or that decompose or deposit on the hot wall to form particles.

4. Heat Source Symmetry and Design

There are basically four heat source symmetries: the *line* source, the *square* source, the pseudo-*hexagonal* source and the pseudo-*ring*. Most systems use line-symmetrical components, i.e. linear double-ended tubes, usually in some cross-lamp array (*square* symmetry source) of lamp banks below and/or above the chamber [74] (see Section IV).

Figure 5 gives an example of a system that uses both "point"-symmetrical and line symmetrical W–halogen lamps [75, 76], comprising a square-symmetrical heat source. In order to compensate for the missing gold reflector coating at the location of the feedthrough for the rotary drive shaft assembly, a heat concentrator has been installed. It includes four identical

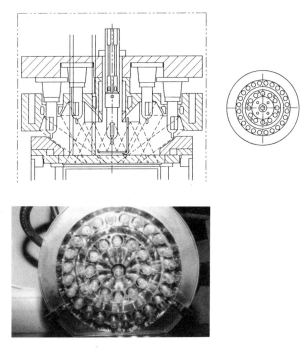

FIGURE 6. Schematic representation of a pseudocircular multizone heating module, with individual flux rings and individual fiber optical pyrometers, used in the Stanford Rapid Thermal Multiprocessor. The center lamp is 2 kW, all others are 1 kW lamps. (From [22] and [77].)

heat focusing devices, each consisting of a W-halogen bulb in a converging reflector housing.

The tungsten-halogen bulb is also used to form concentrically arranged optical flux rings and hexagons for uniform flood heating. An example is the Texas Instruments lamp [22, 77] shown in Fig. 6. Another is the hexagonal lamp panel, prototyped by G-Squared [78] and now being further developed by Applied Materials. It contains up to 108 water-cooled, cylindrical light pipes, each with a W-halogen bulb lamp inside. One lamp is in the center and the others are in six hexagonal zones around it. Note that, although these symmetries are conformal to the wafer, it is very important that the individual lamps cooperate such that the heat intensity distributions of adjacent lamps overlap to give an overall even distribution at the wafer plane. This is accomplished by using different lamp altitudes and special reflector geometry (Fig. 6) or diffuse light pipe reflectors [78], combined with differential lamp powering and/or wafer rotation.

It is obvious that the pseudo-ring or hexagonal symmetry is the most complicated and immature in full uniformity control, yet the most promising, as compared to line and square sources, which are not compatible to the wafer shape.

A *fully ring-shaped* symmetry is offered by the resistance-heated, silicon carbide bell jar [71, 72], which serves as the heat source as well as the reaction chamber (see Section III B).

5. Reflector Designs

Reflector designs have the same general symmetries as the lamps and/or the chamber: line, square, and pseudo-circular.

An example of a reflector with improved uniformity is a computer-designed reflector with segmented lenticular shape that creates a series of virtual line sources from a single line source arc lamp [79, 80]. Another design is that of a curved reflector around the wafer to compensate for edge losses [81] (see also Section V).

Often the reflector is conformal to, or comprising, the chamber walls. In that case most designs make use of a gradient in the reflectivity of the reflector. One way is to make the reflector more reflective at the wafer edge, say by applying an infrared absorbing material on the center part of the top reflector. This example already implies that most reflector optimization is done for the stationary part of the heat cycle, where the edge effects are fundamentally different from those in the ramp-up part (see Section V). For the transient parts, radiation may be adjusted by a dynamically controlled wafer-to-reflector motion, as proposed by Kakoschke [81], although adaptive multizone control is a better solution.

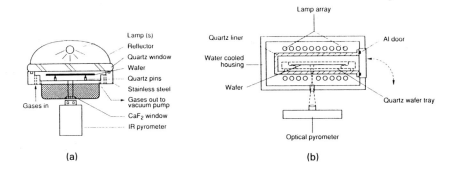

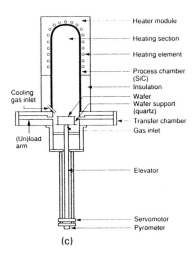

FIGURE 7. Schematic illustrations of (a) a cold-wall, (b) warm-wall and (c) a hot-wall chamber. (From [41].)

B. CHAMBER DESIGNS

Commercial RTP systems employ three generic chamber designs, namely the *cold wall, warm wall*, and *hot wall* [41]. Figure 7 shows their typical configurations. Each design affects the temperature measurement and the temperature uniformity in its own way, because of the specifics of the optical and dimensional properties of the walls, windows, and so forth. The three designs are discussed below, with focus on thermal and material aspects.

1. Cold-Wall Chamber

The cold-wall reactor is made from water-cooled metals such as stainless steel, aluminum, or other alloys in cylindrical or rectangular shapes. Reflectivity is accomplished by electropolishing the metal, or by coating it with reflective materials such as gold or aluminum. Often the metal surface is passivated by anodizing it, lining it with quartz inserts [41], or by coating it with a thin quartz layer [61].

The chamber top face is an air- or water-cooled quartz window plate that transmits the optical flux into the chamber. The cold walls have the advantage of minimum thermal memory effects that could further complicate temperature measurement.

Figure 8 shows the effect of active cooling of the quartz top plate on the cooling rate. Clearly, active cooling reduces the thermal memory of the plate. Water cooling of the quartz plate is more effective than air cooling. Especially with high-throughput processing the top plate can be heated up by the many wafers. Thus, parasitic deposition on hot spots may occur, which causes particles and decreases the temperature uniformity.

But water cooling also has disadvantages. These are possible formation of bubbles, contamination in the water, and, most importantly, the less

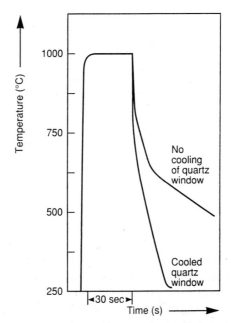

FIGURE 8. The effect of active cooling of a quartz top plate on the wafer cooling rate (Source: MPT Corp.).

efficient coupling of the heat source to the wafer, due to the optical absorption above 1.4 μm by the water column. This problem can be alleviated by using coolants that do not absorb in the near infrared and have enough heat capacity, for example, Fomblin oil.

Other window plate materials are also used, like sapphire. It has about 20 times more thermal conductivity and will create less thermal memory than quartz.

On the other hand, depending on the process, too cold a wall temperature may also give problems in terms of unwanted deposition or condensation of reactants. A tunable wall temperature, say from room temperature to 100 °C, may be an effective compromise here, and so are the use of quartz liners along the metal walls and the use of slower ramp rates up to deposition temperature. Thus, as an example, one can avoid the inclusion of cubic SiC [82] at the Si-epi interface upon the growth from $SiCl_2H_2$. Note that other factors can also play an important role in the recontamination of highly reactive carbon adsorbates and the like. Examples are the precleaning steps and the base vacuum.

2. Warm-Wall Chamber

The warm-wall reactor consists of a quartz envelope that encloses the wafer. The envelope is rectangular, cylindrical, ellipsoidal, or a bell jar [83]. The chamber is surrounded by a reflective water-cooled metal housing. Normally, the quartz is only cooled by forced-air cooling or suction, which is less "active" than water cooling. Thus, quartz reactors can easily reach 200–400 °C, and are referred to as warm-wall chambers. Inside, the wafer is exposed on all sides to the quartz. During the entire thermal cycle the warm wall is at a markedly lower temperature than the wafer.

A quartz reactor wall has several advantages. One is the elimination of possible metal contamination from the metal chamber housing. Another is the extra degree of freedom in choosing the reflector materials. Furthermore, the warm wall suppresses chemical memory effects in the system because of desorption of condensates. However, thermal memory effects may be present that can only partly be eliminated by using preheat cycles. Furthermore, repeated cleaning with etching agents can degrade the transmissive properties of quartz.

3. Hot-Wall Chamber

The hot-wall reactor is the single-wafer analogue of the hot-wall batch reactor, with the exception that during processing the wall temperature is always higher than that of the wafer. This design applies to the continuous resistively heated bell jar. The wall is made from silicon carbide [71, 72] or quartz [72] and serves as a uniform heat radiator. More details appear in Section III A 4.

4. Gas Flow Patterns

So far, we have treated all RTP features in terms of optical, radiative nature only. It should be noted, however, that the continuous hybridization of RTP with CVD and the trend towards lower process temperatures lead to an increased importance of convection and thermal conduction [84, 85] with respect to the radiation term, which is $\propto T^4$ (see Eq. 9.1).

The simplest way to achieve gas flow uniformity (in direction and magnitude) is by linear, laminar gas flow from the inlet side to the outlet side. The redistribution of the gas flow entering through a pipe is done by inserting special quartz baffles perpendicular at the inlet side to offset the Poiseuille distribution.

A good way to improve uniformity of gas supply in CVD is the use of shower heads. Although complex, it offers improved uniformity in epitaxial growth. Usually, the gas enters from a quartz or metal grid above the wafer front side and leaves the chamber at the wafer back side. Often this leaves only the wafer back side for heating and temperature sensing. Note that in this configuration parasitic deposition onto the shower head often occurs because of the heating of the shower head by the wafer front side. This, and the resulting temperature nonuniformity, can be reduced by coating the wafer side of the shower head with a highly reflective coating. The particle problem can be alleviated when the configuration is inverted, that is, when a geometry with the wafer face down is used.

Another option is the use of annular and tangential gas inlets and outlets, which is more complicated but gives a centrosymmetric symmetry conformal to the wafer. Note that tangent inlets have the risk of introducing chemical memory effects.

C. Temperature Sensing, Calibration, and Control

Current RTP systems employ two classes of temperature sensors: *thermoelectric* detectors and *photon* (or *quantum*) detectors. Detailed information on these detectors is given elsewhere [86–88]. This subsection will highlight only the main characteristics of these sensors in the scope of the spectral aspects that play a role in the sensor selection. The thermopile detector will receive some more attention, since it is the "workhorse" in RTP.

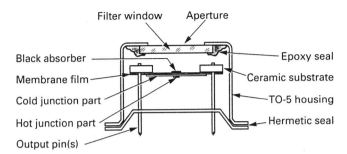

FIGURE 9. Cross section of a thermopile detector, sealed in a TO-5 housing.

1. Thermal Detectors

The thermoelectric detectors mainly used in RTP are the thermocouple and the thermopile. The cantilevered *thermocouple* is the classical sensor in RTP. The basic type is a bimetal junction of Ni-Cr/Ni-Mn-Al, W-5%Re/W-26%Re or Pt-Rh alloy combinations. Its operation is based on the Seebeck effect: a temperature gradient ΔT along the wires (note: *not* across the junction) generates a voltage $\Delta V = \alpha_S \Delta T$ over each lead end, with α_S being the Seebeck coefficient.

Modern *thermopiles* also operate on the Seebeck principle by noncontact absorption of the infrared photons emitted by the hot object. Figure 9 shows a cross section of a typical thermopile. The active detector part is a ceramic heat sink, supporting a 1–3 μm thick membrane of polyethylene terephthalate, etc. Thin film structures are deposited on the hot junction area of this membrane by modern integrated multilayer technology. To increase the absorption efficiency, the other side of the free-standing membrane area is blackened with paintblack or goldblack.

The layers in the hot junction are from materials with highly positive α_S, e.g. Sb, and highly negative α_S, e.g. Bi, in an array of many (up to 200) junctions [89, 90]. The semimetal combination $Bi_{0.9}Sb_{0.1}$/Sb is reported to give the maximum sensitivity for metal thermopiles, exceeding 100 μV/K [89]. Silicon-based thermopiles can give up to twice this value [91, 92].

The shape and the size of the aperture of the thermopile sensor, like every pyrometer, should be such that the hot junctions receive maximum signal and minimum optical noise. Thus the aperture should be conformal to the shape of the junction array. Usually this shape is rectangular or circular [89] (*cf.* Fig. 10). The aperture size should be small enough to screen off perturbing radiation. Figure 11 shows that this perturbing radiation can be divided into zero-order (transmitted) radiation, first-order radiation (missing the

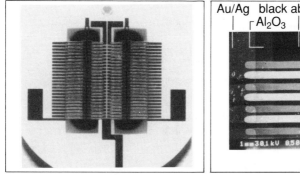
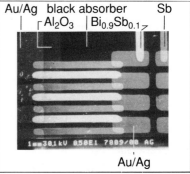

a.   b.

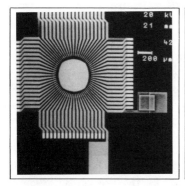
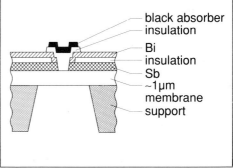

c.   d.

FIGURE 10. Thermopile structures. (After Roozeboom [42].) (a) Rectangular dual element, used by AG Associates (one cold reference half is covered by metal strips); (b) scanning electron microscope (SEM) magnification of upper right part of (a); (c) SEM example of circular single element (black absorber layer removed); (d) cross sectional view of (c).

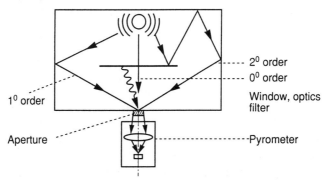

FIGURE 11. Zero-, first-, and second-order perturbing radiation, interfering with the pyrometer reading in an RTP system.

wafer) and second-order radiation, reflected by the wafer and the chamber before reaching the pyrometer. In order to make a thermopile less sensitive to drifting background temperature, it is often water cooled and made in a dual-element version [89] (see Fig. 10). Only one half receives the wafer radiation, while the other cold half is screened off.

Photodetectors utilize a quantum phenomenon in which the infrared photons are absorbed by semiconductor layers. Upon their absorption the photons create carriers that are detected in several ways, as by a photoconductive or photovoltaic diode [86–88]. The detectivity and sensitivity spectra of these sensors were given in Fig. 2c. Note that all types mentioned there are actually used in RTP (see Section IV).

Infrared pyrometry by thermopile and photon sensors dominates over thermocouple control. The main disadvantages of the cantilevered thermocouple are its irreproducible thermal contact, its noninert contact with the wafer, and its degradation upon intense thermal cycling. Thus the thermocouple is used mainly for calibration purposes. One solution is the use of quartz-sheathed thermocouples [75] not only for calibration but also for processing at any temperature, e.g. RTCVD [15–17]. Other inert embedding techniques are described in paragraph III C 2.

Thermopile detectors are favored over photon detectors because they are cheaper, more compact, and sensitive in a broader spectral range (see Fig. 2). On the other hand, they have a slower response with a time constant $\tau \propto 10$ ms [89]. Photon detectors are more sensitive in narrower spectral ranges and have much shorter response times of $\tau \propto 10\,\mu$s [89].

2. Temperature Calibration

The optical pyrometers described above give only a relative signal and need calibration to measure absolute temperature. The calibration of a pyrometer in an RTP system is done at two distinct levels. First is a *system-level* calibration, obtained by matching the pyrometer reading to that of a thermocouple in contact with a sacrificial reference wafer. For frequent system recalibration one prefers indirect methods such as temperature sensitive reactions. If the emissivity is likely to show small variations for the individual wafers within a batch, a *wafer-level* calibration is needed. Here, one measures the extrinsic emissivity indirectly by an *ex situ* reflectivity measurement prior to processing, or one measures the effective emissivity *in situ* before or during processing.

a. System-Level Calibration. The classical means of system-level calibration uses a cantilevered thermocouple, usually chromel–alumel. The method has a few disadvantages such as irreproducible contact, degradation upon repeated high-temperature cycling, heat loss along the leads, and noninert contact causing chemical reaction with the wafer. Some of these problems can be reduced more or less successfully by physically embedding the thermocouple in the wafer. One example of mounting is drilling a cavity in a reference wafer and pasting the thermocouple with heat-conducting ceramic cement [93]. Another example is tungsten inert gas or e-beam welding of W–Re based thermocouples [94] or drilling a hole in a thick (6 mm) silicon wafer, where the hole behaves as a large blackbody cavity enclosing the thermocouple [95]. Instead of a thermocouple one can also use an optical fiber embedded in a susceptor [96].

The cantilevered or embedded thermocouple is normally used only for one-time calibration. For frequent recalibration and checks on the uniformity, accuracy, and reproducibility, one uses indirect test reactions as "thermometers" [97, 98]. In that case one must know the extent of such reactions as a function of temperature. An example is the partial activation of implantations, as measured by their sheet resistance R_s. For certain temperature ranges one can design reactions, from which one can derive the temperature from the derivative dR_s/dT_{average}. Suitable reactions are the rate of oxide growth for the high-temperature range (1000–1200 °C), and partial activation of arsenic, phosphorus, and boron implantations for the medium temperatures (600–1000 °C). These indirect measurements give better reproducibility than the direct methods and complete mappings of the temperature. Therefore they are preferred to direct single or multiple thermocouple measurements. Moreover, they are also preferred to other indirect measurements such as high-speed calorimetry or scanning pyrometry [99]. Also, they enable an easy comparison between the performance of different systems, as in round-robins [100–102].

Indirect temperature measurement and calibration are more difficult in the low-temperature regime (400–600 °C). Yet qualitative indicators exist such as the visualization of isothermal contour lines by melting and recrystallization of the appropriate eutectics. An example is the eutectic alloy composition of Ge/Al layers, with a melting point around 420 °C [103], on top of a silicon wafer [104].

b. Wafer-Level Calibration. Calibration at the wafer level implies some method of emissivity correction. This can occur at the wafer *batch* level, or at the individual *wafer* level.

One example of an *in situ*, batch-level calibration is using an opaque, graphite susceptor coated with silicon carbide [50–52, 105]. The susceptor holds a sacrificial wafer in thermal contact and an embedded thermocouple close to the wafer. The thermocouple is irradiated only by the SiC-coated graphite, which acts as a grey body. Using this compensation technique, one could reduce the variation in thickness of nominally 24.5 mm thick SiO_2 films from 7.5 to 0.4 nm, when growing on Si wafers from four vendors with completely different wafer backside textures.

Most RTP equipment manufacturers offer some emissivity adjustment option by measuring the reflectivity and/or transmissivity of the wafer back side, or are developing it.

The emissivity is then obtained from Kirchhoff's law (Eqs. 9.5 and 9.6).

Ex situ *wafer-level calibration.* The simplest compensation is by measuring the reflection of a single wavelength laser beam, either in a preload station (*ex situ*) or in the reaction chamber (*in situ*). Note that the photons emitted by the laser have a wavelength below the 1.2 μm Si band gap. This implies that the laser light is not transmitted and can only be absorbed or reflected (*cf.* Fig. 2b).

Another way is to use a certain wavelength band, such as a LED or a blackbody source. Peak Systems, for example, measure at room temperature the specular reflection by the wafers that are irradiated in a preload station by a blackbody source, kept around 400 °C [106–108]. The measurement is automatically done prior to processing by a 4.65–5.15 μm pyrometer. This wavelength range corresponds to λ_{peak} of the black radiator. Here again, drastic improvements were claimed in the accuracy and reproducibility for wafers with very different back sides [106].

Note that some of these methods measure at best the extrinsic emissivity for incident wavelengths that do not completely correspond to the process pyrometer wavelength band. Thus, these methods may cause temperature errors due to the possible variations of the emissivity as a function of wavelength. The rest of this section will treat a few effective emissivity measurement techniques.

In situ *wafer-level calibration.* Dilhac *et al.* proposed a method for *measuring the reflection* of the light emitted by the system's lower lamp row [109]. Excessive wafer heating above 100 °C should be avoided, so that one measures only reflectivity without any emissivity.

A strong point is that the measurement is done by the process pyrometer with its own wavelength band (here 3.7–4.0 μm), thus eliminating the differences related to geometric effects (optical and dimensional effects of the chamber walls, etc.). A weak point is the low wafer temperature, where the reflectivity is a strong function of extrinsic parameters such as dopant level, frontside layers, etc. (see paragraph II A 2). For this reason real-time, *in situ* precalibration at process temperature is to be preferred (see below).

Ratio pyrometry (or multicolor pyrometry) involves the use of two or more wavelengths. For a detailed discussion on multicolor pyrometry one is referred to Refs. 110–112. Two-color pyrometry is based on Eqs. 9.2 and 9.4. If the energies E_1 and E_2 in two narrow bands around λ_1 and λ_2 are measured, the ratio R_{12} becomes

$$R_{12} = \frac{E_1 \, d\lambda_1}{E_2 \, d\lambda_2} = \frac{\varepsilon_1 f(\lambda_1, T) \, d\lambda_1}{\varepsilon_2 f(\lambda_2, T) \, d\lambda_2} \tag{9.7}$$

This ratio becomes independent of the emissivity, provided that $\varepsilon_1/\varepsilon_2$ is a constant over the wavelength range selected. So ideally the temperature is obtained regardless of the emissivity, and this pyrometer can be used with varying or unknown emissivity. In practice, however, errors are introduced because of the variations in the monochromatic emissivities as a function of wavelength. So in practice the wavelengths must be close together. On the other hand, the sensitivity increases as the wavelengths move further apart. The problem in getting useful accuracy was pointed out by Coates [112], who concluded that dual-color pyrometry gives no more accurate temperature value than single-color pyrometry using an estimated emissivity. Moreover, the range and the response time of commercial dual-color pyrometers are somewhat limited: 400–800 °C and 200 ms, respectively [113].

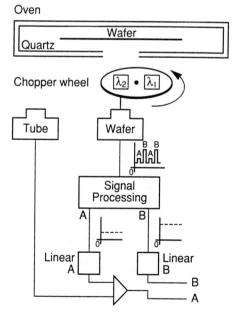

FIGURE 12. Diagram of a recent two-color pyrometry option. (After Mordo et al. [115].)

One form of dual-color pyrometry, closely resembling an earlier published ratio pyrometry method [114], was recently introduced by Mordo et al. [115, 116]. This method was developed for *in situ*, fine correction only after a rough precalibration on wafer batch-level with a series of "master" wafers on a graphite susceptor, embedding a thermocouple [50–52, 105]. These wafers have different backside layers (oxide, nitride, polysilicon), one of which is varying in thickness. When the precalibrated emissivities ε_1 and ε_2 at wavelengths λ_1 and λ_2 are plotted against the layer thickness, they each show a distinct oscillating modulation.

The actual *in situ* correction after the precalibration is illustrated in Figs. 12 and 13. A dual-color photodetector of the PbSe-type is used. It has a response time $\tau = 20\,\mu s$, which is small enough to read two adjacent wavelengths (approx. 3.3–3.5 and 4.3–4.7 μm) at a sampling rate of $\propto 8$ kHz. The signals are separated into two channels by a chopper wheel

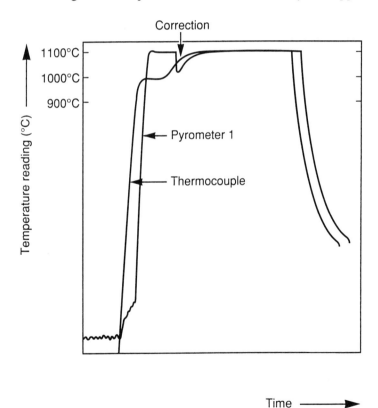

FIGURE 13. Signal correction in two-color pyrometry option of Fig. 12. (After Mordo et al. [115].)

that contains filters of alternating wavelengths. One channel measures and controls the temperature and the other monitors the emissivity. The correlation between the two signals, as obtained from the precalibration, is emissivity independent, so any change in emissivity causes a change in this empirical relationship. Figure 13 gives a nice illustration of what happens upon the insertion of a wafer with a slightly higher emissivity (i.e., thicker backside oxide layer) than the set emissivity belonging to a previous wafer with lower emissivity (thinner oxide). Both wafers are applied with a thermocouple on top. Initially, the first channel controls an apparently equal temperature. Because of the higher emissivity, however, the actual thermocouple reading is too low. Next, the second channel registers an increased wafer emissivity. A software algorithm corrects for this change, which takes about 3 sec prior to the actual processing of each wafer.

Note that the *in situ* correction is usually performed at process temperature or just below it. Note also that the choice of the two wavelengths is such that the combination has minimum interference by warm system parts, such as the quartz reactor (see Fig. 14). Yet the interference is not zero.

A powerful real-time technique is the *ripple amplitude technique* [117], shown in Figs. 15 and 16. Here one uses optical fiber thermometry (OFT)

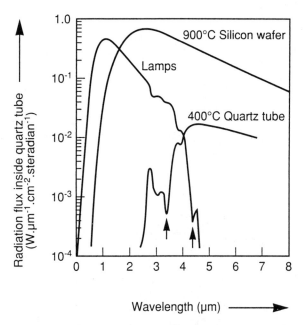

FIGURE 14. Radiation flux characteristics, showing the optima for dual-color pyrometry at 3.3–3.5 and 4.3–4.7 μm, with minimum interfering radiation (Source: AG Associates).

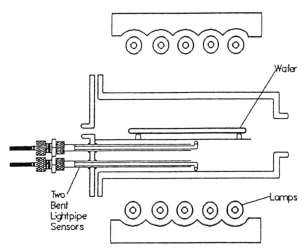

FIGURE 15. Diagram of the setup of the "ripple amplitude" emissivity correction (Source: Accufiber).

with two sapphire light pipes. One views the lamps, the other the wafer back side. With their large numerical apertures (52° cone angle) the sensors are relatively insensitive to wafer roughness, and thus sensitive enough to monitor the 120-Hz modulated ripple amplitude of the 60-Hz lamp light. The first sensor measures the ripple amplitude ΔI_l of the primary lamp light flux $I_l(t)$, and the second sensor measures the amplitude ΔI_w of the total photon flux $I_w(t)$ from the wafer (see Fig. 16). The flux $I_w(t)$ has a reflective

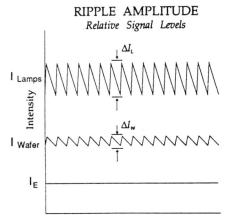

FIGURE 16. Illustration of the principle of the "ripple amplitude" emissivity correction. (From Schietinger et al. [117].)

component $R_w(t)$ superposed on a constant component I_E emitted by the hot wafer. The wafer's reflectivity is obtained by

$$\rho = \Delta I_w / \Delta I_l \qquad (9.8)$$

Next, the wafer emissivity component I_E is obtained by subtracting the wafer reflectivity from the total wafer photon flux, according to Kirchhoff's law, Eq. 9.6:

$$I_E = I_w - I_l(\Delta I_w / \Delta I_l) \qquad (9.9)$$

The emissivity is fed back into a control loop. Note that the Si photodiode is fast enough to detect the 120-Hz ripples. Its detection wavelength at 0.95 μm matches closely to λ_{peak} of the lamp emission. On one hand this ensures a maximum reflection signal, and the 0.95 μm implies that the Si emissivity is no longer sensitive to temperature, thanks to the short wavelength (see Fig. 2). On the other hand, this method remains very sensitive to small detection errors in the primary photon flux from the lamps, and cannot be used for medium temperatures ($\leq 700\,°C$), that is, for low emission/reflection (I_E/I_w) ratios. Some optimization is possible here, for example by using wavelength selective interference mirrors coated on the lamps [41]. These mirrors reject the 0.94–0.96 μm emission by the lamps, and keep the pyrometer from picking up interfering lamp light. Recently, further reductions of the low-end temperature down to 350 °C and of the wafer emissivity below 100 °C were claimed by broadening the wavelength band to 0.5–1 μm [118]. Some problems remain. One is that of excessive

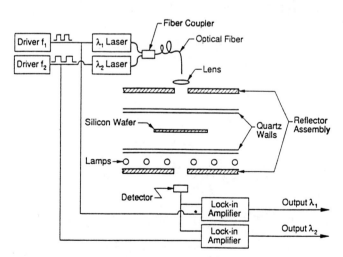

FIGURE 17. *In situ*, real-time temperature measurement by infrared transmission. (From Sturm and Reaves [18].)

heating by reflection and radiation from the hot wafer. Moslehi *et al.* [22, 119] used fibers with water-cooled tips to reduce this. Another problem to be solved is that of parasitic deposition on the fibers.

While the above fiber optics method is *reflection*-based, and especially suited for temperatures above 800 °C, Sturm *et al.* [18, 19, 120] have developed a *transmission*-based, and thus complementary, optical fiber thermometry (OFT) method, which is suited for medium temperatures. A diagram is shown in Fig. 17. Here the 1.3 or 1.55 µm light of a solid-state

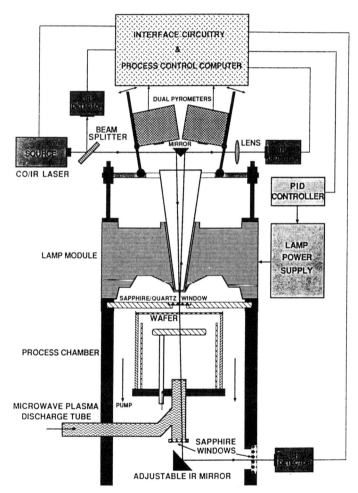

FIGURE 18. Dual-pyrometer temperature measurement with *in situ*, real-time emissivity measurement and compensation by simultaneous transmission and reflection measurement. (From Moslehi [121].)

laser is pulsed into probe fibers, directed toward the wafer top. Because these wavelengths are above the Si band gap, they can only be absorbed by free carriers or transmitted. The free-carrier absorption is temperature dependent and is used to find the temperature. Another fiber below the wafer measures this transmission from 400 to 800 °C. The transmitted fraction is stripped by a lock-in amplifier and normalized by dividing by the room temperature transmission. Note that this method has its limitations. The transmissive properties depend on the sample thickness and composition. Thus each sample needs precalibration. The principle of *combining reflection and transmission* OFT in one RTP chamber has been prototyped by Moslehi [121, 122] (see Fig. 18). A pulsed laser beam with 5.35-μm wavelength is sent into a beam splitter for incident power measurement. The beam is further led by a conical reflective light pipe, or a fiber, through the lamp shroud and through a sapphire window to reach the back side of the wafer in the process chamber. Infrared detectors can measure the transmitted and the reflected signals, yielding the effective emissivity by Eq. 9.5. The emissivity is fed to a processor for real-time compensation of the pyrometer signal(s). The two pyrometers can be operated in several ways. With wavelength bands around 3.8 and 5.1 μm, they can be used as commutating temperature sensors, having a measuring range of 300–1250 °C, or as a ratio pyrometer. Another option is to use one pyrometer at 8–14 μm for measurement of the window plate temperature and to correct for it.

The above option is complicated and not fully mature, yet is a very powerful option.

Another method, recently introduced [123, 124], is that of *wafer diameter extensometry*. Originally designed for off-line calibration, it is now also used for *in situ*, real-time temperature control. Figure 19 shows the

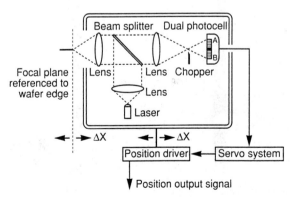

FIGURE 19. Principle of wafer expansion thermometry after Snow [123] (see text) (Source: Peak Systems).

principle. A laser beam is focused through a lens system on the edge of a wafer. The wafer will expand upon heating, thus defocusing its edge and bringing the signals of the dual photocell out of balance. The photocell feeds back the imbalance to the servo system until balance is reached again. The method uses the fact that the thermal expansion of Si is not affected by thin overlayers: $\alpha_{Si} = 4 \times 10^{-6}/°C$. For a 200-mm wafer the displacement is $dL/dT = 0.8\,\mu m/°C$. The temperature resolution of this autofocus technique is claimed to be 0.1–0.3 °C, depending on the wafer diameter and the process temperature, over a range of 100–1300 °C. Note that the edge rounding of the wafer may give some focusing problems. This may be (partially) circumvented by focusing on a small, polished quartz rod pressed against the wafer edge. This in turn may introduce some local heatsink effect. Note also that this method presumes a fully isothermal temperature profile across the wafer, which generally is not the case.

IV. Survey of Commercial Rapid Thermal Processing Equipment

Having discussed the thermophysical aspects of RTP and the individual components of RTP systems, we can now highlight the characteristics of commercial RTP systems. We will focus here on the design aspects. A good design will exclude perturbing zero-, first- and second-order radiation, such that only the Lambertian (diffuse) radiation from the hot wafer reaches the temperature sensor. Table I summarizes the general characteristics of RTP systems from 22 vendors, of which some 14 are currently active. Note that these and other vendors' systems evolve continuously. For that reason Table I is by no means a definitive list.

The table shows that most systems contain either an array of ten or more tungsten–halogen lamps each generating 1–6 kW, or a single arc lamp generating 35 kW or more. The total power, thus installed, is usually sufficient. From the Stefan–Boltzmann law, Eq. 9.1, we can calculate that, for example, the power required to maintain a 200-mm silicon wafer at 1200 °C amounts to 12 kW, assuming that all the thermal energy radiates freely away without backreflection onto the wafer surfaces.

The table shows also that most systems employ thermopile or photon detectors, which come from standard pyrometer manufacturers like Ircon, etc. Only a few are upgraded, such as the InAs detector, which is upgraded to an electronically chopped, dual-channel detector.

A general measure to minimize heat source noise is the use of small apertures and optics on or in front of the pyrometer. This helps to avoid stray radiation.

Table I. Survey of commercial RTP equipment.*

Manufacturer	Latest model	Heat source[a]	Reactor chamber and shape[a]	Standard pyrometer[a,b]	Emissivity correction[c]	Added features[a,c]	Refs.
Addax S.A.[d]	VEGA-908	W-halogen T, B, S	Quartz Rectangular	Thermopile detectors (dual) B 3.7–4.0 μm (<800 °C) 2.0–2.6 μm (>800 °C) Optics/no chopper		Extra side lamps Optional: graphite susceptor for III-Vs, with forced cooling Optional: glove box	7, 109, 154, 181
AET Thermal	JK 6000	W-halogen T	Quartz lined stainless steel Cylindrical, with air-cooled quartz plate	Thermopile detectors (dual) B 3.7–4.0 μm (<800 °C) 2.0–2.6 μm (>800 °C) Optics/no chopper		Cryopump for UHV, wafer face down and cluster module Optional: graphite susceptor sandwich for III–V compounds	
AG Associates	Heatpulse 4108	W-halogen T, B	Quartz Rectangular	Thermopile detector (twin) B 4.3–4.7 μm No optics/no chopper	Graphite precalibration susceptor, with optional two-color (3.4, 4.5 μm) PbSe photon detector for *in situ* fine correction	Cross lamps, poly-Si guard ring and cluster tool Optional: graphite susceptor sandwich for III–V compounds	44, 50, 52, 55, 58, 74, 105, 115, 116, 125, 126, 141, 149, 160
Applied Materials*	MAC	W-halogen T, (B)	Quartz Double-dome: upper and lower quartz domes flanged together	Thermopile detectors (dual) B 4.8–52 μm Optics/no chopper	NA: rotating SiC/ graphite susceptor	Dual point (center edge) pyrometry; gas inlets in dome center, gas outlets in flange; cluster module to their sputtering tool	162, 182, 183

382

Company	Model	Lamps	Chamber	Temperature sensor	Wafer rotation	Notable features	Refs.
ASM Epitaxy	Paragon One, Model R1	W-halogen T, B	Quartz Rectangular	None, but 4 quartz-sheathed thermocouples (1 master in susceptor, 3 slaves in guard ring)	NA: rotating SiC/graphite susceptor	Cross-lamp array, SiC coated graphite guard ring around susceptor, and tube etch clean cycle	15–17, 75, 76, 139, 160
AST Elektronik	SHS-1000G	W-halogen T, B	Hydroxylated quartz and quartz Rectangular	PbS photon detector T, S 2.7–2.8 μm Optics/chopper	Optional ex situ room temperature reflectometry at 2.7–2.8 μm	Guard-rings Cluster module, and wafer-to-reflector movement[d]	104, 133, 159, 160
CVC Products Inc.[e]	2017-CVD	W-halogen T, (B)	Stainless steel Rectangular, with water- or aircooled quartz plate	Thermopile detector T 3.7–4.0 μm Optics/no chopper	Optional: other thermopiles	Cluster module Optional: rotating graphite susceptor and dual point (center/edge) pyrometry[d]	44, 117
DaiNippon Screen	LAW-815-A	W-halogen T, B	Quartz Rectangular	Thermopile detector (twin) B, S 5.5–10.5 μm Optics/no chopper		Cross lamp array Cooled pyrometer windows	130
High Temperature Engineering Corp.	Reliance-800	Continuous resistive heat source around wafer	SiC Bell jar with high aspect ratio	Si photon detector B 0.94–0.96 μm Optics/no chopper		Continuous uniform heat source	44, 71, 72, 101, 102

Table I. Survey of commercial RTP equipment.* Continued

Manufacturer	Latest model	Heat source[a]	Reactor chamber and shape[a]	Standard pyrometer[a,b] Emissivity correction[c]		Added features[a,c]	Refs.
Jipelec	FUV-4	W–halogen B	Stainless steel Cylindrical, with water or aircooled quartz plate	Thermopile detector 4.8–5.2 μm Optics/no chopper	T Reflective cavity effect Optional: 8–14 μm thermopile detector (optics/no chopper)	High-vacuum system with tangent annular gas inlets, UV-lamps (T), and microwave plasma preclean station	24, 44, 167, 182, 184
Koyo-Lindberg	RLA-6100 II	W–halogen T, B	Quartz Rectangular (1 atm) or cylindrical (vacuum)	Thermopile detector 5–8 μm Optics/no chopper	B	Cross-lamp array	101
Modular Process Technology Corp.[f]	CVD-6000	W–halogen T or B	Stainless steel Rectangular, with water-cooled quartz plate	Thermopile detector 4.5–5.0 μm Optics/no chopper	T, B, or S *In situ*, real-time reflectometry at 4.5–5.0 μm[d]	Cross-lamp array, high-vacuum modular system, and tunable wall temperature (25–90 °C) Optional: microwave plasma injection	44

Peak Systems	Horizon	Ac Xe arc lamp T	Stainless steel or Hastelloy Cylindrical, with quartz plate	InAs photon detector (dual-channel) 2.6–3.5 μm Optics/no chopper	B	Integrated *ex situ* room temperature reflectometry at 4.65–5.15 μm *In situ*, real-time optical monitoring of wafer diameter expansion	Si-specific lamp, and cluster module	11, 44, 69, 70, 79, 80, 106–108, 123, 124, 185
Rapro CVD**	Integra One	W-halogen T	Quartz lined stainless steel Cylindrical, with aircooled quartz plate	Thermopile detector 4.8–5.2 μm Optics/no chopper	B	Integrated *in situ*, real-time ellipsometric reflectometry (LEAP), and OFT[d]	Multichamber, load-lock system, rotating quartz holder, shower head, cluster tool Optional: dual point (center/edge) pyrometry[d]	12, 13, 82, 129, 160
Ulvac/ Sinku-Riko	RTA-6RS	W-halogen T	Quartz Bell jar	HgCdTe photon detector 5–9 μm Optics/no chopper	B			83
BCT Spectrum Inc.*	Vision	W-halogen T	Stainless steel and nickel Cylindrical, with air-cooled quartz plate	None, but 2 thermocouples through quartz plate against inverted wafer		NA	Multichamber, load-lock system, double circular bulb lamp array, shower head and etch clean cycle	8, 23

Table I. Survey of commercial RTP equipment.* Continued

Manufacturer	Latest model	Heat source[a]	Reactor chamber and shape[a]	Standard pyrometer[a,b]	Emissivity correction[c]	Added features[a,c]	Refs.
Eaton Corp.*	ROA-400	Dc Ar arc lamp T	Anodized Al Rectangular, with water-cooled quartz plate	Ge photon detector 1.4–1.5 μm Optics/no chopper	B	Si-specific lamp	58, 66–68, 99, 132
Tamarack Scientific Company*	M185	W-halogen B	Stainless steel Rectangular (high aspect ratio), with quartz plate	Thermopile detector 4.8–5.2 μm Optics/no chopper	T	Reflective cavity effect *In situ* ellipsometric end-point detection through cool quartz side windows	9, 61–63
Varian–Extrion*	RTP-8000	W-halogen T	Al or st. steel Rectangular, with air-cooled quartz plate	Thermopile detector 4.8–5.2 μm Optics/no chopper	B	Optional: 8–14 μm thermopile detector (optics/no chopper)	97, 184

* BCT Spectrum, Eaton, G-Squared Semiconductor Corp., LEISK, Tamarack, TEL/Thermco, and Varian discontinued RTP. Applied Materials have been developing prototypes with a star-like array of 16 lamp tubes with, alternatingly, parabolic back reflectors focusing on the wafer center and flat, dispersive back reflectors heating the wafer edge [183], as well as systems with a hexagonal lamp panel [78].
** Rapro CVD, a division of AG Associates.
[a] T, B, S = top, bottom and/or side geometry.   [b] Detector classification of Dimmock [87].
[c] d = in development; OFT = optical fiber thermometry; LEAP = Laser Emissivity Analyzing Pyrometry.
[d] Formerly Addax Technologies.   [e] Formerly Process Products Corporation.   [f] Formerly Nanosil.

Note that some manufacturers (still) employ thermocouple control, especially in dedicated low-temperature CVD systems, but also for high-temperature applications.

From Table I and Fig. 2 we can see three distinct ranges for the pyrometer pass bands: the middle, far, and near infrared. Each of these ranges will be discussed below.

A. THE MIDDLE INFRARED (3-6 μm)

Most systems use pyrometers in the 3-6 μm range. Figure 2 shows that the filters, narrowing the optical bands actually used, are outside the spectra emitted by the heat sources. Thus they circumvent the problems related to the transparency of silicon at low temperatures, which causes the pyrometer to receive interfering radiation from the heat source. The upper limit of most middle-infrared pyrometers is determined by the transparency of quartz, either from the reactor or from (thinned) windows. Its cutoff is at 4-5 μm wavelength, depending on the thickness of the quartz (see Fig. 2a).

Examples: Peak Systems use a xenon arc lamp emitting up to 1.4 μm and an InAs photodetector operating at 2.6-3.5 μm. Thus the pyrometer is immune to the radiation (zero- to second-order) from the arc lamp.

In the case of tungsten–halogen lamps most systems are operated with the quartz of the lamp envelope, the reactor tube, or the window plate acting as filters: beyond 4.5 μm essentially no radiation reaches the wafer. Thus pyrometers with a bandpass filter below or around 4.5 μm are chosen. CVC's single pyrometer operates at 3.7-4.0 μm; MPT's at 4.5-5.0 μm. Applied, Jipelec, Rapro, Tamarack, and Varian use 4.8-5.2 μm.

AG Associates use a twin pyrometer operating at 4.3-4.7 μm [125]. One pyrometer measures the radiation of the wafer through a nominally 0.15-mm thin quartz window. This is thin enough to transmit most of the radiation. Simultaneously, the second pyrometer measures the radiation from the 3-mm thick quartz reactor bottom only, that is, no wafer. The "true" temperature is then approached by electronic correction. Note that the thickness of the thin view window can pose reproducibility problems in the temperature reading. Temperature errors up to 100 °C have been reported due to the varying thickness of the thin view window [126].

It should be noted that in the middle infrared the temperature registration can still be perturbed. One problem is the presence of *warm system parts* (e.g., quartz parts at 400 °C, see Section III A 2) that can radiate false wavelengths into the sensor. For that reason companies like Addax and AET use a dual, commutating pyrometer system. They have a 3.7-4.0 μm pyrometer for the lower process temperatures and a 2.0-2.6 μm pyrometer for higher temperatures, where it has a higher signal-to-noise ratio than the former.

Another problem, only recognized recently, is that the *ambient gas* can absorb wafer radiation in the same wavelength range. An example was given by Chang *et al.* [127], who studied rapid thermal oxidation of a thermocouple-instrumented silicon wafer with N_2O. This reagent has two strong absorption peaks at 4.49 and 4.52 μm [128], whereas their pyrometer has its center wavelength at 4.5 μm. Thus, they observed deviations up to +50-200 °C in the actual thermocouple reading, with respect to the apparent pyrometer reading, as well as heavy oscillations in the temperature during ramp-up.

B. THE FAR INFRARED ($\geq 6\mu$m)

The trend towards lower process temperatures runs parallel to the increasing number of systems with far-infrared detectors. Quartz is no longer transparent here, so window materials such as calcium fluoride [62, 129] and barium fluoride are used, typically located in the reactor bottom. The fluorides have spectral cutoff around 9 and 12 μm, respectively [89].

Using long wavelengths has some advantages. One is that the problem of the transparency of silicon at low temperatures is reduced [49] (*cf.* Fig. 2b). Figure 20 shows that reproducible measurement of temperatures as low as 100 °C [24] is possible for a cold-wall system with water-cooled quartz top plate, when using an 8-14 μm pyrometer with a special, small aperture.

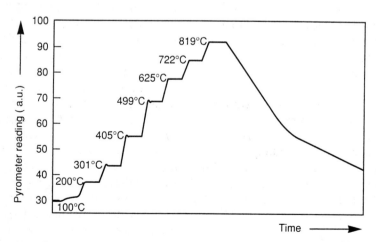

FIGURE 20. Example of low temperature control by using an 8-14 μm pyrometer in a cold-wall system with water-cooled quartz top plate (Source: Jipelec).

Another advantage is that at longer pyrometer wavelengths the oscillation frequency of the emissivity as a function of the thickness of growing overlayers is reduced [39]. The oscillation amplitude increases also, but this can be suppressed by using highly reflective chamber walls [39] (see Section V).

A disadvantage of using long wavelengths is the low signal-to-noise ratio of the detection. Consequently, all reactor components should be kept "cold", including the quartz parts and fluoride windows. If not, these warm system parts interfere with the detection of the wafer radiation: quartz lamp envelopes can reach 400 °C [34], and reactor tubes 100–200 °C ($\rightarrow \lambda_{peak} \cong$ 5–8 μm). The same holds for fluoride windows when attached directly onto the quartz. DaiNippon Screen has reduced this problem by using a twin pyrometer [130] (see Fig. 21). One pyrometer (not shown) views the wafer back side. The other, mounted aside, views the background radiation from the warm quartz walls of the lamps and the reactor via an extra, tilted chip that is placed in the reference sample holder. The BaF$_2$ windows are mounted on *protruded* quartz tubes and are further cooled by nitrogen gas, flowing into the pyrometers.

Examples: most Japanese systems employ far-infrared temperature sensors. Jipelec offers 8–14 μm pyrometry as an option, and so does Varian.

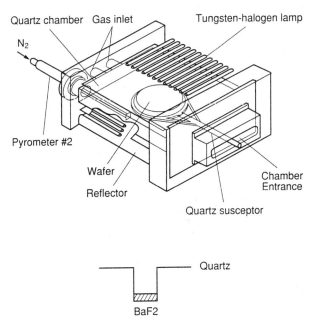

FIGURE 21. Diagram of DaiNippon Screen's RTP system. Note that the bottom pyrometer is not drawn. (Source: DaiNippon Screen).

C. The Near Infrared (0.8–3 μm)

Another trend is toward the near infrared. The part below 1.2 μm is most attractive, because silicon is opaque here. Practically all heat sources have their λ_{peak} centered around this range, so optical noise should be absolutely eliminated. This can be done by selective narrow-band pyrometers in combination with screening these wavelengths from reaching the wafer and the pyrometer [41].

A clear advantage of shorter wavelengths is that the temperature error ΔT due to an emissivity error $\Delta \varepsilon$ is small:

$$\Delta T = \frac{T^2 \lambda \Delta \varepsilon}{c_2 \varepsilon} \tag{9.10}$$

Equation 9.10 can be derived from the Planck equation, Eq. 9.2, by assuming that $e^{c_2/\lambda T} \gg 1$ [131].

Examples: HTE use a 0.94–0.96 μm narrow-band pyrometer, where silicon is opaque. Eaton selected a pyrometer with a 1.4–1.5 μm bandpass filter [132]. This range coincides with a strong absorption peak by the water wall in their argon arc lamp. AST Elektronik make their reactor selectively absorbing by making the top and the sides of the reactor from synthetic, hydroxylated quartz [133], which has a strong absorption at 2.6–2.8 μm (see Fig. 2). The reactor bottom is from normal quartz. Perturbing stray light from the lamps is further reduced by a hydroxylated quartz plate below the reactor, in front of the bottom lamps. This bottom plate contains a small quartz view window to allow the pyrometer to sense the wafer radiation at 2.6–2.8 μm.

V. Temperature Nonuniformity, System Modeling, and Effective Emissivity

The design of the RTP system and, to some extent, that of the integrated circuit patterns on the wafers plays a crucial role in the steady-state and the transient temperature uniformity. Poor designs can lead to significant temperature differences across the wafer, thus causing stress. Above the elastic limit this stress is relieved by plastic deformation, which leads to crystallographic slip [59] and disruption of circuits. Today this is the most serious problem in RTP.

The accomplishment of sufficient temperature uniformity is not a matter of uniform primary radiation flux to the entire wafer; the issue is much more complicated.

Nonuniformity is primarily due to edge loss and patterned structures. *Patterned* wafers (re)act differently and in a more complicated way than *bare*, or *nonpatterned* wafers with planar layers. The latter will have "only" a radial temperature gradient due to the different local energy balances between primary light flux and energy loss at the edges with respect to the center. A perfectly uniform temperature profile is achieved only when, for all parts of a wafer, the primary radiation flux and the reabsorbed heat radiation are balanced.

This section is mainly based on the excellent *modeling* and testing of *system design*, done by Kakoschke *et al.* [57, 58, 134] and Vandenabeele and Maex [45, 54, 56]. Their findings will be related to earlier modeling work on reflective [135] and blackbody cavities [73].

A. NONPATTERNED WAFERS

For nonpatterned wafers Kakoschke and Bussmann [58] have shown and explained that the dynamic temperature nonuniformity ($\Delta T = |T_{center} - T_{edge}|$) increases with higher ramp rates and shorter steady states, and that ΔT was the lowest in a non- or low-reflective chamber, i.e., in a "black" environment. The reflectivity of the chamber has been modeled with Monte Carlo simulations by Kakoschke *et al.* [57, 136, 137] to show its ambivalent function. Figure 22 shows what happens upon irradiation of a wafer in a

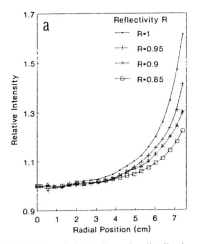 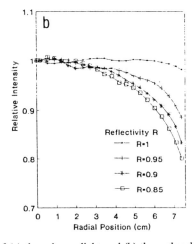

FIGURE 22. Relative intensity distributions of (a) the primary light and (b) the reabsorbed heat radiation over the wafer position with different "photon box" reflectivities. (From Kakoschke [57].)

reflecting box. Uniform primary radiation flux on the wafer gives a uniform temperature profile around the wafer center where it is all absorbed. Compared to the center part of the wafer, the edges, having higher surface/mass ratio than the wafer center, will absorb more heat during ramp-up and radiate more in the steady state. Thus, while a uniform primary light flux calls for a low-reflective, or even black chamber (Fig. 22a) to smooth out this effect, the opposite holds for uniform energy loss (Fig. 22b). Consequently, any improvement in primary heat flux from the lamps (e.g., laterally open reflectors) will counteract the back-reflected heat flux, because of the *synergy* between the two fluxes.

For that reason the reflectivity of the reflectors in lamp systems should be optimized such that a fairly uniform temperature profile is created at the very end of the ramp-up stage. It is at this stage of a thermal cycle that nonuniformity-induced slip is most critical. Mostly the optimization is done for the steady-state part, by making or shaping the reflectors to be more reflective at the edges, thus compensating for nonuniform back-reflection.

Further optimization can be achieved by decoupling the above synergy between temperature distribution and energy loss. One way to do this is to optimize the wafer–reflector distance, i.e. the aspect ratio of the photon box. The effect was simulated by Kakoschke *et al.* [57, 136, 137], and is shown in Fig. 23. The figure shows that the synergy effect becomes noticeable at 10 cm distance between top and bottom reflectors, that is, at a wafer–reflector distance <5 cm. So by moving the reflectors further apart one redistributes the primary and the back-reflected flux on the wafer.

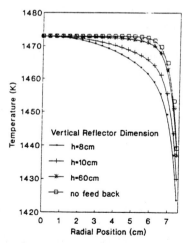

FIGURE 23. Calculated temperature distributions for various heights of a "photon box" with 100% reflecting reflectors. (From Kakoschke [57].)

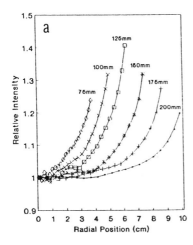 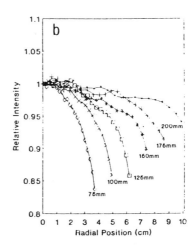

FIGURE 24. Relative intensity distributions of (a) the primary light and (b) the reabsorbed heat radiation for several wafer diameters. (From Kakoschke [57].)

Another very effective way to improve temperature uniformity is the use of *guard rings* around the wafer [59], usually Si or SiC coated graphite. Figure 24 shows the calculated distributions of the primary light flux and the reabsorbed heat radiation for wafer diameters ranging from 75 to 200 mm [57]. It is clear that the extension of the wafer diameter results in a more uniform temperature distribution. More importantly, the nonuniform part of these distributions is outside the wafer on the ring.

Note that guard rings may give problems upon high-throughput processing in the absence of adaptive edge temperature control. In that case the ring, especially a thick one, is not allowed to cool down enough. Consequently, it starts emitting an excess of heat to the wafer edge.

A more ideal solution is the use of dynamic compensation of the transient nonuniformities. This can be done by lateral edge loss compensation, by using an independently controlled subsidiary heat source around the wafer [81], or by active multizone control, preferably with guard rings.

B. PATTERNED WAFERS

On top of the uniformity problems, described above, patterned wafers (re)act in an even more complicated manner. If we envisage a wafer after LOCOS isolation, the center part of the wafer (the part illuminated during photolithography) will be covered with thick field oxide. This region will have a higher absorptivity than the peripheral regions, nonilluminated

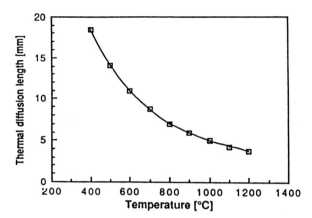

FIGURE 25. Lateral thermal diffusion length as a function of temperature for a 0.6-mm thick silicon wafer. (From Maex [45].)

during photolithography, that remain uncovered. Consequently the central field oxide part will heat up faster than the edge region. Figure 25 shows the lateral heat diffusion length in silicon as a function of temperature. It shows that discontinuities in the optical wafer properties in principal are felt over distances of the order of 1–2 cm [6, 45].

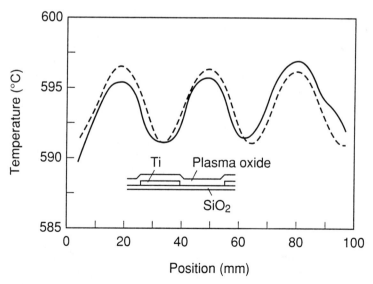

FIGURE 26. Calculated (- - - -) and measured (——) temperature across a 100-mm wafer with alternating pattern, after 10 s at 600 °C (200 K/s ramp rate) in an Eaton ROA-400 system. (From [58].)

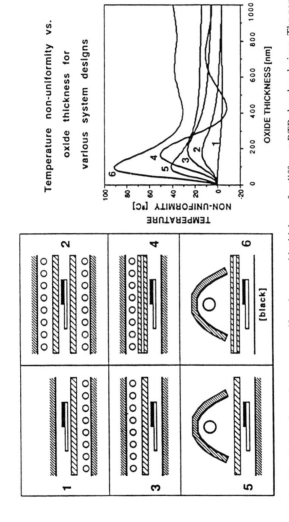

FIGURE 27. Simulation of steady-state noruniformity vs. oxide thickness for different RTP chamber designs. The nonuniformity is the temperature difference between the covered and the noncovered silicon halves. (From Vandenabeele and Maex [56].)

This will have an effect on the local temperatures, depending on the feature size of the patterns. For large-scale patterns above 1 cm the differences in heat absorption will not be averaged out due to the finite heat diffusion in Si. An example is given in Fig. 26. For a 1.5-cm structure size a temperature variation of 6 °C was observed [58]. The nonuniformity decreases for feature sizes below 1 cm, and is negligible below 1 mm, where the differences are leveled out completely.

As to the *layout design* over the wafer, we can conclude that one should avoid a design with large adjacent regions with different optical properties. It is obvious that the above not only holds for static differences in optical properties, such as oxide patterns, but also for dynamic differences. An example was given by Eichhammer et al. [6], who studied the self-aligned silicidation of large-size ULSI Ti and Co patterns. During silicidation, which starts from an initially uniform surface layer, the optical properties change. This results in temperature differences up to 20 °C.

If patterns are present or being formed, the effect of temperature nonuniformity also depends strongly on the *system design*, as shown by a recent round-robin [102]. Experimental results have also been modeled in terms of the optical parameters of the wafer and the chamber [54, 56] (see Fig. 27). It is evident that a reflective chamber reduces the effective energy loss, and thus the pattern-induced nonuniformity. To suppress nonuniformities during the *steady state* it is best to illuminate only the unpatterned side during the ramping stage and to compensate the energy loss during the stationary state by double-sided heating [54, 56].

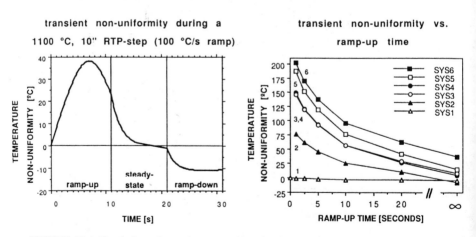

FIGURE 28. Simulation of transient nonuniformity for wafers, half covered with 0.2 μm oxide, in the RTP systems of Fig. 27. Left: time dependent in system 2; right: after completion of the ramp-up time. (From [56].)

Dynamic nonuniformity can be reduced by increasing the contribution of backside heating with respect to frontside heating [54, 56]. This is shown in Fig. 28. Here it is clear that transient nonuniformity is also suppressed by increasing ramp-up times.

C. EFFECTIVE EMISSIVITY AND CAVITY DESIGN

The above modeling work on system design suggested the use of multiple reflections by remote, highly reflective walls. This relates to the *reflective cavity*, published by Sheets [61–63] and shown in Fig. 29. If a wafer is placed in such a cavity, the radiation from one spot on the wafer is emitted diffusely (Lambertian) and thus reflected numerously onto the entire wafer. This way the reflective cavity works as a kaleidoscope, as treated in Section II. Thus the effective emissivity is integrated over the wafer and increased. Especially at larger distances between the top and bottom reflectors the local temperature differences are reduced, which renders the wafer more "black" [135]. Note, however, that there is a practical upper limit to this distance. The reflectivity of the reflectors is less than unity. This means that any increase in the reflector aspect ratio requires more installed lamp power to offset the energy loss at the walls. Also, the reactor volume increases and thus the possible chemical memory effects.

Blackbody cavities were modeled long ago by Gouffé [73]. He calculated that the apparent emissivity of a body can increase by imposing the correct geometry to it, in particular high aspect ratios. For the effective emissivity

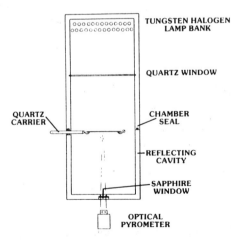

FIGURE 29. Cross-section of a reflective RTP cavity. (From Sheets [61].)

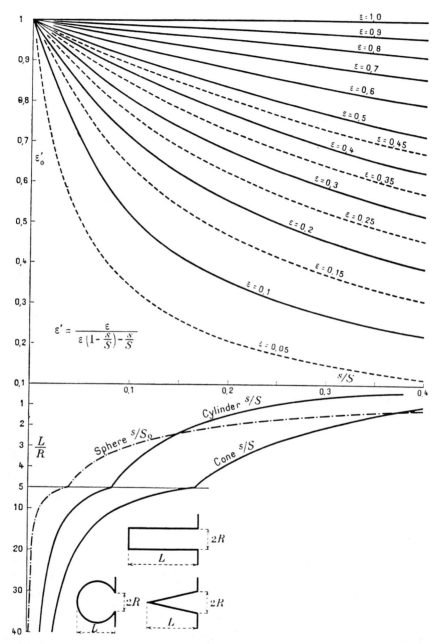

FIGURE 30. Effective emissivities of cylindrical, conical, and spherical cavities. (From Gouffé [73].)

ε' of a body, Gouffé gives

$$\varepsilon' = \frac{\varepsilon}{\varepsilon(1 - s/S) + s/S} \quad (9.11)$$

where ε is the intrinsic emissivity of the material of the cavity and s and S are its aperture area and its interior surface area, respectively (see Fig. 30). The figure shows the emissivities of cavities with simple geometric shapes. In the lower part, the value of s/S as a function of the aspect ratio is read. The effective emissivity ε' is found via this value by reading up to the intrinsic emissivity of the cavity material. Thus, we can derive that for the cylindrical silicon carbide bell jar chamber [71, 72] with its aspect ratio $L/R = 5$ and ε = 0.7 the effective emissivity is increased to ε' = 0.96. When a wafer is placed at this position in such a cavity it can be considered as a continuation of the cavity wall, with the same intrinsic emissivity $\varepsilon_{Si} = \varepsilon_{SiC} = 0.7$. Thus, also the wafer's effective emissivity is increased towards unity (black), which practically eliminates the effects of edges, patterns, backside roughness, and so on.

Note that the blackbody cavity theory [73] holds also in practice very well for the resistively heated bell jar chamber. The reflective cavity theory [135] for lamp-heated RTP chambers holds less well in practice, because of the cold walls. Yet, ideally it also leads to an increased emissivity for larger cavity aspect ratios. The ideal cavity concept requires that the cross section of the top reflector should extend from one wafer edge to the other over the wafer front side. The space between the edges and the reflector should be minimum, such that no (bottom) lamp light enters and that no Lambertian radiation escapes the cavity. As the wafer cannot make physical contact to the cold walls, this enclosure should be approached by reflectors around the wafer edge, such that the wafer "sees" itself (i.e. same effective emissivity, approaching unity).

In case one has no tools to handle hot wafers (such as Bernoulli pickup), one disadvantage of a highly reflective chamber could be in the cool-down transient part of a temperature cycle. All of a sudden one needs a highly absorbing wall to enable rapid cooling. To this end, it has been suggested that reflectors be developed with high reflectivity for wavelengths below 3 μm, and high absorptivity for those above 3 μm, or active wafer cooling [40]. An example that focuses on rapid uniform cooling is that of an RTP system with a separate rapid quenching chamber coupled to a conventional quartz process chamber [138]. After thermal processing, the specimen was withdrawn into the quenching chamber. Quenching ambients ranged from controlled vacuum to liquid nitrogen. Thus a blockwave-shaped temperature cycle was claimed without slip formation. A disadvantage of a chamber with a highly reflective bottom, in combination with a bottom

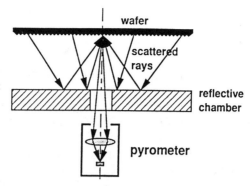

FIGURE 31. Dependence of effective emissivity on wafer backside roughness in a chamber bottom. (From [56].)

pyrometer, is that its readout becomes more sensitive to wafer backside roughness [56] (see Fig. 31). A rougher back side can scatter more reflected light into the pyrometer, and thus increase the effective emissivity. This problem may be suppressed by using a smaller pyrometer view angle or by making the chamber bottom less reflective. From a wafer point of view the use of $\lambda/4$ antireflective back side coatings or epi wafers has been suggested [54], but this is less practical for production.

D. MISCELLANEOUS DESIGN ASPECTS TO IMPROVE PROCESS UNIFORMITY

It should be noted that all the dynamic temperature uniformity improvement treated above is also achieved by using a classical CVD solution: a rotating graphite susceptor [139] behaves as a grey body and enables good uniformity control, albeit at slightly lower heating and cooling rates of about 50 K/s. With this type of reactor, shown in Fig. 5 and having guard rings as well, the atmospheric growth around 600 °C of high-quality epitaxial silicon and strained SiGe has been reported [15–17]. More will be said on this in Section VII.

Susceptors are also used in III–V processing. For GaAs IC implant annealing and other processes, a nonrotary graphite susceptor has been developed that consists of two parts, sandwiching the wafer [140–143].

Especially with RTCVD, using highly heat-conducting, reactive carrier gas such as H_2 with SiH_2Cl_2, convective cooling can be as important as radiation losses, or even more [84, 85], and should be compensated for. This can be done by admixing low specific heat, inert gas, such as Ar, to the reactant mixture. Further optimization is obtained by good gas flow dynamics design [84, 85], such as annular gas inlets and outlets.

VI. Noncontact *In Situ* Real-Time Process Control Options

As stated in Section III C, temperature control in RTP is dominated by pyrometry-based techniques. Yet the search for reproducible *in situ* control, irrespective of the presence of overlayers, continues.

Novel process control techniques can be divided into *temperature measurement* and *end-point detection* techniques. So far, most of these techniques have their practical limitations when compared to regular pyrometry. One is their price, another is that some options call for a sacrificial unexposed die area. Also, quite a few options suffer from parasitic deposition onto the probes (fibers or view windows).

This section describes the options of noninvasive thermometry for RTP, next to pyrometry. An extensive treatment is beyond the scope of this chapter. A recent background article is found elsewhere [113]. Although our focus is on nonpyrometric options, some work on emissivity end-point control by pyrometry is included here.

A. NOVEL TEMPERATURE MEASUREMENT TECHNIQUES

Noncontact techniques for temperature measurement studied so far are Raman scattering, laser interferometry, optical diffraction thermometry, and ellipsometry [113]. Each technique is highlighted below.

1. Raman Scattering

The temperature measurement is based on the inelastic scattering of photons from a probe laser, resulting in two peaks: the Stokes peak, due to the creation of a phonon, and the less intense anti-Stokes peak, due to the absorption of a phonon. The wafer temperature is derived from the ratio of the two peaks [144, 145]. The method has more disadvantages than advantages: it is expensive, it is not very precise ($\pm 50\ °C$), its sensitivity decreases and its noise increases with temperature, and it is not valid during deposition of absorbing films.

2. Laser Interferometry

Several methods exist. One reported recently [146] is by infrared laser interferometry, which detects small changes in the *refractive index* and (some) thermal expansion of semiconductors. However, laboratory experiments so far have been only successful from 300 to 900 K.

A second method reported by Lee *et al.* [147] is based on the propagation rate of *acoustic waves* through a semiconductor as a function of temperature. Here, a pulsed nitrogen laser generates elastic perturbations, which appear as fundamental acoustic Lamb waves that propagate through the wafer, faster at lower temperatures. The wave detection is done by interferometry with

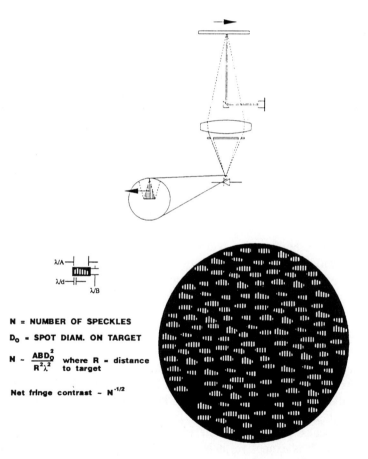

FIGURE 32. Principle of laser extensometry and interference fringe image (see text). (From Voorhes and Hall [148].)

another laser, probing at a distance from the wave generation spot. So far, the method is too immature to draw conclusions about its feasibility.

A third method is based on *extensometry* [148] and illustrated in Fig. 32. The wafer reflects the incident laser light of two adjacent beams. The diffusely reflected light is collected optically such that a photodetector shows an image of speckles, each with interference fringe patterns within. These patterns are used to detect the expansion of the wafer. Changes in temperature cause a displacement of the fringes. The temperature is derived from these displacements at two spots and the thermal expansion coefficient of silicon. An accuracy and repeatability of 0.5 °C were claimed up to 1300 °C process temperature. Yet the problem of sensitivity to system vibrations during operation will have to be addressed.

3. Optical Diffraction Thermometry

This method is based on the temperature-dependent change in the diffraction angle of laser light, incident on a grating etched onto the wafer surface [149, 150]. A temperature increase causes a thermal expansion of the substrate, and thus of the pitch of the grating. This in turn, changes the diffraction angle under laser illumination.

At present the method looks immature. Only a short temperature range was prototyped so far: 20–700 °C. Moreover, the etching of a grating requires processing time and die area.

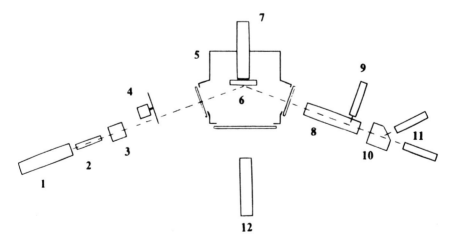

FIGURE 33. Laser emissivity analyzing pyrometry setup. (1) Laser, (2) beam expander, (3) polarizer, (4) chopper, (5) sample chamber, (6) sample, (7) heat source, (8) imaging optics, (9) radiance detector, (10) analyzer, (11) detectors, (12) pyrometer. (From Hansen *et al.* [151].)

4. Ellipsometry

This technique, combined with concurrent pyrometry, is also called laser emissivity analyzing pyrometry. LEAP, recently studied for opaque metals [151], is now also investigated for RTP. As usual, a pyrometer measures the spectral temperature (see Fig. 33). The ellipsometer measures the reflectivity of the wafer by analyzing the polarization state of a reflected laser beam. Thus, with Kirchhoff's law the emissivity is known. If the pyrometer pass band and the laser beam have equal wavelength, this yields the correct thermodynamic temperature via Eq. 9.4. Most information is obtained by rotating analyzer ellipsometry with multiple angle-of-incidence geometry [152, 153].

B. END-POINT DETECTION TECHNIQUES

From the above it may be obvious that temperature measurement as a means for process control has, and will always have, its limitations. For that reason, especially in a production environment, one is investigating other methodologies for *in situ* end-point detection. End-point detection does not solve the problem of temperature nonuniformity, yet it may improve the repeatability of a process. Candidates are ellipsometry, reflectometry, interferometry, and possibly emissivity measurement by either regular pyrometry or the ripple amplitude technique (Section III). A few illustrations of these techniques follow.

1. Ellipsometry

This *in situ* real-time method is based on the very sensitive response of the basic ellipsometric parameters Ψ and Δ upon increasing layer thickness. Figure 34 shows for a growing oxide layer how Ψ and Δ respond in a cyclic way [152, 153]. On the basis of this phenomenon, algorithms have been developed for reproducible, <10 nm oxide thickness control [9]. The accuracy was within 0.1 nm, even with absolute temperature errors as high as 100 °C, and irrespective of film growth rate.

More information on the potential and limitations of monochromatic and spectroscopic ellipsometry for *in situ* layer thickness control is found in a recent review [152]. It must be stressed that the use of ellipsometry for *in situ*, real-time process control relies strongly upon the optical modeling and the constants, used for the (multi)layer structure, and on the experimental

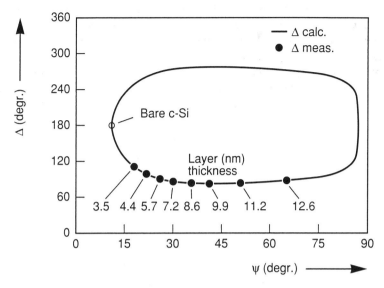

FIGURE 34. Monochromatic ellipsometry results for several samples with increasing SiO_2 layer thickness. (After J. Jans, R. Hollering and M. Erman in *Analysis of Microelectronic Materials and Devices* (M. Grasserbauer and H. W. Werner, eds), pp. 681–717. Reprinted by permission of John Wiley & Sons, Ltd.)

conditions. For example, cooled view windows are essential in avoiding problems of stress-induced optical birefringence.

2. Reflectometry

Another end-point detection technique, used successfully to monitor the subsequent stages of platinum silicide formation from a Pt film on a silicon wafer, is reflectometry [7]. A He–Ne laser beam is reflected from the wafer surface into a photocell with a 632.8 ± 0.1 nm band-pass filter. The reflected signal and that of a thermocouple contacting the wafer back side were recorded during annealing at 580 °C. The film reflectivity dropped in two stages because of the solid state reaction between platinum and silicon, first into Pt_2Si, and next into PtSi. The reflectivity signal remained constant after completion of the second stage, also upon switching off the lamps.

3. Light Interferometry

In a small extension of the above method, incident light is not only reflected but also refracted by growing surface layers with different refractive indices.

Because of optical interference the phase shift in the reflected light will show an oscillating behavior with the thickness of the layer. As an example, this method was used in the solid phase epitaxial growth of an amorphous silicon film on Si by Dilhac et al. [154].

Figure 35 illustrates an *optical fiber* version of the same technique, which is applied on an industrial scale in LPCVD of interference filter layers on car lamps [155, 156]. One uses an optical fiber probe, which detects the

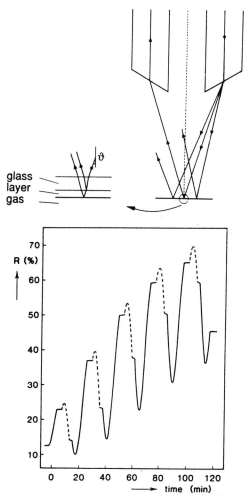

FIGURE 35. Fiber optical light interferometry. (After P. J. Severin and A. P. Severijns, *J. Electrochem. Soc.* **137**, 1306–1309, 1990. Reprinted by permission of the publisher, The Electrochemical Society, Inc.)

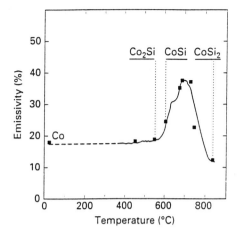

FIGURE 36. *In situ* measurements of the emissivity during $CoSi_2$ formation on identically processed wafers. The emissivity measurement is based on a 5 Hz modulation of the lamp power. (From [157].)

phase shift of reflected light by a Si_3N_4/SiO_2 multilayer stack. The input and output fiber rods are optically coupled by beveling them. Thus, one can conduct phase-locked CVD up to 800 °C.

4. Emissivity Measurement

With the optimum resolution and optics applied onto a pyrometer, it seems possible today to use emissivity as an *in situ*, real-time process monitor. Recent publications indicate that this can be done by straightforward pyrometry. One example is the solid state reaction of a cobalt film on silicon, as investigated by Vandenabeele *et al.* [157]. Here the two-stage formation via CoSi into $CoSi_2$ could be detected (see Fig. 36).

Similar results were claimed for the optical fiber "ripple amplitude" technique [118].

VII. Recent Developments and Future Trends in Rapid Thermal Processing

Even when RTP will overcome the problems of temperature repeatability and uniformity, a major argument against single-wafer processing in today's stand-alone systems is the *inferior throughput*, compared to multiwafer processing. Recent estimations on the average process time per wafer are of the order of 1 min in a conventional furnace and of a few minutes for a single-wafer RTP reactor [158, 159]. The low throughput is, amongst others, due to the limited, asymptotic wafer cooling rate (*cf.* Fig. 20). Typically, from 1000 °C the cool-down time to 700 °C is 3 sec, to 400 °C it is 30 sec, and to 100 °C it is 180 sec [40]. Moreover, a typical throughput number in stand-alone RTP equipment for transport only (no process) is 50 wafers/hour.

Today, wafer fabs are still dominated by stand-alone batch equipment for these throughput and possibly historical reasons. Only when the balance between *cost per wafer* and process specification is in favor will RTP find wider acceptance.

Two factors in favor of RTP are *yield* and, especially in low-volume application specific integrated circuit (ASIC) manufacturing, the reduction in total *turnaround time* (by a factor of 6, see [158]). Developments to improve cycle time and yield, and, to some extent, throughput are

- modular clustering of equipment with
- multi-chamber, single-process modules,
- single-chamber, multi-process modules,
- multivariable process control, and to some extent
- rapid gas switching.

Each of these trends will be discussed below. Especially the modular cluster integration looks successful here. As an example a 4- to 6-module cluster is predicted to enable a throughput of 30 wafers/hour and more, depending on the degree of sophistication of wafer transport, etc. [159]. For example, Bernoulli wafer pick-up enables the transport of wafers at high temperatures, up to 900 °C.

It should be noted that these developments are paralleled by the continuing development of conventional batch furnaces. For example, the problems of horizontal furnaces are being reduced by the introduction of *vertical batch* furnaces. The same holds for horizontal, lamp-heated RTP furnaces for minibatches with a few wafers. These systems have reduced heat loss upon unloading, smaller footprint, and some compatibility with multichamber systems. Here, the trade-off will be that between higher throughput and poorer uniformity.

A. Equipment Architecture and Design

Several RTP companies are working on multichamber cluster tool applications, mostly within the new MESC[2] equipment architecture and interface standards [160]. Their objective is modular process integration, with better

---

[2] MESC is the Modular Equipment Standards Committee of SEMI (Semiconductor Equipment and Materials International), founded March 30, 1989.

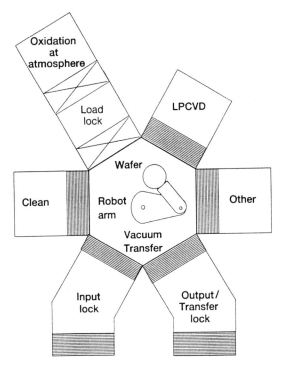

FIGURE 37. Example of rapid thermal processing modules in a radial multichamber cluster. (From [40].)

particle exclusion, interprocess ambient control, higher throughput, flexible operation, etc. A good review of the ins and outs of multichamber RTP integration was recently published by Rosser *et al.* [40]. The developments vary from integrating several RTP modules (e.g. rapid thermal cleaning, rapid thermal oxidation, rapid thermal CVD of polysilicon for MOS gate stack formation [12, 40, 161]), to integrating one RTP module with other equipment (such as metal deposition with a pre-cleaning etch and subsequent rapid thermal silicidation for local interconnect formation [159, 162]). An example of a multichamber cluster is illustrated by Fig. 37.

The critical issues in a multichamber system are the wafer *transfer system* and the transfer *ambient*. The former has to be absolutely reliable. The latter should, ideally, be an ambient that keeps a silicon surface or an interface ultraclean, as in vacuum or a glove box. Both requirements will add to the complexity and the cost of such a system. Also here, throughput interferes. This can happen when, for example, a wafer is transferred from a high temperature module (e.g., oxidation) to a low temperature module (e.g., poly-Si CVD) while still being hot. Its thermal memory may disturb the gas flow dynamics or a preprocess emissivity recalibration step. This illustrates nicely how RTP clustering will be used only when, in return for a lower or equal throughput, a qualitatively superior product is obtained in a shorter cycle time [158].

This might be the case, since each individual chamber can be optimized for customer and reaction-specific (i.e., narrow temperature windows) use, in contrast to the suboptimized multiprocess use, discussed below.

In order to make throughput more acceptable, much effort is given to *in situ beam processing*. This relates to the other driving factor behind RTP, besides shorter process times, being *reduced substrate temperatures*. Concurrent, alternative energy sources are used, such as microwave plasmas and photon (laser and UV) enhancement, to further reduce the thermal budget [163]. Generally, this is done with the regular lamp illumination at the wafer backside, and the microwave injection, etc., at the wafer frontside.

Another development is of a single-chamber, multiprocess system [164, 165]. Compared to the multichamber cluster, the advantages here are the absence of interchamber transfer, which yields the shortest turnaround times, and the best possible interprocess ambient control. Another advantage is the price. More details on the economic impact of single-wafer multiprocessors are found elsewhere [158, 165]. When one can afford multiprocessing in a single chamber, that is the better, more economic choice. A good example, demonstrated recently, is that of an *in situ* preclean, using dry HF, directly followed by epitaxial growth from SiH_2Cl_2 [166]. The film, thus grown, had better quality than the one without preclean. In general, however, a disadvantage of this option is the risk of memory effects or crosscontamination.

Another increasing trend is the hybridization of CVD systems (CVD, MOCVD and UHV-CVD) with RTP, into RTCVD. These systems are often cylindrical, load-locked, cold-wall, metal systems with cooled quartz windows and susceptors. Usually, they operate at reduced pressure, which minimizes the effects of gas flow on the deposition kinetics, and thus uniformity. Yet low pressures require thicker window plates, which absorb more radiation, causing parasitic deposition. Additional problems arise at reduced pressures. Nonuniform mass transport by gas depletion or accumulation, resulting in down-stream nonuniformity, or radial nonuniformity in case of annular gas inlets, is reported for the selective CVD of $TiSi_2$ [167, 168].

Another trend is that wafers are processed face down, in order to minimize particle related problems. More details on the reactor and process design in RTCVD are found elsewhere [169].

B. SCALAR VS. MULTIVARIABLE PROCESS CONTROL

Temperature nonuniformity, induced by edge loss and patterns, can result in yield loss. There are several hardware and software solutions to reduce the nonuniformity to some extent, such as the special reflector and guard ring designs, as discussed in Section V, or multizone heating.

A more flexible solution to edge loss is the use of *multivariable* instead of *scalar* process control. These terms are adapted from Norman [170–172]. A scalar system has only one controlled output (lamp power) and one or no sensed input (pyrometer signal). A multivariable or multizone system has multiple controlled outputs and/or multiple sensed inputs in its control loop. Figure 38 illustrates scalar vs. multivariable control of a multilamp system with three concentric lamp zones [171]. The scalar system has fixed ratios between the powers $P_j(t)$ supplied to the zones j, whereas the zones in the multivariable system are independently controlled.

Sensor signals may include other variables than those related to wafer temperature. Examples are lamp current, temperature of chamber components, gas pressure and composition, wafer surface conditions (grain size, overlayer thickness), and so forth. An extensive list is given by Moslehi [161]. Knowledge about all these variables adds to the control, yet also to the complexity, of a system. We will focus here on systems that are multivariable from the viewpoint of temperature control, since basically all the other sensors are not really needed for a correct process control.

Multivariable control by dynamic, independent powering of lamp zones is shown to have great potential. Note here that the development of the

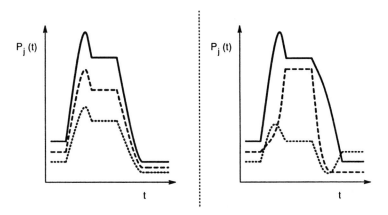

FIGURE 38. Principle of scalar and multivariable control of multizone RTP systems. Scalar control implies dependent control, multivariable control independent control. $P_j(t)$ is the power to each zone j as a function of time t. (From Norman [171].)

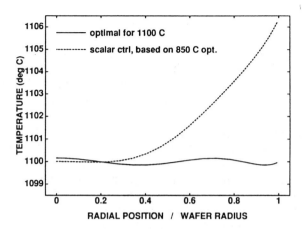

FIGURE 39. Scalar vs. multivariable control at 1100 °C in a system optimized at 850 °C, see text. (From S. A. Norman [172].)

proper software and hardware for the optimum power settings and for the dynamic control is not an easy task [173], yet is very well known in control theory. So far, much work is mainly based on modeling, amongst others by Norman [170–172] and Gyurcsik *et al.* [174]. One example, by Norman [171], illustrates the potential of multivariable control for the three-zone lamp of the Stanford Rapid Thermal Multiprocessor, shown in Fig. 6, Section III. Figure 39 shows the optimum steady-state temperature profiles for a setpoint of 1100 °C with both scalar and multivariable control. It is obvious that scalar control has inferior temperature uniformity. The lamp settings were optimized for 850 °C with a sensor measuring the wafer center. Consequently the rescaling to 1100 °C results in deviations up to +6 °C.

Multivariable control also gets more attention in practical application. Figure 5 shows a commercial RTCVD system with four thermocouples. One in the center is for the master reading, the other three in the guard ring serve as slaves in the multivariable control of individual lamp zones. Figure 40 shows another commercial CVD system [175] in development, with a dual-point pyrometry option for 200-mm wafers. One fixed pyrometer views the wafer center. The other, which may be fixed or scanning, views the edge of the rotating wafer. The seven lamps are multivariably controlled by the pyrometers.

Practically all other commercial systems, however, still use scalar control with multizone heating by crossed lamp banks. The slaves are ratioed to the master by a set of empirical software algorithms, but not independently controlled.

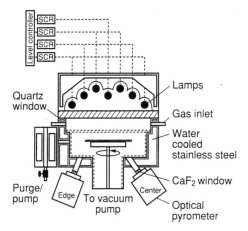

FIGURE 40. Dual-point pyrometry system with rotating wafer and independent 4-zone lamp control. From the outer to the inner zone the powers of the linear lamps are 3, 5, 6 and 6 kW. (Source: Rapro CVD.)

C. THERMAL SWITCHING VS. RAPID GAS SWITCHING

Most publications on RTP have stressed the rapid temperature ramping of the technique. It has even resulted in the term "thermal switch", to indicate that reaction can be initiated and terminated by rapidly ramping the temperature up or down. Using the thermal switch technique, the growth of Si/SiGe superlattices has been reported [18–21, 120], with thermal switching to 700 (Si) and to 625 °C (SiGe), while gases are switched when the substrate is cooled in between. The growth technique is referred to as limited reaction processing (LRP), and was claimed to limit the thermal budget of the wafer to a minimum. It can be used for other structures as well, such as dopant modulation structures [176] and III/V superlattices [177].

It has been shown that rapid *gas switching*, if designed correctly, can result in extremely sharp interfaces [178, 179]. In the case of well-constructed switching valves and a small reaction chamber, memory effects are minimized. Thus gas switching is sufficient, and the "simultaneous" thermal switch is not necessary. Moreover, by keeping the temperature constant one avoids the temperature switch-induced nonuniformities, and possible slip.

As an example, strained Si/SiGe was grown in an atmospheric RTCVD reactor, at temperatures as low as 625 °C [15–17]. The same or even better interface abruptness was obtained, as compared to LRP or MBE [180]. Figure 41 shows a high-resolution TEM cross section of a Si-capped

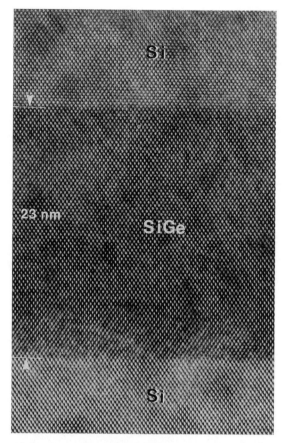

FIGURE 41. High-resolution TEM image of a Si-capped $Si_{0.75}Ge_{0.25}$ layer, grown by atmospheric pressure CVD at 625 °C. (From [15].)

$Si_{0.75}Ge_{0.25}$ layer. From the photograph one can see that the transition at the upper interface is monolayer sharp, and that defect-free epitaxial material was grown. Thus it is possible to use gas switching only. As this is even possible without vacuum, one may conclude that the atmospheric RTCVD of strained SiGe layers is close to being a full production technique.

VIII. Technology Roadmap and Concluding Remarks

RTP has certainly undergone a revolutionary development since its start in the late 1960s. Yet, in full production it has not yet had an indispensable role in today's IC manufacturing with design features down to, say 0.5 μm. The technology roadmap in Table II shows that the potential of RTP grows, the further one moves into the subhalf micrometer technology. What the roadmap shows, above all, is that the developments in RTP will be steady

Table II. RTP Technology Roadmap

	1985		1990		1995		2000
Design rule (μm)	2	1.3	.8	.5	.3	.2	.15
DRAM equiv.	256k	1M	4M	16M	64M	256M	1G
Wafer size (in.)	5	6	6-8	8	8 (-10)		12
t_{ox} (nm)	35	25	20	15	12	10	8
L_{eff} (μm)	2	1.3	.8	.5	.3	.2	.15
x_j (μm)	.3	.25	.2	.15	.12	.1	.08
Repeatability (°C)	±15		±5		±2		±1
Uniformity (°C)	±7-10		±3-5		±2-3		<±2
Accuracy (°C)	±20+		±5-10		±3-5		<±3
Slip	No visible				No measurable		
Temperature	T.C. witness sample	T.C. and pyro high temp.		Broad range pyrom.	Indirect measure of mat. property		Direct measure of temperature
Operational control	Open loop	Closed loop to program recipe					Closed loop to program recipe of secondary end point
Processing environment	Warm wall quartz			Cold wall passivated metal			Cold wall non-metallic
Ambient pressure	Atmospheric						Low pressure processing
Applications	Implant monitor Ti-silicide Ti-nitride			Implant anneal, CVD, Co-silicide, low temp.epi, oxidation, alloying			CVD metal selective epi, low temperature epi, S.O.I.
Particles	.1/cm² >.5 μm		.01/cm² >.3 μm		.005/cm² >.2 μm		.001/cm² >.1 μm
Configuration	Single chamber broad temp.			Multichamber sequential process cluster broad temp.			Application specific narrow temp. window
Computer sophistication	Micro-controller		Micro-processor w. floppy disks and diagn.		Micro-processor w. hard disk SECS		Micro-processor w. hard disk, network controller SECS and extensive diagnostics

ones, but without any major breakthrough. Below, a few statements follow, regarding the developments in RTP during the rest of this decade, which leads us into the 1-Gbit era.

Lamps

- Tungsten–halogen lamps will remain the mainstay in RTP. Their geometry will be in multivariably controlled arrays.

Chambers

- Focus will be on single-wafer RTP chambers having reflectors with high reflectivity and optimum aspect ratio, multizone heating and *in situ*, real-time multivariable sensing.
- In the short term the single-process multichamber cluster tool is more likely than the multiprocess single chamber.
- There will be a continuing merger with CVD. Examples here are the use of slip guard rings, rotating susceptors, quartz-coated metal or nonmetallic walls, but also *in situ* etch cleaning.

Sensors

- Optical (thermopile) pyrometers will remain the mainstay.
- Improved optics and aperture design, combined with special lamp filtering and powering will enable emissivity-independent control. As soon as one is able to measure *in situ*, real-time emissivity, end-point control will be automatically achieved.
- More end-point detection techniques will be used, especially in a production environment.

Processes

- Very soon all batch processes will be possible in RTP. Yet, only when the process is absolutely needed in RTP will it be done, because of technical and cost reasons.
- Atmospheric processes will find earlier and more use than vacuum processes.
- Existing CVD processes in conventional batch CVD furnaces are not necessarily applicable to RTP.

- Single-sided (backside) ramping and double-sided stationary heating will reduce pattern-induced nonuniformity
- Beam-assisted (laser, UV, microwave, etc.) processing may be used to further reduce substrate temperatures.
- Nonorthodox process applications outside the semiconductor field will grow rapidly. These applications will pose new, unique equipment obstacles; one example is the crystallization of α-Si to poly-Si TFT on large glass substrates (30 $\times$ 40 cm^2) for displays under thermal flux annealing conditions where the substrate is scanned between two xenon long-arc lamps.

General

- There is not yet any consensus among the equipment manufacturers and users on the optimum reactor design.
- Modeling is a powerful development tool in optimizing the design of both the reactor and the process.
- The trend towards lower process temperatures will reduce slip-related problems. The need for good temperature control remains, but only for the steady-state phase.

Acknowledgments

The author wishes to thank all companies mentioned in this chapter for reviewing the latest developments on their (sub)systems, and for the permission to publish these. W. B. de Boer and A. T. Vink are acknowledged for critically reading the manuscript. Finally, J. W. Smits and the late A. R. Miedema are acknowledged for the permission to publish this chapter.

References

1. J. M. Fairfield and G. H. Schwuttke, *Solid St. Electron.* **11**, 1175–1176 (1968).
2. M. M. Moslehi, J. W. Kuehne, L. Velo, D. Yin, R. L. Yeakley, S. Huang, B. Jucha, and T. Breedijk, *Soc. Photo-Opt. Instrum. Eng. Symp. Proc.* **1595**, 132–145 (1991).
3. M. M. Moslehi, R. A. Chapman, M. Wong, A. Paranjpe, H. N. Najm, J. W. Kuehne, R. L. Yeakley, and C. J. Davis, *IEEE Trans. Electr. Dev.* **39**, 4–32 (1992).
4. R. Chapman, J. W. Kuehne, P. S.-H. Ying, W. F. Richardson, A. R. Peterson, A. P. Lane, I. C. Chen, L. Velo, C. H. Blanton, M. M. Moslehi, and J. L. Paterson, *Proc. Int. Electr. Dev. Mtg.*, Washington, DC, Dec. 8–11, 1991, pp. 101–104.
5. B. Lojek, *Mater. Res. Soc. Symp. Proc.* **224**, 33–38 (1991); see also Chapter 8.

6. W. Eichhammer, P. Vandenabeele, and K. Maex, *Mater. Res. Soc. Symp. Proc.* **224**, 487-492 (1991).
7. J.-M. Dilhac, C. Ganibal, and T. Castan, *Appl. Phys. Lett.* **55**, 2225-2226 (1989).
8. T. H. Wu, R. S. Rosler, B. C. Lamartine, R. B. Gregory, and H. G. Tompkins, *J. Vac. Sc. Technol. B* **6**, 1707-1713 (1988).
9. C. T. Yu, K. H. Isaak, and R. E. Sheets, *J. Electrochem. Soc.* **137**, 530C (1990); *Semicond. Intern.* **14**(6), 166-169 (1991).
10. Y. Uoochi, A. Tabuchi, and Y. Furumura, *J. Electrochem. Soc.* **137**, 3923-3925 (1990).
11. J. C. Liao and T. I. Kamins, *J. Appl. Phys.* **67**, 3848-3852 (1990).
12. A. Kermani, K. E. Johnsgard, and F. Wong, *Solid State Technol.* **34**(5), 71-73 (1991).
13. J. W. Osenbach, Y. H. Ku, and A. Kermani, *Mater. Res. Soc. Symp. Proc.* **198**, 33-38 (1990).
14. T. Y. Hsieh, K. H. Jung, and D. L. Kwong, *J. Electrochem. Soc.* **138**, 1188-1207 (1991).
15. W. B. de Boer and D. J. Meyer, *Appl. Phys. Lett.* **58**, 1286-1288 (1991).
16. T. I. Kamins and D. J. Meyer, *Appl. Phys. Lett.* **59**, 178-180 (1991); *ibid.* **61**, 90-92 (1992).
17. P. D. Agnello, T. O. Sedgwick, D. J. Meyer, and A. P. Ferro, *Proc. 37th Amer. Vac. Soc. Symp.*, Toronto, 1990, pp. 46-49.
18. J. C. Sturm and C. M. Reaves, *Soc. Photo-Opt. Instrum. Eng. Symp. Proc.* **1393**, 309-315 (1990).
19. J. C. Sturm and C. M. Reaves, *IEEE Trans. Electr. Dev.* **39**, 81-88 (1992).
20. J. F. Gibbons, C. M. Gronet, and K. E. Williams, *Appl. Phys. Lett.* **47**, 721-723 (1985).
21. J. L. Hoyt, C. A. King, D. B. Noble, C. M. Gronet, J. F. Gibbons, M. P. Scott, S. S. Laderman, S. J. Rosner, K. Nauka, J. Turner, and T. I. Kamins, *Thin Solid Films* **184**, 93-106 (1990); see also Chapter 2.
22. M. M. Moslehi, J. Kuehne, R. Yeakley, L. Velo, H. Najm, B. Dostalik, D. Yin, and C. J. Davis, *Mater. Res. Soc. Symp. Proc.* **224**, 143-157 (1991).
23. R. S. Rosler, J. Mendonca, and M. J. Rice, Jr., *J. Vac. Sc. Technol. B* **6**, 1721-1727 (1988).
24. J. L. Regolini, D. Dutartre, D. Bensahel, and J. Penelon, *Solid St. Technol.* **34**(2), 47-48 (1991).
25. A. Katz, A. Feingold, S. J. Pearton, S. Nakahara, M. Ellington, U. K. Chakrabarti, M. Geva, and E. Lane, *J. Appl. Phys.* **70**, 3666-3677 (1991).
26. P. D. Agnello, T. O. Sedgwick, M. S. Goorsky, J. Ott, T. S. Kuan, and G. Scilla, *Appl. Phys. Lett.* **59**, 1479-1481 (1991).
27. T. O. Sedgwick, P. D. Agnello, M. Berkenblit, and T. S. Kuan, *J. Electrochem. Soc.* **138**, 3042-3047 (1991).
28. A. Katz, A. Feingold, S. J. Pearton, C. R. Abernathy, M. Geva, and K. S. Jones, *J. Vac. Sci. Technol. B* **9**, 2466-2472 (1991).
29. R. Pascual, M. Sayer, C. V. R. Vasant Kumar, and L. Zou, *J. Appl. Phys.* **70**, 2348-2352 (1991).
30. T. Suzuki, *J. Appl. Phys.* **69**, 4756-4760 (1991).
31. J. E. Fair, *Solid State Technol.* **35**(8), 47-52 (1992).
32. J. M. Salzer, *Solid State Technol.* **35**(5), 62-63 (1992).
33. K. C. Saraswat, M. M. Moslehi, D. D. Grossman, S. Wood, P. Wright, and L. Booth, *Mater. Res. Soc. Symp. Proc.* **146**, 3-13 (1989).
34. S. R. Wilson, R. B. Gregory, and W. M. Paulson, *Mater. Res. Soc. Symp. Proc.* **52**, 181-190 (1986).
35. R. A. Powell and M. L. Manion, *Mater. Res. Soc. Symp. Proc.* **52**, 441-480 (1986).
36. R. Singh, *J. Appl. Phys.* **63**, R59-R114 (1988).

37. M. J. Hart and A. G. Evans, *Semicond. Sci. Technol.* **3**, 421–436 (1988).
38. C. Hill, in *Laser and Electron Beam Solid Interactions and Materials Processing* (J. F. Gibbons, L. D. Hess, and T. W. Sigmon, eds.), pp. 361–374. Elsevier North Holland, New York, 1981.
39. C. Hill, S. Jones, and D. Boys, in *Reduced Thermal Processing for ULSI* (R. A. Levy, ed.), pp. 143–180. Plenum Press, New York, 1989.
40. P. J. Rosser, P. B. Moynagh, and K. B. Affolter, *Soc. Photo-Opt. Instrum. Eng. Symp. Proc.* **1393**, 49–66 (1990).
41. F. Roozeboom and N. Parekh, *J. Vac. Sc. Technol. B* **8**, 1249–1259 (1990).
42. F. Roozeboom, *Mater. Res. Soc. Symp. Proc.* **224**, 9–16 (1991).
43. F. Roozeboom, *Semicond Intern.* **14**(10), 74 (1991).
44. W. DeHart, *Microelectr. Manufact. Technol.* **14**(7), 44–48 (1991).
45. K. Maex, *Microelectr. Eng.* **15**, 467–474 (1991).
46. J. J. Wortman, J. R. Hauser, M. C. Öztürk, and F. Y. Sorrell, in *ULSI Science and Technology* (J. M. Andrews and G. K. Celler, eds.), pp. 528–540. The Electrochemical Society, Pennington, NJ, 1991.
47. A. J. LaRocca in *The Infrared Handbook* (W. L. Wolfe and G. J. Zissis, eds.), rev. 2nd ed. pp. 2.1–2.97. Environmental Res. Inst. of Michigan, Ann Arbor, MI, 1989.
48. J. F. Snell, in *Handbook of Optics* (W. G. Driscoll and W. Vaughan, eds.), pp. 1.1–1.30. McGraw-Hill, New York, 1978.
49. T. Sato, *Jpn, J. Appl. Phys.* **6**, 339–347 (1967).
50. J. Nulman, *Soc. Photo-Opt. Instrum. Eng. Symp. Proc.* **1189**, 72–82 (1989).
51. J. Nulman, B. Cohen, W. Blonigan, S. Antonio, R. Meinecke, and A. Gat, *Mater. Res. Soc. Symp. Proc.* **146**, 461–466 (1989).
52. J. Nulman, S. Antonio, and W. Blonigan, *Appl. Phys. Lett.* **56**, 2513–2515 (1990).
53. P. Vandenabeele, K. Maex, and R. de Keersmaecker, *Mater. Res. Soc. Symp. Proc.* **146**, 149–160 (1989).
54. P. Vandenabeele and K. Maex, *Soc. Photo-Opt. Instrum. Eng. Symp. Proc.* **1189**, 89–103 (1989).
55. D. W. Pettibone, J. R. Suarez, and A. Gat, *Mater. Res. Soc. Symp. Proc.* **52**, 209–216 (1986).
56. P. Vandenabeele and K. Maex, *Mater. Res. Soc. Symp. Proc.* **224**, 185–196 (1991).
57. R. Kakoschke, *Mater. Res. Soc. Symp. Proc.* **224**, 159–170 (1991).
58. R. Kakoschke and E. Bussmann, *Mater. Res. Soc. Symp. Proc.* **146**, 473–482 (1989).
59. H. A. Lord, *IEEE Trans. Semicond. Manufact.* **1**, 105–114 (1988).
60. M. M. Chen, J. B. Berkowitz-Mattuck, and P. E. Glaser, *Appl. Opt.* **2**, 265–271 (1963).
61. R. E. Sheets, *Nucl. Instrum. Meth. Phys. Res. B* **6**, 219–223 (1985).
62. R. E. Sheets, *Mater. Res. Soc. Symp. Proc.* **52**, 191–197 (1986).
63. R. E. Sheets, US Patent 4 649 261 (10 March 1987); US Patent 4 698 486 (6 Oct. 1987).
64. J. R. Coaton and J. R. Fitzpatrick, *IEE Proc.* **127A**, 142–148 (1980).
65. M. Rehmet, *IEE Proc.* **127A**, 190–195 (1980).
66. D. M. Camm, A. Kjørvel, N. P. Halpin, and A. J. D. Housden, Eur. Patent 186 879 (9 July 1986); US Patent 4 700 102 (13 Oct. 1987).
67. J. C. Gelpey and P. O. Stump, *Microelectron. Manufact. Test.* **6**, 22–27 (1983).
68. J. C. Gelpey and P. O. Stump, *Nucl. Instrum. Meth. Phys. Res. B* **6**, 316–320 (1985).
69. P. J. Walsh and A. Kermani, *J. Appl. Phys.* **61**, 4484–4491 (1987).
70. T. J. Stultz, US Patent 4 820 906 (11 April 1989).
71. C. Lee and G. Chizinsky, *Solid State Technol.* **32**(1), 43–44 (1989).
72. C. Lee, US Patent 4 857 689 (15 Aug. 1989).
73. A. Gouffé, *Revue d'Optique* **24**, 1–10 (1945).

74. A. S. Gat and E. R. Westerberg, US Patent 4 680 451 (14 July 1987); Eur. Patent 290 692 (17 Nov. 1988).
75. M. Robinson and A. E. Ozias, US Patent 4 836 138 (6 June 1989).
76. Anonymous, *Semicond. Intern.* **11**(6), 338 (1988).
77. P. P. Apte, S. Wood, L. Booth, K. C. Saraswat, and M. M. Moslehi, *Mater. Res. Soc. Symp. Proc.* **224**, 209-214 (1991).
78. C. M. Gronet and J. F. Gibbons, patent application WO 91/10873 (25 July 1991).
79. J. L. Crowley, T. J. DeBolski, A. Kermani, and S. E. Lassig, US Patent 4 755 654 (5 July 1988).
80. Anonymous, *Solid State Technol.* **33**(8), 39-41 (1990).
81. R. Kakoschke, Eur. Patent 345 443 (13 Dec. 1989); US Patent 4 981 815 (1 Jan. 1991).
82. K.-B. Kim, P. Maillot, A. E. Morgan, A. Kermani, and Y.-H. Ku, *J. Appl. Phys.* **67**, 2176-2179 (1990).
83. Y. Miyai, K. Yoneda, H. Oishi, H. Uchida and M. Inoue, *J. Electrochem. Soc.* **135**, 150-155 (1988).
84. S. A. Campbell, K.-H. Ahn, K. L. Knutson, B. Y. H. Liu, and J. D. Leighton, *IEEE Trans. Semicond. Manuf.* **4**, 14-20 (1991).
85. K. L. Knutson, S. A. Campbell, and J. D. Leighton, *Mater. Res. Soc. Symp. Proc.* **224**, 203-208 (1991).
86. G. F. Warnke, in *Temperature, its Measurement and Control in Science and Industry* (H. H. Plumb, ed.), Vol. 4, part 1, pp. 503-517. Instrument Society of America, Pittsburgh, 1972.
87. J. O. Dimmock, *J. Electron. Mat.* **1**, 255-309 (1972).
88. A. G. Fischer, *J. Lumin.* **7**, 427-448 (1973).
89. L. R. Wollmann, *Electro-Opt. Syst. Des.* **11**(9), 37-44 (1979).
90. F. Völklein and A. Wiegand, *Sensors and Actuators A* **24**, 1-4 (1990).
91. A. W. van Herwaarden and P. M. Sarro, *Sensors and Actuators* **10**, 321-346 (1986).
92. S. Middelhoek and S. A. Audet, *Silicon Sensors*, pp. 153-199. Academic Press, New York, 1989.
93. S. A. Cohen, T. O. Sedgwick, and J. L. Speidell, *Mater. Res. Soc. Symp. Proc.* **23**, 321-326 (1984).
94. J. L. Hoyt, K. E. Williams, and J. F. Gibbons, US Patent 4 787 551 (29 Nov. 1988).
95. C. J. Russo, *Soc. Photo-Opt. Instrum. Eng. Symp. Proc.* **623**, 133-141 (1986).
96. R. R. Dils, *J. Appl. Phys.* **54**, 1198-1201 (1983); R. R. Dils and A. K. Winslow, US Patent 4 845 647 (4 July 1989).
97. D. T. Hodul and S. Mehta, *Nucl. Instrum. Methods Phys. Res. B* **37/38**, 818-822 (1989).
98. C. B. Yarling and W. A. Keenan, *Microelectron. Manufact. Test.* **12**, 1-14 (1989).
99. J. C. Gelpey, P. O. Stump, J. Blake, A. Michel, and W. Rausch, *Nucl. Instrum. Methods Phys. Res. B* **21**, 612-617 (1987).
100. C. B. Yarling and W. A. Keenan, *Mater. Res. Soc. Symp. Proc.* **146**, 451-460 (1989).
101. C. B. Yarling and W. A. Keenan, *Soc. Photo-Opt. Instrum. Eng. Symp. Proc.* **1189**, 164-173 (1989).
102. P. Vandenabeele and K. Maex, *Soc. Photo-Opt. Instrum. Eng. Symp. Proc.* **1393**, 372-394 (1990).
103. A. J. McAlister and J. L. Murray, *Bull. Alloy Phase Diagrams* **5**, 341-347 (1984).
104. Z. Nènyei and H. Walk, *Pract. Met.* **28**, 305-313 (1991).
105. M. Pecot and J. Nulman, US Patent 4 854 727 (8 Aug. 1989).
106. J. L. Crowley, J. C. Liao and J. C. Gelpey, *Soc. Photo-Opt. Instrum. Eng. Symp. Proc.* **1189**, 64-71 (1989).
107. J. J. Pelletier and T. E. Winter, *Mater. Res. Soc. Symp. Proc.* **146**, 467-471 (1989).

108. J. L. Crowley, A. Kermani, S. E. Lassig, N. H. Johnson, and G. R. Rickords, US Patent 4 969 748 (13 Nov. 1990); US Patent 4 984 902 (15 Jan. 1991).
109. J.-M. Dilhac, C. Ganibal, and N. Nolhier, *Mater. Res. Soc. Symp. Proc.* **224**, 3-8 (1991).
110. P. M. Reynolds, *Brit. J. Appl. Phys.* **15**, 579-589 (1964).
111. J. L. Gardner and T. P. Jones, *J. Phys. E. Sci. Instrum.* **13**, 306-310 (1980).
112. P. B. Coates, *Metrologia* **17**, 103-109 (1981).
113. D. Peyton, H. Kinoshita, G. Q. Lo, and D. L. Kwong, *Soc. Photo-Opt. Instrum. Eng. Symp. Proc.* **1393**, 295-308 (1990).
114. K. Crane and P. J. Beckwidth, US Patent 4 470 710 (11 Sep. 1984).
115. D. Mordo, Y. Wasserman, and A. Gat, *Soc. Photo-Opt. Instrum. Eng. Symp. Proc.* **1595**, 52-60 (1991).
116. A. Gat and D. Mordo, US Patent 5 114 242 (19 May 1992) and Eur. Patent 490 290 (17 June 1992); A. Gat and M. French, US Patent application serial no. 7/624 205, filed 7 Dec. 1990.
117. C. W. Schietinger, B. E. Adams, and C. B. Yarling, *Mater. Res. Soc. Symp. Proc.* **224**, 23-31 (1991); C. W. Schietinger and B. E. Adams, US Patent 5 154 512 (13 Oct. 1992).
118. C. W. Schietinger, private communication.
119. M. M. Moslehi, H. Najm, L. Velo, R. Yeakley, J. Kuehne, B. Dostalik, D. Yin, and C. J. Davis in *ULSI Science and Technology* (J. M. Andrews and G. K. Celler, eds.), pp. 503-527. The Electrochemical Society, Pennington, NJ, 1991.
120. J. C. Sturm, P. V. Schwartz, E. J. Prinz, and H. Manoharan, *J. Vac. Sci. Technol. B* **9**, 2011-2016 (1991).
121. M. M. Moslehi, *Soc. Photo-Opt. Instrum. Eng. Symp. Proc.* **1393**, 280-294 (1990).
122. M. M. Moslehi, US Patent 4 891 499 (2 Jan. 1990). US Patent 4 956 538 (11 Sep. 1990).
123. K. A. Snow, US Patent application serial no. 07/702 991 (20 May 1991); B. W. Peuse, A. Rosekrans, and K. A. Snow, *Soc. Photo-Opt. Instrum. Eng. Symp. Proc.* **1804**, in press (1992).
124. L. Peters, *Semicond. Intern.* **14**,(9), 56-62 (1991).
125. T. E. Thompson and E. R. Westerberg, Eur. Patent 339 458 (2 Nov. 1989).
126. B. Brown, *Proc. 9th European RTP Users Group Meeting*, Harlow (UK), Jan. 29, 1992.
127. J. C. Chang, T. Nguyen, J. S. Nakos, and J. W. Korejwa, *Soc. Photo-Opt. Instrum. Eng. Symp. Proc.* **1595**, 35-38 (1991).
128. A. J. LaRocca in *The Infrared Handbook* (W. L. Wolfe and G. J. Zissis, eds.), rev. 2nd ed. pp. 5.92-5.95. Environmental Res. Inst. of Michigan, Ann Arbor, MI, 1989.
129. F. Wong, C. Y. Chen, and Y. H. Ku, *Mater. Res. Symp. Proc.* **146**, 27-33 (1989).
130. A. Usami, *Denki Kagaku* **57**, 758-765 (1989) (in Japanese).
131. D. P. DeWitt and R. E. Rondeau, *J. Thermophysics* **3**, 153-159 (1989).
132. J. C. Gelpey, P. O. Stump, and J. W. Smith, *Mater. Res. Soc. Symp. Proc.* **52**, 199-207 (1986).
133. H. Walk, German Patent DE 4 012 615 (24 Oct. 1991).
134. R. Kakoschke, *Nucl. Instrum. Meth. Phys. Res. B* **37/38**, 753-759 (1989).
135. R. E. Bedford and C. K. Ma, *J. Opt. Soc. Am.* **64**, 339-349 (1974).
136. R. Kakoschke, E. Bussmann, and H. Föll, *Appl. Phys. A* **50**, 141-150 (1990).
137. R. Kakoschke, E. Bussmann, and H. Föll, *Appl. Phys. A* **52**, 52-59 (1991).
138. A. Katz, M. Albin, and Y. Komem, *J. Vac. Sci. Technol. B* **7**, 130-132 (1989).
139. W. B. de Boer and A. E. Ozias, US Patent No. 4 821 674 (18 April 1989).
140. Anonymous, *Solid State Technol.* **32**(11), 55-56 (1989).
141. S. J. Pearton and R. Caruso, *J. Appl. Phys.* **66**, 663-665 (1989).
142. A. Katz and S. J. Pearton, *J. Vac. Sci. Technol. B* **8**, 1285-1290 (1990).

143. T. E. Kazior, S. K. Brierley, and F. J. Piekarski, *IEEE Trans. Semicond. Manufact.* **4**, 21-25 (1991).
144. T. O. Sedgwick and J. E. Smith, *J. Electrochem. Soc.* **123**, 254-258 (1976).
145. H. W. Lo and A. Compaan, *Phys. Rev. Lett.* **44**, 1604-1607 (1980).
146. V. M. Donnelly and J. A. McCaulley, *J. Vac. Sci. Technol. A* **8**, 84-92 (1990).
147. Y. J. Lee, C. H. Chou, B. T. Khuri-Yakub, and K. C. Saraswat, *Soc. Photo-Opt. Instrum. Eng. Symp. Proc.* **1393**, 366-371 (1990).
148. D. W. Voorhes and D. M. Hall, *Soc. Photo-Opt. Instrum. Eng. Symp. Proc.* **1595**, 61-64 (1991).
149. M. Pichot and M. Guillaume, US Patent 4 525 066 (8 July 1982).
150. A. Durandet, O. Joubert, J. Pelletier, and M. Pichot, *J. Appl. Phys.* **67**, 3862-3866 (1990).
151. G. P. Hansen, S. Krishnan, R. H. Hauge, and J. L. Margrave, *Appl. Optics* **28**, 1885-1896 (1989).
152. J. Jans, R. Hollering, and M. Erman in *Analysis of Microelectronic Materials and Devices* (M. Grasserbauer and H. W. Werner, eds.), pp. 681-717. J. Wiley, Chichester, 1991.
153. H. Z. Massoud, R. K. Sampson, K. A. Conrad, Y.-Z. Hu, and E. A. Irene, in *ULSI Science and Technology* (J. M. Andrews and G. K. Celler, eds.), pp. 541-550. The Electrochemical Society, Pennington, NJ, 1991.
154. J.-M. Dilhac, N. Nolhier, and C. Ganibal, *Appl. Surf. Sci.* **46**, 451-454 (1990).
155. P. J. Severin and A. P. Severijns, *J. Electrochem. Soc.* **137**, 1306-1309 (1990).
156. P. J. W. Severin, *Soc. Photo-Opt. Instrum. Eng. Symp. Proc.* **1266**, 130-135 (1990).
157. P. Vandenabeele, R. J. Schreutelkamp, K. Maex, C. Vermeiren, and W. Coppye, *Mater. Res. Soc. Symp. Proc.* **260**, 653-658 (1992).
158. S. C. Wood and K. C. Saraswat in *ULSI Science and Technology* (J. M. Andrews and G. K. Celler, eds.), pp. 551-565. The Electrochemical Society, Pennington, NJ, 1991.
159. M. E. Bader, R. P. Hall, and G. Strasser, *Solid State Technol.* **33**(5), 149-154 (1990).
160. P. Burggraaf, *Semicond. Intern.* **14**(11), 66-70 (1991).
161. M. M. Moslehi, *Soc. Photo-Opt. Instrum. Eng. Symp. Proc.* **1393**, 280-294 (1990).
162. J. Nulman, *Mater. Res. Soc. Symp. Proc.* **181**, 123-132 (1990); E. Keller, J. Bukhman, S. Gonzales, C. Magnella, J. Nulman, R. Mosely, H. Grunes, and A. Tepman, *Solid State Technol.* **35**(2), 71-74 (1992).
163. R. Singh, S. Sinha, R. P. S. Thakur, and N. J. Hsu, *Mater. Res. Soc. Symp. Proc.* **224**, 197-202 (1991).
164. M. M. Moslehi and K. C. Saraswat, US Patent 4 913 929 (3 Apr. 1990).
165. S. C. Wood, K. C. Saraswat, and J. M. Harrison, *Soc. Photo-Opt. Instrum. Eng. Symp. Proc.* **1393**, 36-48 (1990).
166. P. P. Apte and K. C. Saraswat, *IEEE Trans. Semicond. Manuf.* **5**, 180-188 (1992).
167. A. Bouteville, C. Attuyt, and J. C. Remy, *Appl. Surf. Sci.* **53**, 11-17 (1991).
168. J. L. Regolini, E. Mastromatteo, M. Gauneau, J. Mercier, D. Dutartre, G. Bomchil, C. Bernard, R. Madar, and D. Bensahel, *Appl. Surf. Sci.* **53**, 18-23 (1991).
169. M. C. Öztürk, F. Y. Sorrell, J. J. Wortman, F. S. Johnson, and D. T. Grider, *IEEE Trans. Semicond. Manuf.* **4**, 155-165 (1991).
170. S. A. Norman, *IEEE Trans. Electr. Dev.* **39**, 205-207 (1992).
171. S. A. Norman, Technical Report 91-SAN-1, Information Systems Lab., Stanford University, June 1991.
172. S. A. Norman, PhD thesis, Stanford University, July 1992.
173. J. R. Hauser and R. S. Gyurcsik, *Soc. Photo-Opt. Instrum, Eng. Symp. Proc.* **1392**, 340-351 (1990).

174. R. S. Gyurcsik, T. J. Riley, and F. Y. Sorrell, *IEEE Trans. Semicond. Manufact.* **4**, 9-13 (1991).
175. A. Kermani, M. F. Robertson, Y.-H. Ku, and F. Wong, US Patent 5 002 630 (26 Mar. 1991); F. Wong and Y.-H. Ku, patent application WO 90/14158 (29 Nov. 1990).
176. C. M. Gronet, J. C. Sturm, K. E. Williams, and J. F. Gibbons, *Appl. Phys. Lett.* **48**, 1012-1014 (1986).
177. S. Reynolds, D. W. Vook, and J. F. Gibbons, *Appl. Phys. Lett.* **49**, 1720-1722 (1986).
178. P. J. Roksnoer, J. W. F. M. Maes, A. T. Vink, C. J. Vriezema, and P. C. Zalm, *Appl. Phys. Lett.* **58**, 711-713 (1991).
179. P. J. Roksnoer, J. W. F. M. Maes, A. T. Vink, C. J. Vriezema, and P. C. Zalm, *Appl. Phys. Lett.* **59**, 3297-3299 (1991).
180. P. C. Zalm, C. J. Vriezema, D. J. Gravesteijn, G. F. A. van der Walle, and W. B. de Boer, *Surf. Interf. Anal.* **17**, 556-566 (1991).
181. P. Molle, S. Deleonibus, and F. Martin, *J. Electrochem. Soc.* **138**, 3752-3738 (1991).
182. R. N. Anderson, J. G. Martin, D. Meyer, D. West, R. Bowman, and D. V. Adams, US Patent 5 108 792 (28 April 1992).
183. R. N. Anderson, T. E. Deacon, and D. K. Carlson, Eur. Patent 476 307 (25 March 1992).
184. D. Bensahel, J. L. Regolini, and J. Mercier, *Appl. Phys. Lett.* **55**, 1549-1551 (1989).
185. Anonymous, *Solid State Technol.* **32**(1), 47-48 (1989).
186. D. Aitken, S. Mehta, N. Parisi, C. J. Russo, and V. Schwartz, *Nucl. Instrum. Meth. Phys. Res. B* **21**, 622-626 (1987).

Index

Absorption, 314
  by ambient gas, 388
  band-to-band, 354
  free carrier, 354, 361, 380
Absorptivity, 354
Accuracy, 372, 415
Alloys, silicon-germanium
  applications, 33
  bandgap reduction, 36
  growth rate, 34
  misfit dislocations, 37
  oxygen, 34, 40
  strain, 31, 35
Architecture, 408
Autodoping, 19
  doping transition widths, 19, 27, 28

Bipolar transistor
  doping profile, 16
  processing (see Processing, bipolar transistor)

Cavity, 359, 397
  blackbody, 397
  design, 397
  greybody, 362
  reflective, 359, 397
Chamber design, 365
  aspect ratio in, 358, 397, 399, 416
  cold wall, 366, 415
  hot wall, 367
  quartz, 81
  stainless steel, 81
  warm wall, 367, 415
Chemical vapor deposition, 352, 368, 410, 416
  chamber walls, 116
  RTCVD, 14
    polycrystalline Si-Ge, 98

  polycrystalline silicon, 93
  silicon dioxide, 86
  silicon nitride, 92
  temperature measurement, 114
  temperature uniformity, 111
  thin-film deposition
    activation energy, 84
    methods, 80
    temperature regimes, 84
  titanium silicide, 103
  tungsten, 107
  UHVCVD, 22, 80
Cluster tool, 408
Cycle time, 410

Defects
  amorphous layers, 128, 155, 190, 201, 205, 217
  buried, 147
  critical energy deposition, 129, 139
  interfaces, 128, 131
  ion-beam-induced recrystallization, 131
  light ions, 129
  phase transformation, 128, 144
  regrowth, 132
  threshold damage density, 129
  density, 18
  ion implantation-induced, 123, 174
    annealing kinetics, 155
    defect formation kinetics, 133
    end-of-range dislocations, 140, 157, 161, 178, 197
    extended defects, 134
    hairpin dislocations, 144
    loops, 133, 135, 141, 147, 149, 158
    low dose, 127, 135, 156, 175
    microtwins, 146, 155
    point defects, 133, 158
    projected range, 149, 180
    recoil model, 141

426   Index

removal, 177, 179, 203
screening of point defects, 197, 199
solid phase epitaxial regrowth, 132, 144
solubility effects, 159
subthreshold, 176
theory, 125
laser annealing, 125
misfit dislocations
    effect on device performance, 39
    formation, 37, 39
    RTA induced, 38
    spacing, 37
point defects
    self-interstitials, 133, 175, 180, 192, 197
    supersaturation, 201, 207
    vacancies, 133, 175
silicide damage, 281
slip, 19, 50, 318, 323, 390, 392, 415, 417
stress, 50
    extrinsic, 318
    intrinsic, 318
    shear, critical value, 321, 323
    thermal, 318
    warpage, 318
Devices
    bipolar transistors, 325
        concentrations, 329
        polysilicon emitter, 326
    guard ring, 362, 393, 416
    MOS transistors, 340
    silicided
        bipolar, 228, 235
        contact resistance (*see* series resistance)
        hot electron effects, 285
        latchup, 287
        performance, 229
        polycide, 229, 282
        salicide, 232
        series resistance, 231, 233, 275, 283
Dielectrics, 44
    breakdown, 59, 67
Diffusion
    dopants, 206
        As, 207
        B, 172, 183, 185, 202, 208, 216
        P, 181, 184, 199, 208
        Sb, 206
    grain boundary (silicides), 246
    silicide as diffusion source, 235, 287
    silicides, 281, 288

transient, 124, 170, 180
    activation energy, 209
    amorphous layers, 190, 201, 205, 217
    high dose implant, 188
    low dose implant, 181
    models, 192, 210
    simulation, 191
    time constant, 185, 207
Dopant activation
    amorphous layer effect, 217
    As, 219
    B, 175, 215
    measurement of, 191
    P, 218
    precipitates, 149, 151
    sheet resistance, 214, 217, 220
    theory, 213

Electrical properties of RTO/RTP oxides,
        62, 337, 342
    breakdown, 59, 67
        time dependence, 68
        time zero, 67
    carrier trapping, 59, 61, 69
    charge-to-breakdown, 58
    current-voltage characteristics, 66
    devices
        lifetime, 70
        performance, 61, 70
    electron mobility, 71
    fixed charge, 58, 61
    flatband voltage, 58, 62
    Fowler–Nordheim tunneling, 66
    gate delay, 71
    gate-induced drain leakage, 71
    hot-carrier reliability, 70, 339
    interface traps, 59, 62
    MOSCAP capacitance, 63
    time-to-breakdown, 58
Emissivity, 314, 318, 353
    effective, 357, 390, 396, 397
    endpoint detection, 407
    error, 390
    extrinsic, 357
    intrinsic, 353
    non-patterned wafer, 391
    patterned wafer, 393
    rough wafer backside, 400
Endpoint detection, 404
    by ellipsometry, 404

Index

by emissivity, 407
by light interferometry, 405
by reflectometry, 405
Energy, loss compensation, 316
Epitaxy
  cost, 20
  doping profile, 27
  growth kinetics, 17, 25
  heteroepitaxy, 31
  in ICs, 15
  low temperature, 18
  MBE, 22
  multilayer deposition, 20
  pattern shift, 19
  quality, 29
  reactor design, 23
  selective growth, 20
  surface segregation, 28
Equipment
  commercial, 381
  furnaces, batch
    minibatch, 408
    vertical, 408
  issues in RTO, 46
    factorial experiment design, 50
    heating sources, 46, 51
    in-situ ellipsometry, 48
    manufacturability, 50
    process chamber, 46, 51
    pyrometry, 48
    slip, 50
    system requirements, 46
    temperature measurement and control, 38, 46
    thermal stress, 50
    thermocouples, 48
    ultracleaning, 46
    uniformity, 50, 372, 415
  susceptor, 400
  thin-film deposition
    chamber design, 81
    deposition on chamber walls, 116
    lamps, 81
    temperature measurement, 114
    temperature uniformity, 111
Extensometry, 380, 403

Gases
  flow of, 368, 401
  switching of, 414

Heat source, 51, 360
  arc lamp, 354, 361
  intensity control, 51
  silicon carbide bell jar, 355, 361
  spectral (color) temperature, 354, 360
  tungsten halogen lamp, 354, 360
  water cooled
    arc lamp, 361
    metal, 366
    quartz plates, 366
Heterojunction bipolar transistors
  $Si/Si(1-x)Ge(x)$, 413
    collector current equations, 35
    first CVD grown, 32
    ring oscillator delays, 37
    small geometry devices, 36
    strain, 31, 35
Historical survey, 351

Integrated circuits, application specific, 408
Ion implantation
  amorphous layers, 178, 196
    junction leakage, 201
    self-amorphization, 206
    shallow layers, 200
  damage, 174
    end-of-range, 178, 197
    low dose, 175
    projected range, 180
    removal, 177, 179, 203
    screening of point defects, 197, 199
    subthreshold, 176
  implant through metal, 235, 287

Junctions
  annealing, 171
  formation, 169
  ion implanted, 170
  requirements, 170
  sheet resistance, 214, 217, 220
  silicided, 276, 287 (*see also* Metal-silicon reaction)
    damage, 281
    dopant diffusion, 281, 288
    leakage, 276, 288
    thermal budget, 171

Kaleidoscope, 358, 397
Kirchhoff's law, 358

Lambertain emission, 354, 397
Laser annealing, 351
Light pipe, 358
Limited reaction processing, 13, 397
   doping profiles, 27
      C and O, 30
   in-situ grown diodes, 31
   material quality, 29
   reduced time-temperature exposure, 21
Manufacturing
   budget crisis, 1
   cost reduction, 2
   parameters, 3
      ambient, 5
      contamination, 6
      electrical, 9
      mechanical, 6
      thermal, 4
   RTP, 2
   technology roadmap, 4, 415
Memory effects
   chemical, 367, 397
   thermal, 367
Metalorganic CVD, 23
Metal-silicon (silicide) reaction
   atmosphere effects, 257, 270
   edge effect (*see* linewidth effect)
   grain boundary diffusion, 246
   growth kinetics, 251, 257
   impurities, 255
   linewidth effect, 270
   nucleation, 254, 257, 269
   resistivity, 252, 255
   silicon consumption, 260
   stress, 262
   surface layers, 255
   thickness dependence, 261
Minority carrier lifetime, 18, 29, 31
Modeling
   equipment, 390
   process, 291
      rapid thermal reoxidation of nitrided oxides, 61
      RTO growth kinetics, 55
      transient diffusion, 192, 210
Multichamber processing, 408
Multivariable control, 411

Nitrided oxides
   rapid thermal reoxidation, 61
      charge turnaround, 65
      hot-carrier degradation, 61
      interface traps, 65
      models, 61
      oxide fixed charge, 65
      silicon nitride (*see* Silicon nitride deposition)

Optical fiber thermometry, 376, 379
   by interferometry, 406
   reflection-based, 379
   ripple amplitude, 376, 407
   transmission-based, 379

Particles, 415
Pattern, 393
Photonic contribution, 312
Planck's radiation law, 354
Point defects
   process control, 158
   self-interstitials, 133, 158, 175, 180, 192, 197
   sources, 133
   supersaturation, 201, 207
   vacancies, 133, 175
Polysilicon deposition
   activation energy, 93
   amorphous-crystalline transition, 94
   deposition rate, 95
   diffusion source, 109 (*see also* Polysilicon emitter)
   grain structure, 97
   Si-Ge deposition
      gate electrode, 98
      in-situ doped, 101
      resistivity, 101
      selectivity, 99
      stoichiometry, 99
      surface morphology, 101
   surface roughness, 95
   ultraviolet reflectance, 95
Polysilicon emitter, 108, 326
   arsenic-doped, 334
   double polysilicon, 326
   single polysilicon, 332
Precleaning, 367, 409
Processing
   beam-assisted, 410, 417
   bipolar transistor, 325
      collector, 328, 331

Index

doping concentrations, 329
polysilicon emitter, 326, 332, 334
control, 411
  scalar vs. multivariable, 411
isothermal, 316
  cold wall, 312
  hot wall, 324
MOS
  gate dielectric, 342
  gate oxide integrity, 337
  hot carrier generation, 339
  raised electrodes, 340
Process modeling, 291
Pyrometry
  detectivity, 355
  dual color, 275
  dual point, 412
  perturbing radiation in, 360, 369, 387, 390
  radiation absorption, by ambient gas, 388
  ratio, 374
  sensitivity, 355, 369
  spectral features, 355
    far infrared, 388
    middle infrared, 387
    near infrared, 390

Radiant energy, 315
Radiation
  blackbody, 354
  greybody, 354, 362
  Planck's radiation law, 354
  primary, 315
  pyrometry, radiation effects, 360, 369, 387, 390
  secondary, 315
  spectral features, infrared, 355, 387, 390
Rapid thermal annealing, oxides, 45, 48
  charge-to-breakdown, 58
  dielectric breakdown, 59
  flatband voltage, 58
  interface traps, 59
  oxide fixed charge, 58
  oxide trapping, 59
Rapid thermal oxidation
  experimental observations, 52
  growth kinetics, 51
    activation energy, 53, 55
    doping, 54
    HCl, 53

heat source, 57
linear growth, 52
models, 55
one-step oxidation, 52
parabolic growth, 56
temperature control, 51
temperature uniformity, 54
thermal stress, 54
two-step oxidation, 52
wafer orientation, 53
Reflectivity, 314, 344
  of chamber, 375, 391
  measurement, 373, 379, 405
  ultraviolet, of polysilicon, 95
  of wafer, 358, 378, 405
Reflector, 316, 364
Repeatability, 415
Reproducibility, 372
Round-robin, 372
RTP, oxides, 44, 58
  rapid thermal reflow, 45
  rapid thermal reoxidation of nitrided oxide, 61, 65
RTA, 45, 58
RTB, 45
RTC, 45, 68
RTCVD, 45, 49
RTN, 45, 60, 63, 66
RTO, 45

Seebeck effect, 369
Silicide materials
  bridging, 244
  formation techniques, 238
    buried implantation, 249
    chemical vapor deposition, 250
    codeposition, 241
    ion beam-induced mixing, 249
    thermal reaction, 242
  lateral overgrowth (*see* bridging)
  properties, 237, 239
    agglomeration (*see* stability)
    conductivity, 264
    dopant redistribution, 273
    stability, 264
    thermal grooving (*see* stability)
Silicide processes
  implant through metal, 235, 287
  local interconnection, 236, 289
  polycide, 229

SADS (*see* silicide as diffusion source)
salicide, 231, 242, 244
silicide as diffusion source, 235, 287
titanium silicide deposition, 103
Silicided devices (*see* Devices, silicided)
Silicided junctions (*see* Junctions, silicided)
Silicon dioxide deposition
  film growth (*see* Rapid thermal oxidation)
  gate dielectric, 85
    dichlorosilane/nitrous oxide, 86, 107
    silane/nitrous oxide, 86
  isolation, 88
    annealing, 90
    deposition rate, 90
    etch rate, 91
    step coverage, 89
    sticking coefficient, 89
Silicon nitride deposition, 92
  activation energy, 92
  deposition rate, 92
  in-situ processing, 107
  stoichiometry, 92
Single-wafer processing, 313
Slip (*see* Defects)
Spectrometry
  secondary ion mass spectrometry
    depth resolution, 27
    sensitivity limits, 18
    thermal desorption, 27
Stefan–Boltzmann law, 353
Superlattice of Si-SiGe, 32, 35, 413

Technology roadmap, 4, 414
Temperature
  calibration,
    ex situ, 373
    in situ, 372
    system level, 371
    wafer level, 372
  distribution, 313, 316
  heating sources, 46, 51, 354, 360
  optical fiber thermometry, 376, 379
  uniformity, 54, 314, 390, 397
  wafer temperature measurement
    by acoustic waves, 402
    by ellipsometry, 404
    for epitaxy, 23
    by indirect reaction, 372
    by interferometry, 404
    IR transmission, 24, 379
    measurement and control, 46, 48, 313
    by optical diffraction, 403
    pyrometry, 48, 355, 360, 369, 375, 387, 390
    by Raman scattering, 401
    temperature distribution, 313, 316
    by thermal expansion, 381
    thermocouples, 48, 369, 371
Thermal budget, 171, 311
Thermal detector, 369
  photodetector, 371, 381
  thermocouple, 369, 371
  thermopile, 369, 381
Thermal expansion, 381, 402
Thermal nitridation, 45, 60
  charge build-up, 61
  device degradation, 61
  dielectric breakdown, 68
  dielectric constant, 63
  trapping, 61
  tunneling, 66
Thermal switch, 413
Thermophysics, 352
Thin-film deposition
  activation energy, 84
  methods
    atmospheric pressure CVD, 80
    low pressure CVD, 80
    ultra-high vacuum CVD, 80
  temperature regimes
    mass transport, 84
    surface reaction, 84
Throughput, 407
Titanium silicide deposition, 103
Transmission
  optical media, 355
  silicon, 379
Transmissivity, 358
Tungsten deposition, 107

Uniformity, 372, 415

Wafers, Bernoulli pick-up, 399, 408
Warpage, 318
Wien's displacement law, 354

Yield, 318, 408

ISBN 0-12-247690-5